Study Guide, Volume 1

for

Paul A. Tipler and Eugene Mosca's

Physics for Scientists and Engineers

Fifth Edition

Todd Ruskell
Colorado School of Mines

Gene Mosca
U.S. Naval Academy

W. H. Freeman and Company
New York

Printed in the United States of America

ISBN: 0-7167-8332-0

First printing 2003

CONTENTS

To the Student

This Study Guide was written to help you master Chapters 1 through 20 of Paul Tipler and Eugene Mosca's *Physics for Scientists and Engineers,* Fifth Edition. Each chapter of the Study Guide contains the following sections:

I.   Key Ideas. A brief overview of the important concepts presented in the chapter.

II. Physical Quantities and Key Equations. A list of the constants, units, and basic equations introduced in the chapter.

III. Potential Pitfalls. Warnings about mistakes that are commonly made.

IV. True or False Questions and Responses. Statements to test whether or not you understand essential definitions and relations. All the false statements are followed by explanations of why they are false. In addition, many of the true statements are followed by explanations of why they are true.

V.   Questions and Answers. Questions that require mostly qualitative reasoning. A complete answer is provided for each question so that you can compare it with your own.

VI. Problems, Solutions, and Answers. With few exceptions, the problems come in pairs; the first of the pair is followed by a detailed solution to help you develop a model, which you can then implement in the second problem. One or more hints and a step-by-step guide accompany most problems. These problems will help you build your understanding of the physical concepts and your ability to apply what you have learned to physical situations.

What Is the Best Way to Study Physics?

Of course there isn't a single answer to that. It is clear, however, that you should begin early in the course to develop the methods that work best for you. The important thing is to find the system that is most comfortable and effective for you, and then stick to it.

In this course you will be introduced to numerous concepts. It is important that you take the time to be sure you understand each of them. You will have mastered a concept when you fully understand its relationships with other concepts. Some concepts will *seem* to contradict other concepts or even your observations of the physical world. Many of the True or False statements and the Questions in this Study Guide are intended to test your understanding of concepts. If you find that your understanding of an idea is incomplete, don't give up; pursue it until it becomes clear. We recommend that you keep a list of the things that you come across in your studies that you do not understand. Then, when you come to understand an idea, remove it from your list. After you complete your study of each chapter, bring your list to your most important resource, your physics instructor, and ask for assistance. If you go to your instructor with a few well-defined questions, you will very likely be able to remove any remaining items from your list.

Like the example problems presented in the textbook, the problem solutions presented in this Study Guide *start with basic concepts,* not with formulas. We encourage you to follow this practice. Physics is a collection of interrelated basic concepts, not a seemingly infinite list of disconnected, highly specific formulas. Don't try to memorize long lists of specific formulas, and then use these formulas as the starting point for solving problems. Instead, focus on the concepts first and be sure that you understand the ideas before you apply the formulas.

Probably the most rewarding (but challenging) aspect of studying physics is learning how to apply the fundamental concepts to specific problems. At some point you are likely to think, "I understand the theory, but I just can't do the problems." If you can't do the problems, however, you probably don't understand the theory. Until the physical concepts and the mathematical equations become your tools to apply at will to specific physical situations, you haven't really learned them. There are two major aspects involved in learning to solve problems: drill and skill. By *drill* we mean going through a lot of problems that involve the direct application of a particular concept until you start to feel familiar with the way it applies to physical situations. Each chapter of the Tipler textbook contains about 35 single-concept problems for you to use as drill. *Do a lot of these!*—at least as many as you need in order to feel comfortable handling them.

By *skill* we mean the ability both to recognize which concepts are involved in more advanced, multi-concept problems, and to apply those concepts to particular situations. The text has several intermediate-level and advanced-level problems that go beyond the direct application of a single concept. As you develop this skill you will master the material and become empowered. As you find that you can deal with more complex problems—even some of the advanced-level ones—you will gain confidence and enjoy applying your new skills. The examples in the textbook and the problems in this Study Guide are designed to provide you with a pathway from the single-concept to the intermediate-level and advanced-level problems.

A typical physics problem describes a physical situation—such as a child swinging on a swing—and asks related questions. For example: If the speed of the child is 5.0 m/s at the bottom of her arc, what is the maximum height the child will reach? Solving such problems requires you to apply the concepts of physics to the physical situation, to generate mathematical relations, and to solve for the desired quantities. The problems presented here and in your textbook are exemplars; that is, they are examples that deserve imitation. When you master the methodology presented in the worked-out examples, you should be able to solve problems about a wide variety of physical situations.

To be successful in solving physics problems, study the techniques used in the worked-out example problems. A good way to test your understanding of a specific solution is to take a sheet of paper, and—without looking at the worked-out solution—reproduce it. If you get stuck and need to refer to the presented solution, do so. But then take a fresh sheet of paper, start from the beginning, and reproduce the entire solution. This may seem tedious at first, but it does pay off.

This is not to suggest that you reproduce solutions by rote memorization, but that you reproduce them by drawing on your understanding of the relationships involved. By reproducing a solution in its entirety, you will verify for yourself that you have mastered a particular example problem. As you repeat this process with other examples, you will build your very own personal base of physics knowledge, a base of knowledge relating occurrences in the world around you—the physical universe—and the concepts of physics. The more complete the knowledge base that you build, the more success you will have in physics.

All the problems in the Study Guide are accompanied by step-by-step suggestions on how to solve them. As previously mentioned, the problems come in pairs, with a detailed solution for the odd-numbered problem. We suggest that you start with an odd-numbered problem and study the problem steps and the worked-out solution. Be sure to note how the solution implements each step of the problem-solving process. Then try the second, "interactive" problem in the pair. When attacking a problem, read the problem statement several times to be sure that you can picture the problem being presented. Then make an illustration of this situation. Now you are ready to solve the problem.

You should budget time to study physics on a regular, preferably daily, basis. Plan your study schedule with your course schedule in mind. One benefit of this approach is that when you study on a regular basis, more information is likely to be transferred to your long-term memory than when you are obliged to cram. Another benefit of studying on a regular basis is that you will get much more from lectures. Because you will have already studied some of the material presented, the lectures will seem more relevant to you. In fact, you should try to familiarize yourself with each chapter before it is covered in class. An effective way to do this is first to read the Key Concepts of that Study Guide chapter. Then thumb through the textbook chapter, reading the headings and examining the illustrations. By orienting yourself to a topic *before* it is covered in class, you will have created a receptive environment for encoding and storing in your memory the material you will be learning.

Another way to enhance your learning is to explain something to a fellow student. It is well known that the best way to learn something is to teach it. That is because in attempting to articulate a concept or procedure, you must first arrange the relevant ideas in a logical sequence. In addition, a dialogue with another person may help you to consider things from a different perspective. After you have studied a section of a chapter, discuss the material with another student and see if you can explain what you have learned.

We wish you success in your studies and encourage you to contact us at truskell@Mines.edu if you find errors, or if you have comments or suggestions.

Acknowledgments from Todd Ruskell

I want to thank Eugene Mosca, the primary author of the Study Guide that accompanied the fourth edition of the textbook. Much of the material in this edition comes either directly or indirectly from his efforts. I also want to thank Paul Tipler and Eugene Mosca for writing a textbook that has been a delight to work with. I would like to thank the publishing staff, especially Brian Donnellan, for his patience with me throughout this project. I am deeply indebted to the many reviewers whose work has made this a better Study Guide than it could otherwise have been. Finally, I wish to thank my wife Susan for her patience and support.

January 2003

Todd Ruskell
Colorado School of Mines

Chapter **1**

Systems of Measurement

I. Key Ideas

Section 1-1. Units. When we measure any physical quantity, we are comparing it to some agreed-on, standard **unit.** Scientists (and others) use a system of units called the **Système Internationale (SI)** in which the standard units for length, time, and mass are the meter, the second, and the kilogram, respectively.

When the names of units are written out, they always begin with a lowercase letter, *even when* the unit is named for a person. Thus, the unit of temperature named for Lord Kelvin is the kelvin. The abbreviation for a unit also begins with a lowercase letter, *except when* the unit is named for a person. Thus, the abbreviation for the meter is m, whereas the abbreviation for the kelvin is K. The exception is the abbreviation for the liter, which is L. Abbreviations of units are not followed by periods.

All written-out prefixes for powers of 10, such as *mega* for 10^6, begin with lowercase letters. For multiples less than or equal to 10^3, the abbreviations of the prefixes are lowercase letters, such as k for *kilo*. For multiples greater than 10^3, however, the prefix abbreviations are uppercase letters, such as M for *mega*. The prefixes for powers of 10, as well as their abbreviations, are listed in Table 1-1 on page 5 of the main text.

You may also encounter other systems of units. The cgs system is based on the *c*entimeter, *g*ram, and *s*econd. The centimeter is defined as 0.01 m. The gram is defined as 0.001 kg. In the United States, a unit of force, the pound, is chosen to be a fundamental unit. The fundamental unit of time in this system is the second, and the fundamental unit of length is the foot. The foot is defined as exactly one-third of a yard, which we can define in terms of the meter:

$$1\,\text{yd} = 0.9144\,\text{m}$$
$$1\,\text{ft} = \tfrac{1}{3}\,\text{yd} = 0.3048\,\text{m}$$

making the inch equal to 2.54 cm.

Section 1-2. Conversion of Units. One unit can be converted to another by multiplying it by a **conversion factor** that equals 1. If we divide each side of the relation

$$60\,\text{sec} = 1\,\text{min}$$

by 1 min, we obtain the conversion factor

$$\frac{60\,\mathrm{s}}{1\,\mathrm{min}} = 1$$

To convert 7.4 min to seconds, we multiply 7.4 min by this conversion factor:

$$7.4\,\cancel{\mathrm{min}} \times \frac{60\,\mathrm{s}}{1\,\cancel{\mathrm{min}}} = 444\,\mathrm{s}$$

In this expression, we think of the "min" units as canceling each other. One may draw lines through units that cancel as shown.

Section 1-3. Dimensions of Physical Quantities. The **dimensions** of a physical quantity express what kind of quantity it is—whether it is a length, a time, a mass, or whatever. For example, the dimensions of velocity are length per unit time. The corresponding units might be miles per hour or meters per second. Whenever we add or subtract quantities, they must have the same dimensions. Also, both sides of an equation must have the same dimensions. Checking that the dimensions are in fact the same is often a useful way of checking for mistakes in setting up equations.

Section 1-4. Scientific Notation. Very often in physics we find ourselves dealing with very large or very small numbers. This is much simpler to do if the numbers are written in **scientific notation,** that is, as a number between 1 and 10 multiplied by the appropriate power of 10. An example is 1.67×10^5 for the number 167,000. When numbers are multiplied (or divided), the powers of 10 are added (or subtracted). Another common way to write large numbers is to use "E-notation." With this notation, 167,000 can be written as 1.67E5 or 1.67e5. Either the upper or lowercase "e" is acceptable. Here, the "e" is shorthand for "times ten to the."

Section 1-5. Significant Figures and Order of Magnitude. The quantities we deal with in physics are not always pure numbers but instead are often (in principle or in fact) the results of measurements. They are thus known to only limited precision. The number of digits we use to express a quantity is, by implication, an expression of its precision. We therefore use only those digits that are **significant figures,** that is, those digits that have meaning. The term *figure* is synonymous with the term *digit,* and physicists commonly use them interchangeably.

The number of significant figures in the result of multiplication or division is no greater than the least number of significant figures in any of the factors.

The result of addition or subtraction of two numbers has no significant figures beyond the last decimal place where both of the original numbers had significant figures.

Many examples in this Study Guide will be done with data to three significant figures, but occasionally we will say, for example, that a table top is 1 m by 2 m rather than taking the time and space to say it is 1.00 m by 2.00 m. Any data you see can be assumed to be known to three significant figures unless otherwise indicated.

The **order of magnitude** of a value is an estimate of the value to the nearest power of 10.

II. Physical Quantities and Key Equations

Physical Quantities

1 inch (in) = 2.54 centimeters (cm)

1 foot (ft) = 0.3048 meters (m)

1 mile (mi) = 1.61 kilometer (km)

1 liter (L) = 1×10^{-3} m

Key Equations

There are no key equations for this chapter.

III. Potential Pitfalls

Without units, expressions of physical quantities do not have meaning. Usually, it is best to use SI units. If you are given problems with quantities in other units, you should usually convert them to SI units before proceeding with the problem. When converting units, remember that all conversion factors must have a magnitude of 1.

This course will require you to solve a large number of problems. For most of these, you will be using a calculator. Do not blindly believe whatever answer your calculator displays. Check your results by estimating the order of magnitude of the calculation on a piece of scratch paper. Do not use all of the digits displayed by the calculator. Write down only the significant digits.

Be careful when entering numbers using "e-notation" in you calculator. In this notation, the value 10 is written as 1e1, not 10e1.

IV. True or False Questions and Responses

True or False

____ 1. The length of the meter depends on the duration of a second.

____ 2. The SI unit of mass is the gram.

____ 3. Two quantities having different dimensions can be multiplied, but they cannot be subtracted.

____ 4. All conversion factors are equal to 1.

____ 5. The second is defined in terms of the frequency of a certain kind of light emitted by cesium atoms.

_____ 6.   The SI units of force, energy, and other physical quantities are defined in terms of the fundamental units of mass, length, and time.

_____ 7.   Two quantities having the same dimensions must be measured in the same units.

Responses to True or False

1.   True. According to the current definition of the meter, if the duration of the second doubled, the length of the meter would also double.

2.   False. It is the kilogram.

3.   True.

4.   True.

5.   True.

6.   True.

7.   False. For example, time can be measured in hours, minutes, seconds, or other units.

V.   Questions and Answers

Questions

1.   Is it possible to define a system of units for measuring physical quantities in which one of the fundamental units is not a unit of length?

2.   What properties should an object, system, or process have for it to be a useful standard of measurement for a physical quantity such as time or length?

3.   A furlong is one-eighth of a mile and a fortnight is two weeks. Of what physical quantity is a furlong per fortnight a unit? What is the SI unit for this quantity? Find the conversion factor between furlongs per fortnight and the corresponding SI unit.

4.   Acceleration a has dimensions L/T^2, and those of velocity v are L/T. The velocity of an object that has accelerated uniformly through a distance d is either $v^2 = 2ad$ or $v^2 = (2ad)^2$. Which one must it be?

5.   If you use a calculator to divide 3411 by 62.0, you will get something like 55.016 129. (Exactly what you will get depends on your calculator.) Of course, you know that not all these figures are significant. How should you write the answer?

6.   If two quantities are to be added, do they have to have the same dimensions? The same units? What if they are to be divided?

Answers

1. Certainly. It could have fundamental units for time and speed. The unit for length could then be defined in terms of these fundamental units.

2. It should be reproducible at a variety of times and places, and it should be precise.

3. The quantity is speed. The SI unit for speed is the meter per second. We determine the conversion factor as follows:

$$\left(\frac{1\ \text{furlong}}{\text{fortnight}}\right)\left(\frac{1\ \text{fortnight}}{14\ \text{days}}\right)\left(\frac{1\ \text{day}}{24\ \text{hr}}\right)\left(\frac{1\ \text{hr}}{3600\ \text{s}}\right)\left(\frac{1\ \text{mi}}{8\ \text{furlongs}}\right)\left(\frac{1610\ \text{m}}{1\ \text{mi}}\right)=1.66\times10^{-4}\ \text{m/s}$$

The conversion factor is

$$\frac{1.66\times10^{-4}\ \text{m/s}}{1\ \text{furlong/fortnight}}$$

4. It must be $v^2 = 2ad$ because both sides of the equation have the dimensions L^2/T^2.

5. You are dividing a number with four significant figures by a number with three significant figures. The result should therefore have three significant figures. It should be written 55.0.

6. To be added, the quantities have to have the same dimensions but not the same units. For example, if you add 4 feet and 3 inches you get 4 feet 3 inches. (Normally, you will want to convert the units of one of the quantities to those of the other before adding them.) Quantities can be divided even if they have different dimensions, as when distance is divided by time to get speed.

VI. Problems, Solutions, and Answers

Example #1. Write the following in scientific notation without prefixes: (*a*) $22\,\mu$m, (*b*) 233.7 Mm, (*c*) 0.4 kK, (*d*) 20.0 pW.

1. Represent each number in scientific notation. Using Table 1-1 on page 5 of the text, we can determine the power of ten for each of the prefixes. Add the exponents.	$22\,\mu\text{m} = \left(2.2\times10^{1}\right)10^{-6}\ \text{m} = 2.2\times10^{-5}\ \text{m}$ $233.7\,\text{Mm} = \left(2.337\times10^{2}\right)10^{6}\ \text{m} = 2.337\times10^{8}\ \text{m}$ $0.4\,\text{kK} = \left(4\times10^{-1}\right)10^{3}\ \text{K} = 4\times10^{2}\ \text{K}$ $20.0\,\text{pW} = \left(2.00\times10^{1}\right)10^{-12}\ \text{W} = 2.00\times10^{-11}\ \text{W}$

Example #2. Write the following in scientific notation without prefixes: (*a*) 330 km, (*b*) 33.7 μm, (*c*) 0.03 K, (*d*) 77.5 GW.

1. Represent each number in scientific notation. Using Table 1-1 on page 5 of the text, we can determine the power of ten for each of the prefixes. Add the exponents.	$330\,\text{km} = 3.3 \times 10^5\,\text{m}$ $33.7\,\mu\text{m} = 33.7 \times 10^{-5}\,\text{m}$ $0.03\,\text{K} = 3 \times 10^{-2}\,\text{K}$ $77.5\,\text{GW} = 7.75 \times 10^{10}\,\text{W}$

Example #3. In the following equation, the distance x is in meters, the time t is in seconds, and the velocity v is in meters per second. What are the dimensions and SI units of the constants A, B, C, D, and E? *Hint:* Both exponents and the arguments of trigonometric functions must be dimensionless. (The argument of $\cos\phi$ is ϕ.)

$$x = Avt + B\sin(Ct) + Dt^{1/2}3^{x/E}$$

1. Since exponents must be dimensionless, the dimensions of E must be the same as those for x.	E has dimensions of length, L. Because x has units of meters, m, so does the quantity E.
2. Since arguments of trigonometric functions must be dimensionless, the dimensions of C must cancel the dimensions of t.	C has dimensions of $1/T$, and units of $1/\text{s}$.
3. The three terms added together on the right must each have the same dimensions as x, or they couldn't be added together to equal x.	vt has dimensions of L already, so A must be dimensionless. Trigonometric functions are dimensionless, so B must have dimensions of L, and units of m. Constants are dimensionless, and $t^{1/2}$ has dimensions of $T^{1/2}$. As a result, D must have dimensions of $L/T^{1/2}$, which in this problem corresponds to units of $\text{m}/\text{s}^{1/2}$.

Example #4. In the following equation, the distance x is in meters, the time t is in seconds, and the velocity v is in meters per second. What are the dimensions and SI units of the constants A, B, C, D, and E?

$$v = Bt\left[\sqrt{Ax} + \cos^2(Ct)\right] - D2^{Et}$$

1. Since exponents must be dimensionless, the quantity Et must be dimensionless.	E has dimensions of $1/T$, and units of $1/\text{s}$.

2. Since arguments of trigonometric functions must be dimensionless, the dimensions of C must cancel the dimensions of t.	C has dimensions of $1/T$, and units of $1/\text{s}$.
3. The three terms added together on the right must each have the same dimensions as v, or they couldn't be added together to equal v.	Trigonometric functions are dimensionless, so the quantity $\sqrt{Ax}$ must also be dimensionless. This gives A dimensions of $1/L$ and units of $1/\text{m}$. Bt must have the same dimensions as v, so B has dimensions of L/T^2, and units of m/s^2. Constants are dimensionless, so D must have dimensions of L/T and units of m/s.

Example #5. Write the following using prefixes and abbreviations for the SI units: (*a*) 25,000 meters per second, (*b*) 0.004 seconds, (*c*) 4,200,000 watts, (*d*) 0.1 kilograms. For example, 0.0001 meters = 100 μm. (The number preceding the prefix should always be less than 1000 and equal to or greater than 1.)

1. Factor the number by 10^{+3} or 10^{-3} until you have a number less than 1000 and equal to or greater than 1. Determine the prefix by looking at Table 1-1 on page 5 of the text.	$25{,}000\,\text{meters per second} = 25\times10^3\,\text{m/s} = 25\,\text{km/s}$ $0.004\,\text{seconds} = 4\times10^{-3}\,\text{s} = 4\,\text{ms}$ $4{,}200{,}000\,\text{watts} = 4.2\times10^6\,\text{W} = 4.2\,\text{MW}$ $0.1\,\text{kilograms} = 100\times10^{-3}\,\text{kg} = 100\,\text{g}$

Example #6. Write the following using prefixes and abbreviations for the SI units: (*a*) 0.000 025 meters per second, (*b*) 0.0445 seconds, (*c*) 0.000 000 032 watts, (*d*) 25,640 kilograms. For example, 0.0001 meters = 100 μm. (The number preceding the prefix should always be less than 1000 and equal to or greater than 1.)

1. Factor the number by 10^{+3} or 10^{-3} until you have a number less than 1000 and equal to or greater than 1. Determine the prefix by looking at Table 1-1 on page 5 of the text.	$0.000\,025\,\text{meters per second} = 25\,\mu\text{m/s}$ $0.0445\,\text{seconds} = 44.5\,\text{ms}$ $0.000\,000\,032\,\text{watts} = 32\,\text{nW}$ $25{,}640\,\text{kilograms} = 25.64\,\text{Mg}$

Example #7. A very fast sprinter can run the 100-m dash in slightly under 10 s. This means his average speed is slightly greater than 10 m/s. Using conversion factors, convert 10 m/s to miles per hour.

1. Multiply the speed by conversion factors until you are left with miles per hour. There are 1.61 km in one mile.	$\left(\dfrac{10\ \cancel{m}}{\cancel{s}}\right)\left(\dfrac{1\ \cancel{km}}{10^3\ \cancel{m}}\right)\left(\dfrac{1\ mi}{1.61\ \cancel{km}}\right)\left(\dfrac{60\ \cancel{s}}{1\ \cancel{min}}\right)\left(\dfrac{60\ \cancel{min}}{1\ hr}\right) = 22.4\ mi/hr$

Example #8.   The speed of light in empty space is 3×10^8 m/s. Use conversion factors to determine the speed of light in feet per nanosecond.

1. Multiply the speed by conversion factors until you are left with miles per hour.	
	3×10^8 m/s $= 1.02$ ft/ns

Chapter **2**

Motion in One Dimension

I. Key Ideas

When describing motion in this chapter, we are restricting ourselves to studying the motion of a single particle. The particle concept is central to all of physics. An object can be treated as a particle if variations in its structure do not occur and it is not rotating. That is, an object is a particle if it is a nonrotating perfectly rigid object. Initially we will consider only motion of particles in one dimension.

Section 2-1. Displacement, Velocity, and Speed. If we make an *x*-axis along the line of motion, with its origin *O* at some reference point, we can specify a particle's position by a single variable *x*. The sign of *x* indicates to which side of the origin *O* the particle is located. We describe motion of the particle by specifying how its position changes with time. When a particle starts at position x_1 and moves to position x_2, the **displacement** Δx of the particle is defined as the net change in its position, that is

$$\Delta x = x_2 - x_1$$ Displacement

The **average velocity** v_{av} of the particle is this displacement divided by the time interval Δt in which the displacement occurs.

$$v_{av} = \frac{\Delta x}{\Delta t}$$ Average velocity

Like position and displacement, average velocity has a direction in space (in one dimension, a sign) as well as a magnitude. The velocity is positive when the particle moves in the positive *x* direction, and negative when it moves in the negative *x* direction.

The **average speed** of a particle is the total distance it has traveled divided by the total time taken to travel that distance.

$$\text{avg. speed} = \frac{\text{total distance}}{\text{total time}} = \frac{s}{t}$$ Average speed

The average speed is not equal to the average velocity. Speed has no direction associated with it. The average speed is also not equal to the magnitude of the average velocity because the distance traveled by a particle can be much different than its displacement.

Instantaneous Velocity. What we usually mean when we say "velocity" is instantaneous velocity. The **instantaneous velocity** is the displacement divided by the time interval, in the limit that the time interval—and thus the displacement—approaches zero. In mathematical language we would say, "The velocity of a particle is the time derivative of its position." On a position-versus-time graph (see Figure 2-4 on page 21 of the text), average velocity is the slope of a line connecting the points representing the initial and final positions of the particle. The instantaneous velocity at a point is the slope of the line tangent to the curve at that point (see Figure 2-5 on page 21 of the text).

$$v = \lim_{\Delta t \to 0} \frac{\Delta x}{\Delta t} = \frac{dx}{dt} \qquad\qquad \text{Instantaneous velocity}$$

The instantaneous speed is simply the magnitude of the instantaneous velocity.

Relative Velocity. Velocity is always measured relative to a **reference frame.** It is often convenient to match a reference frame to a rigid object. For example, a reference frame can be associated with either the ground or a railroad car. It can also be associated with the flowing water in a river under certain circumstances. Suppose that dye markers simultaneously released at three arbitrary points in a river are observed to drift downstream in such a manner that the triangle formed by lines connecting the markers maintains its size and shape. We can then think of the water as rigid. If the dye markers drift in this way, there are no eddy currents, so the structure and arrangement of the water does not change.

Position measurements are always made relative to coordinate systems that are attached to reference frames. For example, suppose you are on a railroad car and are walking toward its front end. Also suppose we measure your position using a coordinate axis that is attached to the car, with the origin O at the car's rear, and with the forward direction taken as positive, as in Figure 2-1. Your position x is then a distance d from the back of the car, and your velocity v is the time rate of change of x. Alternatively, suppose we measure your position using a second coordinate axis, identical with the first but with its origin O' at the front of the car. Your position x' using this coordinate axis is then negative d' from the front of the car, and your velocity v' is the time rate of change of x'.

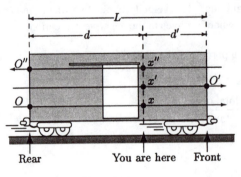

Figure 2-1

The distance between you and the back of the car plus the distance between you and the front of the car equals the length L of the car. That is, $d + d' = L$. Expressing this relation in terms of the positions, we have $x = d$ and $x' = -d'$, so $x - x' = L$. To find the velocities relative to each coordinate system we differentiate each side of this equation to get $v - v' = 0$ or

$$v' = v$$

where L is constant so $dL/dt = 0$. That is, the velocities measured by the two coordinate axes are equal. This is always the case; the velocity of a particle is the same as measured by any two coordinate axes, as long as they are both parallel and attached to the same reference frame.

Suppose we again measure your velocity in the train, this time using a third coordinate axis, one identical with the first except that the *backward* direction is taken as positive. Your position x'' is then negative d from the back of the train so $x = -x''$ and differentiating both sides gives $v = -v''$. In this case it appears that the velocities are not equal even though the two coordinate axes are attached to the train. Although the signs of the velocities are different, they both represent the same speed and direction—toward the *front* of the train—so it is correct to say that the velocity relative to the reference frame (the train) is the same in both cases. The velocities have different signs because they refer to coordinate axes with opposite directions taken as positive. You can avoid this potentially confusing situation by always choosing coordinate axes with the same direction taken as positive.

Often it is the case that the motion of an object is measured relative to two distinct reference frames. For example, your pal swims directly downstream in a river. Let the shore be one reference frame and the river water be the other. Her velocity relative to the shore v_{PS}, equals her velocity relative to the water v_{PW} plus the velocity of the water relative to the shore v_{WS}:

$$v_{PS} = v_{PW} + v_{WS} \qquad\qquad \text{Relative velocity}$$

Suppose we choose two coordinate axes, one fixed to the shore and the other fixed to the water, and that for each downstream is the positive direction. Then, if her velocity relative to the water is +1 m/s and the velocity of the water relative to the shore is +2 m/s, according to our equation, her velocity relative to the shore is +3 m/s.

Section 2-2. Acceleration. The mathematical relationship between acceleration and velocity is the same as the relationship between velocity and position. Thus the **average acceleration** is the change in the instantaneous velocity Δv divided by the time interval Δt in which the velocity changes. The **instantaneous acceleration** is the average acceleration in the limit that the time interval—and thus the change in velocity—approaches zero. That is, the acceleration is the time derivative of the velocity.

$$a_{av} = \frac{\Delta v}{\Delta t} \qquad\qquad \text{Average acceleration}$$

$$a = \lim_{\Delta t \to 0} \frac{\Delta v}{\Delta t} = \frac{dv}{dt} = \frac{d^2 x}{dt^2} \qquad\qquad \text{Instantaneous acceleration}$$

Section 2-3. Motion with Constant Acceleration. The simplest form of accelerated motion is motion in which the acceleration remains constant. When the acceleration is constant, the velocity increases linearly with time, and the average velocity equals the numerical average of the initial velocity v_1 and final velocity v_2; that, is $v_{av} = (1/2)(v_1 + v_2)$. When we combine these observations with the fact that the displacement is the average velocity times the time, we get the following equations.

$$v_{av} = \tfrac{1}{2}(v_1 + v_2)$$
$$v = v_0 + a(\Delta t)$$
$$\Delta x = x - x_0 = v_0(\Delta t) + \tfrac{1}{2} a(\Delta t)^2 \qquad \text{Constant acceleration equations}$$
$$v^2 = v_0^2 + 2a\,\Delta x$$

Usually we set the initial time to be at $t = 0$ seconds. In this case, $\Delta t = t_2 - t_1 = t$ in the above equations, simplifying the writing of the equations.

One of the most common instances of motion with constant acceleration is that of a freely falling object. Near the surface of the earth, all objects fall with a constant acceleration of $g = 9.81\,\text{m/s}^2 = 32.2\,\text{ft/s}^2$, in a direction toward the surface of the earth, as long as air resistance can be neglected.

Section 2-4. Integration. Integration is the inverse of differentiation. Therefore we can express the kinematic relations as

$$\frac{dx}{dt} = v \quad \Leftrightarrow \quad \Delta x = \int_{t_1}^{t_2} v\,dt$$
$$\frac{dv}{dt} = a \quad \Leftrightarrow \quad \Delta v = \int_{t_1}^{t_2} a\,dt \qquad \text{Integral form of the kinematic definitions}$$

It follows that the area under the velocity-versus-time curve is the change in position (the displacement) of the particle, and the area under the acceleration-versus-time curve is the change in velocity.

The average value of the velocity v is then given by

$$v_{av} = \frac{1}{t_2 - t_1} \int_{t_1}^{t_2} v(t)\,dt \qquad \text{Average value of the velocity}$$

However, average velocity is also defined as $v_{av} = \Delta x / \Delta t$ (Section 2-1). By substituting Δx for $\int v\,dt$ (see the first integral form in this section) the integral expression for average velocity reduces to

$$v_{av} = \frac{1}{t_2 - t_1} \Delta x = \frac{\Delta x}{\Delta t}$$

the definition of average velocity.

II.   Physical Quantities and Key Equations

Physical Quantities

Distance and displacement	meters m
Time	seconds s
Velocity and speed	m/s
Acceleration	m/s^2
Acceleration due to gravity	$g = 9.81\,\text{m/s}^2 = 32.2\,\text{ft/s}^2$

Key Equations

Displacement	$\Delta x = x_2 - x_1$
Average velocity	$v_{av} = \dfrac{\Delta x}{\Delta t}$
Average speed	$\text{avg. speed} = \dfrac{s}{t}$
Instantaneous velocity	$v = \lim\limits_{\Delta t \to 0} \dfrac{\Delta x}{\Delta t} = \dfrac{dx}{dt}$
Relative velocity	$v_{PS} = v_{PW} + v_{WS}$
Average acceleration	$a_{av} = \dfrac{\Delta v}{\Delta t}$
Instantaneous acceleration	$a = \lim\limits_{\Delta t \to 0} \dfrac{\Delta v}{\Delta t} = \dfrac{dv}{dt} = \dfrac{d^2x}{dt^2}$

Constant acceleration equations

$$v_{av} = \tfrac{1}{2}(v_1 + v_2) \qquad\qquad v = v_0 + a(\Delta t)$$
$$\Delta x = x - x_0 = v_0(\Delta t) + \tfrac{1}{2}a(\Delta t)^2 \qquad v^2 = v_0^2 + 2a\,\Delta x$$

Integral form of the kinematic definitions

$$\Delta x = \int_{t_1}^{t_2} v\,dt \qquad\qquad \Delta v = \int_{t_1}^{t_2} a\,dt$$

III. Potential Pitfalls

A lot of what is done in physics depends on representing various quantities by algebraic variables. We are using t for the time of an event, for example. Avoid confusing the value of a quantity with a change in, or increment of, that quantity. For instance, 10:30 a.m. on Tuesday is a specific point (instant) in time, but 10.5 hours is a time interval. Similarly, x is the position of a particle at some instant, but the change in position in some time interval, its displacement, is $\Delta x = x_2 - x_1$.

When doing kinematics problems, pay careful attention to signs. Signs indicate direction in space. If you are doing a free-fall problem, for instance, and you call upward the positive direction, then the acceleration is $-g$, because in free-fall the acceleration is downward. In a problem, once you choose the positive direction, all signs must be consistent with that choice. Also, $g = |a_g| = 9.81 \, \text{m/s}^2$, which means it can never be negative.

Speed is the magnitude of the velocity, but average speed is not necessarily the magnitude of the average velocity. Average speed is defined as the *total distance* traveled divided by the time interval, whereas average velocity is defined as the *net displacement* divided by the time interval.

Velocity and acceleration can be either positive or negative. In everyday usage it is common to say "deceleration" for the rate of decrease in speed. However, when a particle is slowing down, the acceleration can be either positive or negative, depending on the choice for the positive direction. *When a particle is slowing down, the acceleration and the velocity are always opposite in direction.* In one-dimensional motion, opposite in direction goes hand-in-hand with opposite in sign.

The equations of motion with constant acceleration are used throughout the text. We recommend that you spend a few minutes now and commit them to memory. Don't forget that they only apply to situations where the acceleration is constant. These equations are developed with reference to the initial position x_0 and initial velocity v_0. In working a problem, it is often convenient to choose the origin at the initial position so that x_0 equals zero.

IV. True or False Questions and Responses

True or False

_____ 1. An object can be treated as a particle if it is not rotating and its internal structure does not vary.

_____ 2. If the average speed of a particle over a certain 3-s interval is 1.5 m/s, the average velocity over the same 3-s interval must be either +1.5 m/s or −1.5 m/s.

_____ 3. The average speed of a particle cannot be negative.

_____ 4. If a car is driven for 1 hr at an average velocity of +60 km/hr, the distance it travels in that same 1-hr interval is 60 km.

_____ 5. An instant is not a time interval, but a point in time like 10:23 a.m.

_____ 6. A particle with a position that is given by $x(t) = At + B$ (A and B are properly dimensioned constants) is moving with a constant velocity.

_____ 7. A particle with a position that is given by $x = Ct^2$ (C is a properly dimensioned constant) has an acceleration given by $4C$.

_____ 8. If over a certain time interval the velocity of a given particle changes from zero to v_f, its average velocity over this interval is $v_f / 2$.

_____ 9. If at a certain instant the acceleration of a particle is positive, its velocity must also be positive.

_____ 10. During a certain time interval, the area under a graph of velocity of a particle versus time is the distance the particle has traveled during that interval.

_____ 11. Relative to a track John runs with a velocity of +8 mi/hr due north and Marie runs with a velocity of –8 mi/hr due south. It follows that John and Marie run with the same velocity relative to the track.

_____ 12. If the velocity of train A relative to train B is +80 km/hr due north then it follows that the velocity of train B relative to train A is –80 km/hr due north (or +80 km/hr due south).

Responses to True or False

1. True.

2. False. The average velocity depends only on the initial and final positions and not on the distance traveled.

3. True.   4. True.

5. True. In technical usage, the terms instant, instantaneously, moment, and momentarily refer to a point in time and not a finite time interval.

6. True. Differentiating will show that the velocity is A.

7. False. The acceleration is $2C$, as differentiating twice will verify.

8. False. The average velocity necessarily equals $v_f/2$ when the acceleration is constant. While it could, the average velocity does not necessarily equal $v_f/2$ when the acceleration varies.

9. Not necessarily. If a particle is moving in the negative direction and is slowing down, its velocity is negative while its acceleration is positive.

10. False. The area under a velocity-versus-time curve equals displacement, not distance.

11. True.   12. True.

V.   Questions and Answers

Questions

1. Figure 2-2 shows the position of a particle versus time. The following questions refer to the intervals between the times shown. (_a_) During which interval(s) does the velocity remain zero? (_b_) During which interval(s) does the velocity remain constant? (_c_) During which interval(s) is the average velocity zero? (_d_) During which interval(s) does the acceleration remain negative?

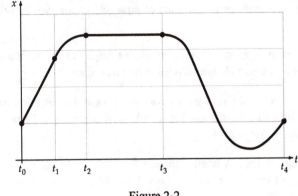

Figure 2-2

2. Starting from rest, a world-class sprinter can run 100 m in 10 s, but he cannot run 30 m in 3 s. Neither can he run 400 m in 40 s. Why not?

3. At some instant a car's velocity is +15 m/s; 1 s later it is +11 m/s. If the car's acceleration is constant, what is its average velocity in this one-second interval? How far does the car go during that interval? Can you answer these questions if you don't know whether or not the acceleration is constant?

4. An elevator is moving upward at a constant speed of v_e when a bolt falls off its undercarriage. Does the bolt immediately descend, or does it continue to ascend until its velocity becomes zero?

5. A bolt falls off its undercarriage when an elevator is (a) moving upward at constant speed v_0, (b) moving downward at constant speed v_0, (c) at rest. Assuming the elevator is at the same height in all three cases and disregarding the effects of air resistance, in which case(s) does the bolt reach the bottom of the shaft at the greatest speed?

6. Is the relation between velocity and position the same as the relation between acceleration and velocity?

Answers

1. (a) The velocity of the particle equals the slope of the tangent to the curve. For $t_2 < t < t_3$ the slope is zero, so for that interval $v = 0$. (b) The slope is constant for $t_0 < t < t_1$ and again for $t_2 < t < t_3$. (c) The average velocity for an interval equals the slope of the straight line that connects the point at opposite ends of the interval. The slope of the line connecting the points at t_0 and t_4 is zero as is that connecting the points at t_2 and t_3. It follows that the average velocity is zero for the interval $t_0 < t < t_4$ and also for the interval $t_2 < t < t_3$. (d) The acceleration is negative when the slope of the tangent to the curve decreases as t increases, which it does throughout the interval from $t_1 < t < t_2$.

2. It takes a sprinter several seconds to reach top speed. Thus in the first three seconds he is traveling slower than 10 m/s most of the time. He can't average 10 m/s over 40 s because a sprinter is unable to sustain top speed for that length of time.

3. If the acceleration is constant, $v_{av} = (1/2)(v_1 + v_2) = 13$ m/s and $\Delta x = v_{av}\Delta t = 13$ m. If you do not know whether or not the acceleration is constant, you cannot determine these quantities.

4. The bolt continues to ascend until its velocity reaches zero.

5. This can best be determined by examining the formula $v^2 = v_0^2 + 2a(x - x_0)$, where a and $x - x_0$ are the same and v_0 differs for the three cases. By examination you can see that the speed v is least for case (c), where v_0 equals zero, and is the same for cases (a) and (b), where the initial velocity squared is the same.

6. Yes. They are both 1st derivative relationships.

VI. Problems, Solutions, and Answers

Example #1. A car travels 40.0 km along a straight road at a speed of 86.0 km/hr and then goes 40.0 km farther at a speed of 50.0 km/hr. What is the car's average velocity for the entire trip?

Picture the Problem. This problem has two segments, each with a different speed. When calculating the total time or total displacement, we will need to add the time or displacement from each of the two segments.

1. The car's average velocity is change in displacement divided by change in time, so we need to calculate those two parameters.	$v_{av} = \dfrac{\Delta x}{\Delta t}$
2. Find Δx.	$\Delta x = 40.0\,\text{km} + 40.0\,\text{km} = 80.0\,\text{km}$
3. Find Δt. We need to calculate the time required to travel each leg of the trip.	$\Delta t_1 = \dfrac{\Delta x_1}{v_1} = \dfrac{40.0\,\text{km}}{86.0\,\text{km/hr}} = 0.465\,\text{hr}$ $\Delta t_2 = \dfrac{\Delta x_2}{v_2} = \dfrac{40.0\,\text{km}}{50.0\,\text{km/hr}} = 0.800\,\text{hr}$ $\Delta t = \Delta t_1 + \Delta t_2 = 0.465\,\text{hr} + 0.800\,\text{hr} = 1.265\,\text{hr}$
4. Use the calculated values of Δx and Δt in the expression from step 1.	$v_{av} = \dfrac{80.0\,\text{km}}{1.265\,\text{hr}} = 63.2\,\text{km/hr}$

Example #2—Interactive. You're bicycling across central Oklahoma on a perfectly straight road. You started at 8:30 a.m. and you've covered 21.0 miles by 11:15 a.m. when your chain breaks. You haven't any spare parts with you, so you have to walk the bike back to the last town at a speed of 2.60 mi/hr. If you get there at 1:30 p.m., what was (a) your displacement, (b) your average velocity, and (c) your average speed for the whole trip?

Picture the Problem. Displacement is the *net* change in position. This problem has two segments, one in which you traveled forward and one backward. **Try it yourself.** Work the problem on your own, in the spaces provided, to get the final answer.

1. Find the displacement during each of the two time intervals.	$\Delta x = 15.2\,\text{mi}$
2. Average velocity is total *displacement* divided by total time.	$v_{av} = 3.04\,\text{mi/hr}$
3. Average speed is total *distance* divided by total time.	avg. speed = 5.37 mi/hr

Example #3. Starting at a time we will call $t = 0$, a man walks east from his office to Wendy's, has lunch, and then walks west to his bank. His trip is shown on the graph of position versus time in Figure 2-3. (*a*) What was his average velocity from his office to the bank? (*b*) How much time did he spend at lunch? (*c*) How far is it from Wendy's to the bank? (*d*) At what point in the trip was he walking fastest?

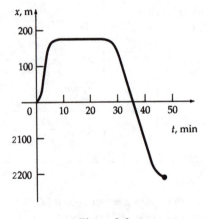

Figure 2-3

Picture the Problem.   The average velocity between any two points is the slope of a straight line drawn between those two points. If the instantaneous slope at any point is zero, then the man must be momentarily at rest. The man ended up at the bank.

1. The average velocity of the man from the office to the bank would be the slope of the line from his starting point (0,0) to the bank, which was at a displacement of –200 m, reached approximately 46 min after he started walking.	(a) $v_{av} = \dfrac{\Delta x}{\Delta t} = \dfrac{(-200\,m) - 0\,m}{46\,min - 0\,min}$ $v_{av} = \dfrac{-200\,m}{46\,min} \times \dfrac{1\,min}{60\,s} = -0.072\,m/s$
2. While the man ate lunch, he had no displacement. This occurred from $t = 10\,min$ to $t = 30\,min$.	(b) 20 min
3. Wendy's appears to be at about $x = 170$ m, and the bank is at $x = -200$ m.	(c) 370 m
4. Greatest slope equals greatest speed.	(d) This appears to be on his way to Wendy's.

Example #4—Interactive. A woman drives west from Nashville toward Memphis on I-40. (Assume that the highway is straight.) At Bucksnort, TN, she develops car trouble and has to be towed back to Nashville for repairs before she can continue her journey. She finally arrives at Memphis at 1700 hours, military time. The entire trip is described on the position-versus-time graph in Figure 2-4. (a) What was her average velocity before her car broke down? (b) What was it for the whole trip? (c) Assume that, without the breakdown, she would have continued at her initial (average) velocity. How much time did the breakdown cost her? (d) During what time interval was her velocity the greatest? **Try it yourself.** Work the problem on your own, in the spaces provided, to get the final answer.

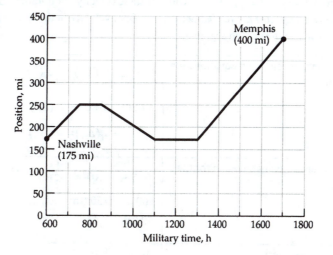

Figure 2-4

1. Average velocity before the car broke down is the slope of the line describing the motion during that time.	(a) $v_{av} = 50\,mi/hr$

2. For the whole trip, we take the slope of the line from the starting point to the ending point.	(b) $v_{av} = 20.5 \, \text{mi/hr}$
3. Assuming initial, constant velocity, find the total length of time of the trip.	
4. Her actual time was 11 hr, so the breakdown cost her the difference between this time and the time calculated in step 3.	(c) $t_{breakdown} = 6.5 \, \text{hr}$
5. Greatest velocity is the greatest slope.	(d) From $t = 1300$ to $t = 1700$ hours

Example #5. Maurice Greene holds the world record for the 100-m sprint with a time of 9.79 s. Assume that he starts from rest and runs with uniform acceleration for the first 50.0 m and thereafter runs at his top (constant) speed. (a) What is his acceleration in the first part of the run? (b) Assuming his acceleration and top speed are the same as in part (a), how long would it take him to run 200.0 m?

Picture the Problem. This problem has two segments, one with constant acceleration (of duration Δt_1, from $t = 0$ to $t = t_1$) and one with constant velocity (of duration Δt_2, from $t = t_1$ to $t = 9.79 \, \text{s}$). The two segments are related because the final velocity in the first segment is equal to the (constant) velocity in the second segment.

1. Write the constant acceleration equations for the first segment.	$x_1 = x_0 + v_0(\Delta t_1) + \dfrac{1}{2} a_1 (\Delta t_1)^2$ $50.0 \, \text{m} = 0 + 0 + \dfrac{1}{2} a_1 (\Delta t_1)^2$ $v_1 = v_0 + a_1(\Delta t_1) = a_1(\Delta t_1)$
2. Write the constant velocity equation for the second segment.	$x_2 = x_1 + v_1(\Delta t_2) + \dfrac{1}{2} a_2 (\Delta t_2)^2$ $100.0 \, \text{m} = 50.0 \, \text{m} + v_1(\Delta t_2) + 0$ $50.0 \, \text{m} = v_1(\Delta t_2)$
3. Relate the two time intervals.	$\Delta t_2 = 9.79 \, \text{s} - \Delta t_1$
4. Substitute the result from step 3 into the results from step 2.	$50.0 \, \text{m} = v_1(9.79 \, \text{s} - \Delta t_1)$

5. Substitute v_1 from step 1 into step 4. You can then solve the remaining two equations, the one you just created, and the second equation in step 1, for the two unknowns, the acceleration during the first segment, and the time duration of the first segment.	$50.0\,\text{m} = a_1(\Delta t_1)(9.79\,\text{s} - \Delta t_1)$ $50.0\,\text{m} = \frac{1}{2}a_1(\Delta t_1)^2 \Rightarrow \Delta t_1 = \sqrt{\frac{100.0\,\text{m}}{a_1}}$ $50.0\,\text{m} = (9.79\,\text{s})a_1\sqrt{\frac{100.0\,\text{m}}{a_1}} - a_1\frac{100.0\,\text{m}}{a_1}$ $150.0\,\text{m} = (97.9\sqrt{\text{m}\cdot\text{s}})\sqrt{a_1}$ $a_1 = 2.34\,\text{m/s}^2$ $\Delta t_1 = 6.53\,\text{s}$ $v_1 = a_1(\Delta t_1) = 15.3\,\text{m/s}$
6. To run 200 m, the sprinter will still run the first 50 m in 6.67 s. The remaining 150 m will be run at the sprinter's top speed, v_1.	$\Delta t_1 = 6.53\,\text{s}$ $150.0\,\text{m} = v_1\Delta t_2$ $\Delta t_2 = \frac{150.0\,\text{m}}{v_1} = \frac{150.0\,\text{m}}{15.3\,\text{m/s}} = 9.82\,\text{s}$ $\Delta t = \Delta t_1 + \Delta t_2 = 16.4\,\text{s}$

Discussion. The men and women's world records for the 200-m sprint are currently held by Michael Johnson, at 19.32 s, and Florence Griffith-Joyner, at 21.34 s. What is wrong with our model? Visit http://www.iaaf.org/ for more information about these sprinting events, including better models for the motion of a sprinter.

Example #6—Interactive. A model rocket starting from rest accelerates straight upward with an acceleration three times that of gravity until it is 300.0 m above the ground. At that point its engine quits. How long after it was fired does the rocket hit the ground? Neglect air resistance.

Picture the Problem. This is a two-segment problem. The first segment has a constant acceleration equal to *3g upward*, and the second segment will have an acceleration of *g* downward. The final velocity for the first segment will be equal to the initial velocity of the second segment. The position of the rocket at the beginning and end of each segment should be labeled. The total time the rocket is in the air will be the sum of the time required for each of the two segments. **Try it yourself.** Work the problem on your own, in the spaces provided, to get the final answer.

1. Write constant acceleration equations for the first segment.	
2. Write constant acceleration equations for the second segment.	
3. Solve the equations in step 1 for v_1 and Δt_1.	

4. Solve the equation from step 2 for Δt_2.	
5. The total time is the sum of the two time intervals.	$\Delta t = 31.7\,\text{s}$

Example #7. A man standing on a cliff 50.0 m above the level of the ground below throws a stone straight up in the air. The stone falls back past the edge of the cliff and strikes the ground 5.00 s after it was thrown. With what initial velocity did the man throw the stone?

Picture the Problem. Because we know the total time the stone was in the air, the net displacement of the stone, and the stone's acceleration (it is equal to g in the downward direction because it is in free-fall after it is thrown), we can solve directly for the initial velocity of the stone. Assuming the origin is at the man's hand, and "up" as positive, the total displacement of the stone will be negative, and so will its acceleration.

1. Write the constant acceleration equation for the ball.	$\Delta x = v_0(\Delta t) + \dfrac{1}{2}a(\Delta t)^2$
2. Solve this equation for v_0.	$v_0 = \dfrac{\Delta x - (1/2)a(\Delta t)^2}{\Delta t}$
3. Plug in the numbers to answer the question.	$v_0 = \dfrac{-50.0\,\text{m} - (1/2)(-9.81\,\text{m/s}^2)(5.00\,\text{s})^2}{5.00\,\text{s}} = 14.5\,\text{m/s}$

Example #8. The speed of sound in air is about 340.0 m/s. If you drop a stone in a well and hear the splash 3.80 s later, how deep is the well?

Picture the Problem. This is a two-segment problem. In the first segment the stone is in free-fall until it hits the water. In the second segment, the sound of the splash travels the same distance back up the well at the speed of sound. The time required for each of these two segments must add up to 3.80 s.

1. Write an equation of motion for the stone.	$x = x_0 + v_0\,\Delta t + \frac{1}{2}a(\Delta t)^2$ $x = \frac{1}{2}(9.81\,\text{m/s}^2)\,\Delta t_{\text{stone}}^2$
2. Write an equation of motion for the sound.	$x = v_{\text{sound}}\,\Delta t_{\text{sound}} = (340.0\,\text{m/s})\Delta t_{\text{sound}}$

3. Relate the two times.	$3.80\,\text{s} = \Delta t_{\text{stone}} + \Delta t_{\text{sound}}$ $\Delta t_{\text{sound}} = 3.80\,\text{s} - \Delta t_{\text{stone}}$
4. Set the distance from step 1 equal to that from step 2, and make the substitution for t_{sound} from step 3.	$\frac{1}{2}(9.81\,\text{m/s}^2)\,\Delta t_{\text{stone}}^2 = (340.0\,\text{m/s})\Delta t_{\text{sound}}$ $\frac{1}{2}(9.81\,\text{m/s}^2)\,\Delta t_{\text{stone}}^2 = (340.0\,\text{m/s})(3.80\,\text{s} - \Delta t_{\text{stone}})$
5. Solve the quadratic equation for the time it takes the stone to fall.	$\Delta t_{\text{stone}} = 3.61\,\text{s}$
6. Now that we know the time, plug it back in to step 1 to find the depth of the well.	$x = \frac{1}{2}(9.81\,\text{m/s}^2)(3.61\,\text{s})^2 = 63.9\,\text{m}$

Example #9. A ball thrown straight up is caught by the thrower. Use the constant-acceleration equations to show that (neglecting air resistance) for a ball thrown straight up, the rise time equals the fall time.

Picture the Problem. We know the velocity of the ball at the top of its flight is zero. Therefore, we can use the velocity equation to solve for the rise time. We can use the position equation to solve for the total time of flight and compare them.

1. Use the constant acceleration equation for the velocity to find the rise time of the ball. If we allow $t_1 = 0$, then $\Delta t = t_{\text{rise}}$.	$v = v_0 + a\,\Delta t$ $0 = v_0 - g t_{\text{rise}}$ $t_{\text{rise}} = v_0 / g$
2. Use the constant-acceleration position equation to solve for the total time, again setting $t_1 = 0$. Δy is zero because the ball starts and ends at the same position.	$\Delta y = v_0 t + \frac{1}{2}at^2$ $0 = v_0 t_{\text{total}} - \frac{1}{2}g t_{\text{total}}^2$ $t = 0\,\text{s},\quad 2v_0/g$

Discussion. Since the total time is twice the rise time, the fall time must be equal to the rise time. This is a universal feature of all thrown objects that start and end at the same height.

Example #10—Interactive. For the same thrown ball in the previous example, use the constant-acceleration equations to show that (neglecting friction) the speed of the ball the instant prior to its being caught is the same as the speed of the ball the instant it leaves the thrower's hand. The ball is caught at the same height that it is thrown.

Picture the Problem. We are not told anything about the time in this constant-acceleration, free-fall problem, so we would like to avoid it if possible. **Try it yourself.** Work the problem on your own, in the spaces provided, to get the final answer.

1. Write the time-independent constant acceleration equation, and see what it tells us.	$v^2 = v_0^2$

Discussion. The fact that the initial and final velocities are the same is a universal feature of all thrown objects that start and end at the same height. It is a result of the symmetrical nature of the problem.

Example #11. The velocity of a particle is given by the equation $v = 3t^2 + 4$, where v is in meters per second and t is in seconds. Determine the displacement during the time interval from $t = 2.00\,\text{s}$ to $t = 5.00\,\text{s}$.

Picture the Problem. The acceleration is not constant, so we cannot use the constant acceleration equations. Instead, we have to use the general integral form for displacement.

1. Write out the integral form for displacement.	$\Delta x = \int_{t_1}^{t_2} v\, dt$
2. Use the given initial and final times and the expression for the velocity to solve for the total displacement.	$\Delta x = \int_{2s}^{5s} (3t^2 + 4)\, dt$ $\Delta x = t^3 + 4t \big|_{t=2}^{t=5} = 125 + 20 - (8 + 8) = 129\,\text{m}$

Example #12—Interactive. The acceleration of a particle is given by the equation $a = 4t - 5$, where a is in meters per second squared and t is in seconds. The velocity at $t = 0$ is $-6.00\,\text{m/s}$. What is the velocity at $t = 4.00\,\text{s}$?

Picture the Problem. The acceleration is not constant, so we cannot use the constant acceleration equations. Instead, we have to use the general integral form for displacement. **Try it yourself.** Work the problem on your own, in the spaces provided, to get the final answer.

1. Write out the integral form for velocity.	
2. Plug in values and solve. Don't forget to include the initial velocity.	$v(4\,\text{s}) = 6.00\,\text{m/s}$

Example #13. You are driving your new car at 18.0 m/s (40 mph) along a straight, level, country road. Suddenly you notice a very large corn picker in front of you, moving in the same direction, at 8.00 m/s. It takes up the entire road, so there is no way for you to pass it. At the instant you hit the brakes, the corn picker is 28.0 m ahead of your car. Assuming constant acceleration, what must your acceleration be to avoid a collision?

Picture the Problem. This is a classic "chase" problem. In these kinds of problems the typical question asked is, "When are the objects at the same place?" This often occurs with additional constraints. To find the acceleration required to avoid a collision, we want to find the parameters for which a collision just barely occurs. This provides us with the threshold acceleration for avoiding the collision. That is, we want both the car's and picker's velocities to be equal at some time, t, and we want the position of both the front of the car and the rear of the picker to be equal at that same time.

1. Draw a sketch of the problem at two different times. Pick the forward direction to be positive. This will make your acceleration negative.	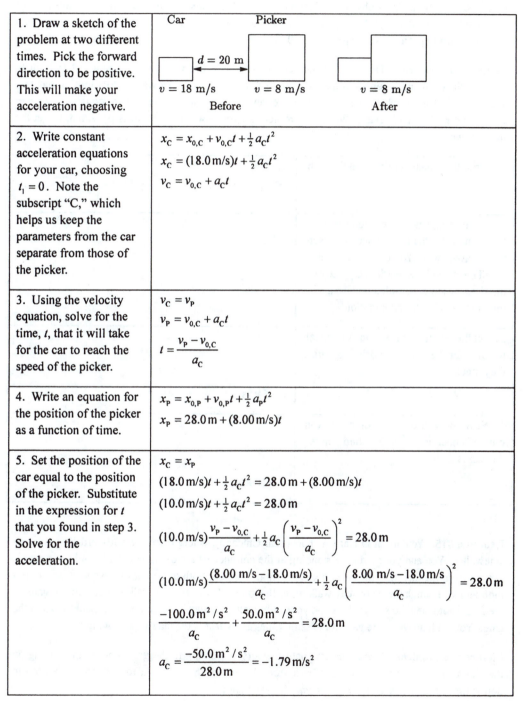
2. Write constant acceleration equations for your car, choosing $t_i = 0$. Note the subscript "C," which helps us keep the parameters from the car separate from those of the picker.	$x_C = x_{0,C} + v_{0,C}t + \frac{1}{2}a_C t^2$ $x_C = (18.0\,\text{m/s})t + \frac{1}{2}a_C t^2$ $v_C = v_{0,C} + a_C t$
3. Using the velocity equation, solve for the time, t, that it will take for the car to reach the speed of the picker.	$v_C = v_P$ $v_P = v_{0,C} + a_C t$ $t = \dfrac{v_P - v_{0,C}}{a_C}$
4. Write an equation for the position of the picker as a function of time.	$x_P = x_{0,P} + v_{0,P}t + \frac{1}{2}a_P t^2$ $x_P = 28.0\,\text{m} + (8.00\,\text{m/s})t$
5. Set the position of the car equal to the position of the picker. Substitute in the expression for t that you found in step 3. Solve for the acceleration.	$x_C = x_P$ $(18.0\,\text{m/s})t + \frac{1}{2}a_C t^2 = 28.0\,\text{m} + (8.00\,\text{m/s})t$ $(10.0\,\text{m/s})t + \frac{1}{2}a_C t^2 = 28.0\,\text{m}$ $(10.0\,\text{m/s})\dfrac{v_P - v_{0,C}}{a_C} + \frac{1}{2}a_C\left(\dfrac{v_P - v_{0,C}}{a_C}\right)^2 = 28.0\,\text{m}$ $(10.0\,\text{m/s})\dfrac{(8.00\,\text{m/s} - 18.0\,\text{m/s})}{a_C} + \frac{1}{2}a_C\left(\dfrac{8.00\,\text{m/s} - 18.0\,\text{m/s}}{a_C}\right)^2 = 28.0\,\text{m}$ $\dfrac{-100.0\,\text{m}^2/\text{s}^2}{a_C} + \dfrac{50.0\,\text{m}^2/\text{s}^2}{a_C} = 28.0\,\text{m}$ $a_C = \dfrac{-50.0\,\text{m}^2/\text{s}^2}{28.0\,\text{m}} = -1.79\,\text{m/s}^2$

Discussion. We solved for the acceleration that would just barely cause a collision. For any negative acceleration greater than 1.79 m/s^2, a collision will be avoided.

Example #14—Interactive. Al sees Bob drive past his house at 20.0 m/s and realizes he must catch Bob to get the day's physics assignment. Al jumps in his car and starts it—this takes 15.0 s —and takes off after Bob. Suppose Al accelerates at a *uniform* rate of 1.40 m/s^2, and Bob continues driving at a constant speed of 20.0 m/s. *Where* and *when* does Al catch up to Bob?

Picture the Problem. This is a classic "chase" problem, but with two segments. We want to know both when and where Bob and Al have the same position. However, Al's motion has two segments: One in which he doesn't move and one in which he has a constant nonzero acceleration. **Try it yourself.** Work the problem on your own, in the spaces provided, to get the final answer.

1. Write the equation of motion for Bob.	
2. Write an equation of motion for Al. Remember, for the first 15 seconds, Bob moves away while Al doesn't move at all. This provides a relationship between the time variable for Bob's motion and the time variable for Al's motion.	
3. Set the positions of Al and Bob equal to solve for the time, Bob drives before they meet.	$t = 54.5\,\text{s}$
4. Now that you know the time, you can plug this back in to step 2 to find where Al passes Bob.	$x = 1090\,\text{m}$

Example #15. You are at O'Hare Airport (near Chicago), where there are 400-yard-long moving sidewalks. You and your twin are standing at the entrance of a moving sidewalk, and your gate is 300 yards in the direction the sidewalk is moving. Suppose both of you have a walking speed of 3 mph and you simultaneously start walking to the gate. However, you walk directly to the gate on the floor, not using the sidewalk, while your twin walks on the sidewalk and then backtracks to the gate. You both arrive at the gate at the same time. How fast is the sidewalk moving?

Picture the Problem. There are three paths: yourself walking to the gate, your twin walking on the sidewalk, and your twin walking back from the end of the sidewalk to the gate. We need to define times and speeds *relative to the gate* for each path.

1. Write an equation for the distance you travel. We will let L be the total length of the sidewalk.	$s_{you} = v(\Delta t)$ $0.75L = v(\Delta t)$
2. Find the speed of your twin, relative to the gate, while she is walking on the sidewalk.	$v_{tg} = v_{ts} + v_{sg} = v + v_{sg}$
3. Now we can write an equation of motion for your twin on the sidewalk.	$s_{twin} = v(\Delta t)$ $L = (v + v_{sg})(\Delta t_1)$
4. Your twin also has to backtrack. Write an equation of motion for that stage.	$0.25L = v(\Delta t_2)$
5. Both you and your twin arrive at the gate at the same time.	$\Delta t = \Delta t_1 + \Delta t_2$
6. Rewrite steps 1, 3, and 4, solving for the time intervals.	$\Delta t = \dfrac{0.75L}{v} \quad ; \quad \Delta t_1 = \dfrac{L}{v + v_{sg}} \quad ; \quad \Delta t_2 = \dfrac{0.25L}{v}$
7. Substitute these expressions into step 5. Rearrange and divide through by L.	$\dfrac{0.75L}{v} = \dfrac{L}{v + v_{sg}} + \dfrac{0.25L}{v}$ $\dfrac{1}{v + v_{sg}} = \dfrac{0.5}{v}$
8. Solving for the speed of the sidewalk relative to the gate, we get:	$v + v_{sg} = \dfrac{v}{0.5} = 2v$ $v_{sg} = v = 3\,\text{mph}$

Discussion: Another way to solve this problem is to realize that at the instant your twin exits the moving sidewalk both you and your twin must be the same distance from the gate. At this instant, relative to the gate you will have moved a distance of $0.5L$, while your twin has traveled a distance L. It follows that while on the sidewalk your twin's speed, relative to the gate, is twice the walking speed. Thus the sidewalk's speed is equal to the walking speed.

To save you from doing extra work, it is a good idea to simplify solutions with variables as much as possible before inserting actual numbers. In this case, you might have initially thought that solving the problem would require converting yards to miles, or vice versa. However, by working through the solution with variables, you find that only the ratios of distances are needed; in the final expression for sidewalk speed, none of the actual distances are required and the unit conversion is not necessary.

Example #16—Interactive. You can swim 1.00 mph in still water. Suppose you jump off a pier and swim against a 0.50 mph current to a second pier 0.25 miles upstream. Upon reaching the second pier you immediately turn around and swim downstream to the first pier. (*a*) How long does the round trip take? (*b*) What total distance do you swim relative to the shore? (*c*) relative to the water?

Picture the Problem. This problem has two segments, one in which you swim upstream, with one velocity relative to the shore, and a second segment in which you swim downstream with a different velocity relative to the shore. **Try it yourself.** Work the problem on your own, in the spaces provided, to get the final answer.

1. Find your velocity relative to the shore while swimming both upstream and downstream.	
2. Find the time it takes to swim 0.25 miles upstream and downstream.	
3. The total time for the trip is the sum of the time it took for each section.	$t = 0.67\,\text{hr} = 40\,\text{min}$
4. Relative to the shore, you swam 0.25 mi upstream and 0.25 miles downstream.	$s_{\text{shore}} = 0.5\,\text{mi}$
5. To find the distance swam relative to the water, remember your swim speed relative to the water is 1.00 mph, and you swam for 0.67 hr.	$s_{\text{water}} = 0.67\,\text{mi}$

Chapter 3

Motion in Two and Three Dimensions

I. Key Ideas

The motion of a particle can be described in terms of its position, velocity, and acceleration, all of which are vectors. Often the motion of an object can be represented by the motion of a single point, and then the object can be treated as a particle. **Vectors** are quantities that have both magnitude and direction. For example, consider a train with a velocity of 80 mi/hr toward the northwest. The speed of the train—how fast it's going—is the magnitude of the velocity vector, while northwest—the direction it's going—is the direction of the velocity vector. The tug of a rope pulling a sled is an example of another vector quantity—force.

Scalars are quantities that have magnitude but no direction, like the temperature of a cup of coffee.

In printed materials vectors are frequently represented by boldface symbols, for example, **A** and **B**. Both here and in the text they will be represented by placing an arrow over a boldfaced, italicized symbol; for example, $\vec{A}$ and $\vec{B}$. The magnitude of $\vec{A}$ is written $|\vec{A}|$, $\|\vec{A}\|$, or simply A. The magnitude of a vector is never negative.

Section 3-1. The Displacement Vector. The **displacement vector** of a moving particle is a quantity that is graphically represented by an arrow with a tail that starts at the initial position of the particle and a head that ends at its final position. The length and direction of this arrow are the magnitude and direction of the displacement vector. The magnitude of the displacement vector has dimensions of length. As illustrated in Figure 3-1, the displacement vector depends only on the initial and final positions of the particle, not on the path taken by the particle as it moves from the initial to the final position.

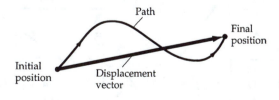

Figure 3-1

If a particle moves from point 1 to point 2, and then from point 2 to point 3, the net displacement of the object is the same as if the particle moved directly from point 1 to point 3. This concept of vector addition is shown in Figure 3-2a. The displacement $\vec{A}$ (from point 1 to point 2) followed by displacement $\vec{B}$ (from point 2 to point 3) is equivalent to displacement $\vec{C}$ (from point 1 to point 3). In the mathematics of vectors, this combining of $\vec{A}$ and $\vec{B}$, which results in $\vec{C}$, is called **vector addition**. Figure 3-2b illustrates the vector addition of $\vec{A}$ and $\vec{B}$ to get $\vec{C}$; that is, $\vec{A} + \vec{B} = \vec{C}$. The prescription to follow in order to add $\vec{B}$ to $\vec{A}$ to get $\vec{C}$ is to draw the arrow representing $\vec{A}$, then draw the arrow representing $\vec{B}$ with its tail at the head of the arrow representing $\vec{A}$. Then draw the arrow representing $\vec{C}$ from the tail of $\vec{A}$ to the head of $\vec{B}$. The arrows, of course, must be drawn to scale, and the direction of the arrows is the same as the direction of the actual displacements.

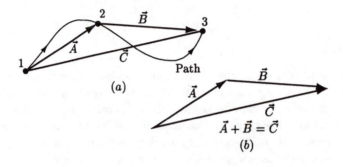

Figure 3-2

Section 3-2. General Properties of Vectors. All vectors satisfy the same rules of arithmetic. If you understand the rules satisfied by displacements, you will understand the rules for all vectors. Besides the vector addition already discussed, there are several other mathematical properties of vectors.

Vector Subtraction. We are all familiar with the subtraction of ordinary numbers. In the equation $a - b = d$, d is the number that must be added to b to get a; that is, $b + d = a$. Analogously, in the vector equation $\vec{A} - \vec{B} = \vec{D}$, $\vec{D}$ is the vector that must be added to $\vec{B}$ to get $\vec{A}$; that is $\vec{B} + \vec{D} = \vec{A}$, as shown in Figure 3-3(a).

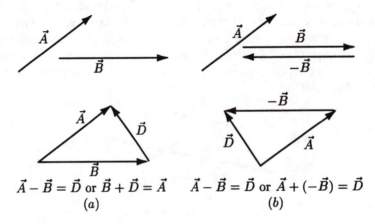

$\vec{A} - \vec{B} = \vec{D}$ or $\vec{B} + \vec{D} = \vec{A}$ $\vec{A} - \vec{B} = \vec{D}$ or $\vec{A} + (-\vec{B}) = \vec{D}$
(a)   (b)

Figure 3-3

The vector $-\vec{B}$ is equal in magnitude and opposite in direction to the vector $\vec{B}$. Figure 3-3(b) illustrates that subtracting $\vec{B}$ from $\vec{A}$ is equivalent to adding $-\vec{B}$ to $\vec{A}$. Note that in both Figure 3-3(a) and Figure 3-3(b) the vectors $\vec{A}$, $\vec{B}$, and $\vec{D}$ have the same lengths and directions. That is, the exact same vector $\vec{A}$ is used in both parts of the figure, even though its position has changed.

Addition of Vectors by Components. When a particle undergoes a displacement $\vec{A}$, the component of the displacement A_x is number of units the particle travels in the $+x$ direction, as illustrated in Figure 3-4. If the particle moves in the $-x$ direction, $A_x < 0$. Similarly, the component of the displacement A_y is number of units the particle travels in the $+y$ direction. The angle θ is usually measured with respect to the x axis.

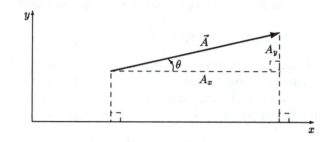

Figure 3-4

$$A_x = A\cos\theta \qquad A = \sqrt{A_x^2 + A_y^2}$$
$$A_y = A\sin\theta \qquad \tan\theta = A_y / A_x$$

Component-vector relations

Components of vectors are not themselves vectors, but are ordinary signed (real) numbers that can be added, subtracted, multiplied, and divided like ordinary numbers. When adding vectors, the x component of the sum equals the sum of the x components of the individual vectors. The corresponding relations are valid for the y and z components. Thus if $\vec{C} = \vec{A} + \vec{B}$, then $C_x = A_x + B_x$, $C_y = A_y + B_y$, and $C_z + A_z + B_z$.

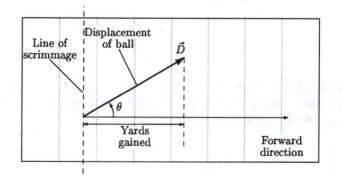

Figure 3-5

For an application using vector components we look to the American game of football. In this game, the yards gained during each play equal the forward component of the displacement $\vec{D}$ of the football during the play (see Figure 3-5); that is, yards gained $= D\cos\theta..$ The forward component of the football's displacement is negative if the play ends with the ball behind the line of scrimmage (an imaginary line across the field through the position of the ball at the start of each play). In a series of plays, the total yards gained is the sum of the yards gained—that is, the sum of the forward components of the ball's displacements—for each play.

Unit vectors. Unit vectors are vectors used to indicate directions; they are dimensionless with a magnitude of unity. A unit vector can be formed by dividing a vector by its magnitude. The unit vectors $\hat{i}$, $\hat{j}$, and $\hat{k}$ are commonly used to indicate the x, y, and z directions, respectively. The vector $A_x\hat{i}$ is the product of the component A_x and the unit vector $\hat{i}$. It is a vector that is parallel to the x axis (or antiparallel if A_x is negative) and has a magnitude $|A_x|$. A general vector $\vec{A}$ can thus be written as the sum of three vectors, each of which is parallel to a coordinate axis:

$$\vec{A} = A_x\hat{i} + A_y\hat{j} + A_z\hat{k}$$

It seems natural that, if a displacement of 3 m to the east is denoted $\vec{A}$, a sequence of two such displacements would be denoted $2\vec{A}$; that is, $\vec{A} + \vec{A} = 2\vec{A}$. The generalization of this process is that the product of a number s (for scalar) and a vector $\vec{A}$ is a vector whose magnitude is $|sA|$ and whose direction is the same as that of $\vec{A}$ if s is positive, and opposite to it if s is negative.

Section 3-3. Position, Velocity, and Acceleration. **Position vectors** are vectors from the origin of a coordinate system to the location of a particle. As shown in Figure 3-6, the position $\vec{r}$ of a particle has components x, y, and possibly z, which are the coordinates of the location of the particle. The $\vec{r}$ vector can be written as

$$\vec{r} = x\hat{i} + y\hat{j} + z\hat{k}$$

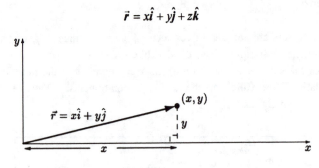

Figure 3-6

The **instantaneous-velocity vector** is the time rate of change of the position vector. Its direction is the direction of motion of the particle. Speed is the magnitude of the velocity. The equations relating the vector velocity, vector position, and time are:

$$\vec{v}_{av} = \frac{\Delta\vec{r}}{\Delta t} \qquad\qquad \vec{v} = \lim_{\Delta t \to 0}\frac{\Delta\vec{r}}{\Delta t} = \frac{d\vec{r}}{dt}$$

or

$$v_{x,av} = \frac{\Delta x}{\Delta t} \qquad\qquad v_x = \lim_{\Delta t \to 0} \frac{\Delta x}{\Delta t} = \frac{dx}{dt}$$

$$v_{y,av} = \frac{\Delta y}{\Delta t} \qquad\qquad v_y = \lim_{\Delta t \to 0} \frac{\Delta y}{\Delta t} = \frac{dy}{dt} \qquad\qquad \text{Velocity}$$

$$v_{z,av} = \frac{\Delta z}{\Delta t} \qquad\qquad v_z = \lim_{\Delta t \to 0} \frac{\Delta z}{\Delta t} = \frac{dz}{dt}$$

Relative Velocity. Consider a person standing on a railroad car that is moving relative to the ground. Consider the position $\vec{r}_{pc}$ of the person relative to the moving car, the position $\vec{r}_{pg}$ of the person relative to the ground, and the position $\vec{r}_{cg}$ of the car relative to the ground. These relative position vectors are illustrated in Figure 3-7, where g, p, and c represent points fixed to the ground, person, and car, respectively.

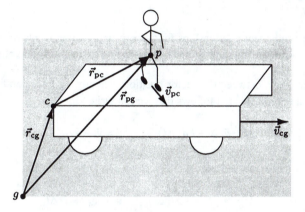

Figure 3-7

The relation between these position vectors is $\vec{r}_{pg} = \vec{r}_{pc} + \vec{r}_{cg}$. Because points g, p, and c are moving, a short time later they will be at slightly different locations. The changes in their position are related by the equation $\Delta\vec{r}_{pg} = \Delta\vec{r}_{pc} + \Delta\vec{r}_{cg}$.

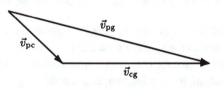

Figure 3-8

By dividing each term of this last equation by the time interval Δt and identifying the terms (in the limit that the time interval Δt approaches zero) as relative velocity vectors (see Figure 3-8), we get

$$\vec{v}_{pg} = \vec{v}_{pc} + \vec{v}_{cg} \qquad\qquad \text{Relative velocity equation}$$

Relative velocity problems can often be solved by using only this equation. However, it is often helpful to recognize that the velocity of the person relative to the ground and the velocity of the ground relative to the person are equal in magnitude and opposite in direction: $\vec{v}_{pg} = -\vec{v}_{gp}$.

The Acceleration Vector. The **instantaneous-acceleration vector** is the time rate of change of the velocity vector. Only when the magnitude and the direction of the velocity are both fixed is the acceleration zero. The equations relating the vector acceleration, vector velocity, and time are

$$\vec{a}_{av} = \frac{\Delta\vec{v}}{\Delta t} \qquad\qquad \vec{a} = \lim_{\Delta t\to 0}\frac{\Delta\vec{v}}{\Delta t} = \frac{d\vec{v}}{dt}$$

or

$$a_{x,av} = \frac{\Delta v_x}{\Delta t} \qquad\qquad a_x = \lim_{\Delta t\to 0}\frac{\Delta v_x}{\Delta t} = \frac{dv_x}{dt}$$

Acceleration

$$a_{y,av} = \frac{\Delta v_y}{\Delta t} \qquad\qquad a_y = \lim_{\Delta t\to 0}\frac{\Delta v_y}{\Delta t} = \frac{dv_y}{dt}$$

$$a_{z,av} = \frac{\Delta v_z}{\Delta t} \qquad\qquad a_z = \lim_{\Delta t\to 0}\frac{\Delta v_z}{\Delta t} = \frac{dv_z}{dt}$$

One can also write component relationships for the x, y, and z components of the acceleration and velocity, similar to those relating the x, y, and z components of position and velocity.

Since the relationships between the vector position, velocity, and acceleration are the same as the one-dimensional relationships, it follows that we can write

$$\Delta\vec{r} = \vec{v}_0\,\Delta t + \tfrac{1}{2}\vec{a}\left(\Delta t\right)^2 \qquad \vec{v} = \vec{v}_0 + \vec{a}\,\Delta t$$

or

$$\Delta x = v_{0x}\,\Delta t + \tfrac{1}{2}a_x\left(\Delta t\right)^2 \qquad v_x = v_{0x} + a_x\,\Delta t$$

$$\Delta y = v_{0y}\,\Delta t + \tfrac{1}{2}a_y\left(\Delta t\right)^2 \qquad v_y = v_{0y} + a_y\,\Delta t$$

for the case of constant acceleration. In the above equations,

$$v_{0x} = v_0\cos\theta \qquad\qquad v_{0y} = v_0\sin\theta$$

and θ is the angle the initial velocity vector makes with the positive x axis.

Section 3-4. Special Case 1: Projectile Motion. Projectiles are objects moving under the influence of only gravity, such as the massive sphere that is thrown by a shotputter (neglecting air resistance). Near the earth's surface the horizontal component of a projectile's acceleration is zero and the vertical component of its acceleration is a constant $g = 9.81\,\text{m/s}^2 \pm 0.03\,\text{m/s}^2$ in the downward direction. On the earth's surface g varies slightly with location, but is usually specified as the constant value of 9.81 m/s^2. Assuming that the positive y direction is upward, the x axis is horizontal, and θ is the angle of the initial velocity vector above the horizontal, the projectile motion equations are:

$$v_{0x} = v_0\cos\theta \qquad\qquad v_{0y} = v_0\sin\theta$$

$$v_x = v_{0x} \qquad\qquad v_y = v_{0y} - gt$$

Projectile motion equations

$$\Delta x = v_{0x}t \qquad\qquad \Delta y = v_{0y}t - \tfrac{1}{2}gt^2$$

Because projectile motion problems involve both horizontal and vertical motion, solving these problems almost always involves a system of at least two equations and two unknowns. However, the fact that the time t is the same for both horizontal and vertical motion can simplify calculations.

Section 3-5. Special Case 2: Circular Motion. Movement along a circular path, or a segment of a circular path, is called **circular motion**. A particle moving at constant speed along a circular path (uniform circular motion) still is accelerating because it has a changing velocity vector since the instantaneous direction of motion is always changing. The direction of this change is always in the **centripetal** direction, meaning toward the center of the circular path. The magnitude of this centripetal acceleration depends on the instantaneous speed of the particle and the radius of the circular motion:

$$a_c = \frac{v^2}{r}$$ Centripetal acceleration

The motion of a particle experiencing uniform circular motion is often described in terms of the time required for one complete revolution, T, called the **period.** The period also depends on the speed of the particle and the radius of the circle:

$$T = \frac{2\pi r}{v}$$ Period

If the speed of the particle is changing while it is moving along a circular path, the particle also experiences a **tangential acceleration**, tangent to the circular path. This tangential acceleration has a magnitude that is simply the time rate of change of the speed of the particle:

$$a_t = \frac{dv}{dt}$$ Tangential acceleration

II. Physical Quantities and Key Equations

Physical Quantities

Acceleration due to gravity $g = |\vec{g}| = 9.81 \, \text{m/s}^2 = 32 \, \text{ft/s}^2$

Key Equations

Component-vector relations $A_x = A\cos\theta$ $A = \sqrt{A_x^2 + A_y^2}$
                                      $A_y = A\sin\theta$ $\tan\theta = A_y / A_x$

Velocity $\vec{v}_{av} = \frac{\Delta \vec{r}}{\Delta t}$ $\vec{v} = \lim\limits_{\Delta t \to 0} \frac{\Delta \vec{r}}{\Delta t} = \frac{d\vec{r}}{dt}$

or

$$v_{x,av} = \frac{\Delta x}{\Delta t} \qquad\qquad v_x = \lim_{\Delta t \to 0} \frac{\Delta x}{\Delta t} = \frac{dx}{dt}$$

$$v_{y,av} = \frac{\Delta y}{\Delta t} \qquad\qquad v_y = \lim_{\Delta t \to 0} \frac{\Delta y}{\Delta t} = \frac{dy}{dt}$$

$$v_{z,av} = \frac{\Delta z}{\Delta t} \qquad\qquad v_z = \lim_{\Delta t \to 0} \frac{\Delta z}{\Delta t} = \frac{dz}{dt}$$

Acceleration $\qquad \vec{a}_{av} = \frac{\Delta \vec{v}}{\Delta t} \qquad\qquad \vec{a} = \lim_{\Delta t \to 0} \frac{\Delta \vec{v}}{\Delta t} = \frac{d\vec{v}}{dt}$

or

$$a_{x,av} = \frac{\Delta v_x}{\Delta t} \qquad\qquad a_x = \lim_{\Delta t \to 0} \frac{\Delta v_x}{\Delta t} = \frac{dv_x}{dt}$$

$$a_{y,av} = \frac{\Delta v_y}{\Delta t} \qquad\qquad a_y = \lim_{\Delta t \to 0} \frac{\Delta v_y}{\Delta t} = \frac{dv_y}{dt}$$

$$a_{z,av} = \frac{\Delta v_z}{\Delta t} \qquad\qquad a_z = \lim_{\Delta t \to 0} \frac{\Delta v_z}{\Delta t} = \frac{dv_z}{dt}$$

Relative velocity equations $\qquad \vec{v}_{AC} = \vec{v}_{AB} + \vec{v}_{BC} \qquad\qquad \vec{v}_{AB} = -\vec{v}_{BA}$

Constant acceleration equations $\qquad \Delta \vec{r} = \vec{v}_0(\Delta t) + \frac{1}{2}\vec{a}(\Delta t^2) \qquad \vec{v} = \vec{v}_0 + \vec{a}(\Delta t)$

or

$$\Delta x = v_{0x}(\Delta t) + \tfrac{1}{2}a_x(\Delta t)^2 \qquad v_x = v_{0x} + a_x(\Delta t)$$

$$\Delta y = v_{0y}(\Delta t) + \tfrac{1}{2}a_y(\Delta t)^2 \qquad v_y = v_{0y} + a_y(\Delta t)$$

$$v_{0x} = v_0 \cos\theta \qquad\qquad v_{0y} = v_0 \sin\theta$$

Centripetal acceleration $\qquad a_c = \frac{v^2}{r}$

Tangential acceleration $\qquad a_t = \frac{dv}{dt}$

Period $\qquad T = \frac{2\pi r}{v}$

III. Potential Pitfalls

Don't forget to use appropriate notation to indicate vectors and unit vectors.

When you add vectors, remember that you cannot simply add their magnitudes. The equation $\vec{C} = \vec{A} + \vec{B}$ does not mean that $C = A + B$. Both of these equations are true only if $\vec{A}$ and $\vec{B}$ are in the exact same direction.

The displacement between two points is defined as the straight-line magnitude and direction between them "as the crow flies." Its magnitude is not necessarily equal to the distance actually traveled by the particle between the points.

It can be challenging to arrive at the equations for relative velocities. To be successful at this, it is best to consistently use the subscript notation described below. With this notation, $\vec{v}_{AB}$ is the velocity of A relative to B, while $\vec{v}_{BA}$ is the velocity of B relative to A. The relation between these velocities is expressed by the equation $\vec{v}_{AB} = -\vec{v}_{BA}$. (This equation formally expresses the observation that if A is moving north at 30 m/s relative to B, then B must be moving south at 30 m/s relative to A.)

Consider reference frames A, B, and C that are in motion relative to each other. An equation for the relative velocities of these frames is $\vec{v}_{AC} = \vec{v}_{AB} + \vec{v}_{BC}$. Note that the left subscript on both sides of this equation is A, the right subscript on both sides is C, and the adjacent interior subscripts are both B. When an additional moving frame D is considered, an equation relating the velocities is $\vec{v}_{AD} = \vec{v}_{AB} + \vec{v}_{BC} + \vec{v}_{CD}$. There are two subscript rules for these equations.

Rule 1. The left exterior subscripts of both sides of the equation must be identical and the right exterior subscripts must be identical. In the equation $\vec{v}_{AD} = \vec{v}_{AB} + \vec{v}_{BC} + \vec{v}_{CD}$, A and D are the exterior subscripts for both sides of the equation.

Rule 2. On either side of the equation, adjacent subscripts of adjacent relative velocity terms must be identical. That is, for the expression $\vec{v}_{AB} + \vec{v}_{BC} + \vec{v}_{CD}$, B is the common adjacent subscript for the first two terms, whereas C is the common subscript for the second and third terms.

The projectile motion equations apply only when a particle is in free-fall, that is, when the only force affecting its motion is gravity. If air resistance affects its motion these equations do not apply.

Consider a projectile thrown straight upward. Don't think that when it reaches the highest point in its trajectory both its velocity and acceleration are zero. Its velocity, which is momentarily zero at its peak, is changing its direction, which means its acceleration is not zero. In fact, its acceleration is the same at the top as it is on the way up and on the way down: 9.81 m/s^2, directed downward.

Consider a projectile thrown both upward and to the side, like a shotput. When such a projectile is at the top of its arc its velocity is not zero. Only the y component of its velocity is zero. The x component of its velocity is constant throughout the motion.

For motion in one dimension, constant speed means that the acceleration is zero. This is not true for motion along a curved path. As long as the direction of the velocity vector is changing, even if the speed is constant, the acceleration vector cannot be zero.

IV. True or False Questions and Responses

True or False

_____ 1.   Three dimensions are necessary to describe the general motion of a particle.

_____ 2.   The displacement of a particle is the change in its position.

_____ 3.   A component of a vector is itself a vector.

_____ 4.   All vector quantities have both a magnitude and a direction, and all follow the same rules for vector addition that consecutive displacements follow.

_____ 5.   The magnitude of a vector quantity is a dimensionless number.

_____ 6.   Both velocity and time must be expressed as vector quantities.

_____ 7.   Vectors may be added by adding corresponding components.

_____ 8.   In Figure 3-9, $\vec{A} + \vec{C} = \vec{B} + \vec{D} + \vec{E}$.

_____ 9.   In Figure 3-9, $\vec{A} + \vec{B} + \vec{C} = \vec{D} - \vec{E}$.

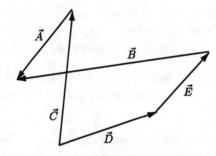

Figure 3-9

_____ 10.   For any two vector quantities $\vec{P}$ and $\vec{Q}$, $\vec{P} - \vec{Q}$ is a vector having the same magnitude as, but a direction opposite to, $\vec{P} + \vec{Q}$.

_____ 11.   When a particle is in motion, the difference between its position vector at time t_2, and its position vector at time t_1 is its displacement during the time interval $t_2 - t_1$.

_____ 12.   The instantaneous-velocity vector is always in the direction of motion.

_____ 13.   If a particle moves in a straight line, its position and velocity vectors are parallel.

_____ 14.   The magnitude of the acceleration vector is equal to the rate at which the _speed_ of the particle changes with respect to time.

_____ 15.   A projectile is a body in motion that cannot be treated as a particle.

_____ 16. The time it takes for a bullet fired horizontally to reach the ground is the same as if it were dropped from rest from the same height.

_____ 17. For a particle undergoing uniform circular motion, the instantaneous-acceleration vector equals zero.

_____ 18. For a particle in uniform circular motion the average acceleration vector for a time interval is in the same direction as the change in velocity vector for that time interval.

_____ 19. For a particle in uniform circular motion the instantaneous-acceleration vector is directed radially outward.

Responses to True or False

1. True

2. True. This is the definition of displacement in fact.

3. False. A component of a vector is the projection of the vector along a given direction. It is a scalar quantity. Thus, if $\vec{A} = \left(6\hat{i} + 8\hat{j} \right)$ ft , its x and y components are 6 ft and 8 ft respectively.

4. True.

5. False. It has no direction, but it may well have dimensions. Thus, if $\vec{A} = \left(6\hat{i} + 8\hat{j} \right)$ ft , the magnitude is $\left| \vec{A} \right| = 10$ ft .

6. False. Time has no direction in space.

7. True.

8. True.

9. False.

10. False. This describes not $\vec{P} - \vec{Q}$, but $-\vec{P} - \vec{Q}$. (See also Question 2.)

11. True.

12. True.

13. False. See Figure 3-10.

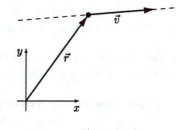

Figure 3-10

14. False. For instance, consider something moving along a curved path at constant speed. As long as the direction of the velocity vector is changing, the acceleration cannot be zero.

15. False. In fact, we have been assuming that projectiles are particles.

16. True. This is because the initial *vertical* component of the velocity is zero in either case, so the vertical motions are the same.

17. False. See the explanation to Question #14.

18. True.

19. False. It is directed radially inward, in the direction of the change in velocity vector for a sufficiently short time interval.

V. Questions and Answers

Questions

1. You throw a baseball from the outfield to a friend at home plate. In general, the distance the ball travels is not equal to the magnitude of its displacement vector. Which of the two is larger?

2. Two displacement vectors $\vec{A}_1$ and $\vec{A}_2$ add to give a resultant of zero. What can we say about the two displacements?

3. Is it possible to drive your car around a curve without accelerating?

4. Suppose you have "sighted in" a rifle so that it will hit whatever the sights are aimed directly at on a target 150 m away over level ground. When shooting at a target 90 m away, should you aim above, below, or directly at the target?

5. Describe briefly what kind of motion a particle is undergoing when (*a*) the position vector changes in magnitude but not in direction; (*b*) the velocity vector changes in magnitude but not in direction; (*c*) the position vector changes in direction but not in magnitude; and (*d*) the velocity vector changes in direction, but not in magnitude.

Answers

1. The magnitude of the displacement is the straight-line distance between the two points, so the actual distance cannot be less than the magnitude of the displacement. They would be equal only if the path of the ball were a straight line. Since the ball must rise and fall some in flight, the distance it travels must be greater than the magnitude of the displacement.

2. The equation $\vec{A}_1 + \vec{A}_2 = 0$ means that $\vec{A}_1 = -\vec{A}_2$, so the displacements are equal and opposite. If these are the two legs of a trip, then the second leg is a return to the starting point.

3. No. The question points out the difference between everyday usage and the scientific meaning of the term "accelerating." Scientifically speaking, acceleration occurs whenever the

magnitude or direction of the velocity vector is changing. Since your velocity is at least changing in direction, you cannot drive around a curve without accelerating.

4.   When you "sight in" a rifle at a particular distance, you are adjusting the sights to compensate for the distance by which the bullet falls below its initial line of flight. The axis of the barrel thus points above the sight line. When the target is closer than the distance for which the rifle was sighted, the bullet will fall a shorter distance. If you aim at the target, the bullet will be high when it gets there, so you should aim *below* the target to compensate.

5.   (*a*) In this case, the particle is moving in a straight line toward or away from the origin. (*b*) The particle is moving in a straight line (since the direction of $\vec{v}$ isn't changing) with variable speed. (*c*) This would mean that the moving particle stays at the same distance from the origin, so it must be traveling in a circle around the origin. (*d*) This indicates that whatever the moving particle is doing, it's doing it at constant speed.

VI. Problems, Solutions, and Answers

Example #1. At 3:00 p.m. you pass mile marker 160 as you are driving due south on Interstate 77. At mile marker 138, you turn off on U.S. 54. At 3:40 p.m. you have gone 14.0 mi due southwest on U.S. 54. What was your average velocity (magnitude and direction) over this 40.0-min interval?

Picture the Problem. Where are you at the end of the 40 minutes? Your average velocity will be the total displacement divided by the time interval.

1. It is usually extremely helpful to sketch out the displacements during each segment of the problem.	
2. Find the x and y components of each individual displacement vector.	$s_{1x} = 0$; $s_{1y} = -22.0\,\text{mi}$ $s_{2x} = -(14.0\,\text{mi})\cos 45° = -9.90\,\text{mi}$ $s_{2y} = -(14.0\,\text{mi})\sin 45° = -9.90\,\text{mi}$
3. The total displacement is the sum of the two vector displacements. The magnitude of the displacement is the square root of the sum of the squares of the components of the displacement.	$s_x = s_{1x} + s_{2x} = -9.90\,\text{mi}$ $s_y = s_{1y} + s_{2y} = -31.9\,\text{mi}$ $s = \sqrt{s_x^2 + s_y^2} = 33.4\,\text{mi}$

4. The direction of the displacement can be found from the x and y components.	$\tan\theta = \dfrac{s_x}{s_y} = \dfrac{-9.90\,\text{mi}}{-31.9\,\text{mi}} = 0.310$ $\theta = 17^\circ$
5. The average velocity is the displacement divided by the time.	$\vec{v}_{av} = \dfrac{\Delta\vec{r}}{\Delta t} = \left(\dfrac{33.4\,\text{mi}}{40.0\,\text{min}}\right)\left(\dfrac{60.0\,\text{min}}{1\,\text{hr}}\right)$ $= 50.1\,\text{mi/hr}$ directed 17° west of south

Example #2—Interactive   A car is traveling along a mountain road. At a certain instant it is moving due west across level ground at a speed of 18.0 m/s; 2.00 s later it is moving north down a steep hill at an angle of 15.0° below the horizontal at 10.0 m/s. Calculate the car's average acceleration over this time interval.

Picture the Problem.   The change in the car's velocity divided by the time is the average acceleration. **Try it yourself.** Work the problem on your own, in the spaces provided, to get the final answer.

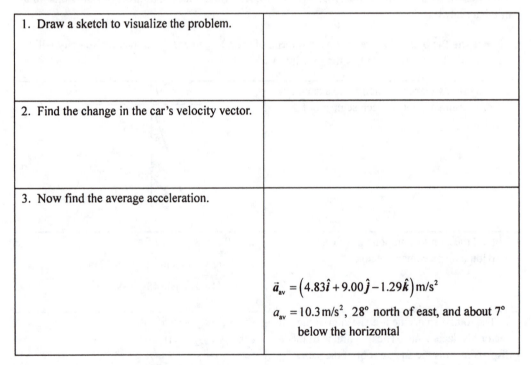

1. Draw a sketch to visualize the problem.	
2. Find the change in the car's velocity vector.	
3. Now find the average acceleration.	$\vec{a}_{av} = \left(4.83\hat{i} + 9.00\hat{j} - 1.29\hat{k}\right)\text{m/s}^2$ $a_{av} = 10.3\,\text{m/s}^2,\ 28^\circ$ north of east, and about 7° below the horizontal

Example #3.   A pilot starts a trip from St. Louis to Memphis, which is 385 km due south. She flies due south at her maximum airspeed of 90.0 m/s but fails to correct for a crosswind of 15.0 m/s directed due east. (*a*) What is her velocity relative to the ground? (*b*) At the time she expects to reach Memphis, where will she actually be? (*c*) How much longer will it take her to get to Memphis?

Picture the Problem. This is a relative velocity problem. Her net velocity will be the sum of the two velocities, and will be responsible for her net displacement, which will not be due south.

1. Draw a sketch to picture the velocity addition.	
2. Find her velocity relative to the ground.	$\vec{v}_{plane,ground} = \vec{v}_{plane,air} + \vec{v}_{air,ground}$ $= \left(-90.0\hat{j} + 15.0\hat{i}\right)$ m/s $v_{plane,ground} = \sqrt{(-90.0)^2 + 15.0^2} = 91.2$ m/s $\tan\theta = \dfrac{v_x}{v_y} = \dfrac{15.0}{90.0}$ $\theta = 9.5°$ east of due south
3. Find how long she would expect the trip to take.	$\Delta t = \dfrac{385\times10^3 \text{ m}}{90.0 \text{ m/s}} = 4280 \text{ s}$
4. After that time, she will be exactly 385 km south of her starting point, but she will also be east of that point because of the crosswind. How far east?	$\Delta x = v_x\left(\Delta t\right) = \left(15.0 \text{ m/s}\right)\left(4280 \text{ s}\right) = 64.2 \text{ km}$
5. To get back to Memphis, she will have to travel 64.2 km straight into the wind, so her ground speed will now be 75 m/s. How long will it take?	$\Delta t = \dfrac{64.2\times10^3 \text{ m}}{75.0 \text{ m/s}} = 856 \text{ s} = 14.3 \text{ min}$

Example #4—Interactive. The current at a certain point where the Wabash River is 80.0 m wide flows at 0.400 m/s. A swimmer sets out for a point directly across the river. (*a*) If the swimmer's maximum speed is 0.750 m/s relative to still water, in what direction should he swim to go directly to his goal? (*b*) How long will it take him to get there?

Picture the Problem. This is a relative velocity problem. Once we find the swimmer's direction, we will use the component of his velocity perpendicular to the river to determine how long it will take to cross the river. **Try it yourself.** Work the problem on your own, in the spaces provided, to get the final answer.

1. Draw a sketch to show the addition of velocities.	
2. Use the relative velocity relationship to find the direction he should swim. The velocity of the swimmer with respect to the ground should be 100% across the river. Lets call this the x direction, and upstream the y direction.	$\theta = 32.2°$ upstream from straight across
3. Now find his across-stream speed, and use that to find the time required.	$\Delta t = 126\,\text{s}$

Example #5. A ball rolls off a tabletop 0.900 m above the floor and lands on the floor 2.60 m away from a point that is directly under the edge of the table. At what speed did it roll off?

Picture the Problem. Remember the horizontal and vertical components of the ball's projectile motion can be treated separately. The vertical motion will provide the time the ball is in the air, which we can then use to find the initial horizontal velocity.

1. Write an expression for the motion of the ball in the y direction, and solve for the time. The initial y velocity is zero as the ball rolls off the table edge. We always use the positive value for time, if we have a quadratic equation to solve. The origin is on the floor, directly below the edge of the table.	$y = y_0 + v_{0y}t + \frac{1}{2}a_y t^2$ $0\,\text{m} = (0.900\,\text{m}) + 0 - \frac{1}{2}(9.80\,\text{m/s}^2)t^2$ $t = 0.428\,\text{s}$
2. Now that we know the time, we can use this in a similar expression for the motion of the ball in the x direction to solve for the initial x velocity. Since the initial y velocity was zero, the initial x velocity is the initial velocity.	$x = x_0 + v_{0x}t + \frac{1}{2}a_x t^2$ $2.60\,\text{m} = 0 + v_{0x}(0.428\,\text{s}) + 0$ $v_{0x} = 6.07\,\text{m/s}$

Example #6—Interactive. A diver leaps from a springboard 6.00 m above the water with a velocity of 6.00 m/s at an angle 30° from the vertical. How far from a point directly beneath the end of the board, and at what speed, does she hit the water?

Picture the Problem. Remember the horizontal and vertical components of the diver's projectile motion can be treated separately. The vertical motion will provide the time the diver is in the air, which we can then use to find the horizontal displacement. **Try it yourself.** Work the problem on your own, in the spaces provided, to get the final answer.

1. Write an expression for the motion of the diver in the y direction, and solve for the time. The initial y velocity must be found from the initial speed and angle.	
2. Use this time in the x equation of motion to find the horizontal displacement.	$x = 5.27\,\text{m}$
3. The final speed will have both x and y components. You have the x component already. Use the velocity equation to find the y component.	$v = \sqrt{v_x^2 + v_y^2} = 12.4\,\text{m/s}$

Example #7. A basketball player takes a shot when he is standing 24.0 ft from the 10.0-ft high basket, as shown in Figure 3-11. If he releases the ball at a point 6.00 ft from the floor and at an angle of 40° above the horizontal, with what speed must he throw the ball for it to hit the hoop?

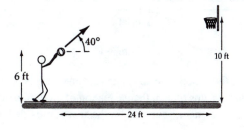

Figure 3-11

Picture the Problem. In this problem, we can use the horizontal motion to solve for the time of flight in terms of the initial speed. We will then put this expression for time into the vertical equation of motion.

1. Write the equation for the horizontal motion and solve for the time to reach the basket in terms of initial speed. The origin is the point of release of the ball.	$x = x_0 + v_{0x}t + \frac{1}{2}a_x t^2$ $24.0\,\text{ft} = 0 + \left(v_0 \cos 40°\right)t + 0$ $t = \dfrac{24.0\,\text{ft}}{v_0 \cos 40°}$

2. Substitute this expression for time into the equation of motion for the vertical direction. Remember to use 32 ft/s^2 for the acceleration due to gravity.	$y = y_0 + v_{0y}t + \frac{1}{2}a_y t^2$ $4.00\,\text{ft} = 0 + \left(v_0 \sin 40°\right)\left(\dfrac{24.0\,\text{ft}}{v_0 \cos 40°}\right) - \frac{1}{2}g\left(\dfrac{24.0\,\text{ft}}{v_0 \cos 40°}\right)^2$ $4.00\,\text{ft} = \left(24.0\,\text{ft}\right)\tan 40° - \frac{1}{2}g\left(\dfrac{24.0\,\text{ft}}{v_0 \cos 40°}\right)^2$ $v_0 = 31.3\,\text{ft/s}$

Example #8—Interactive. General Lee's artillery is taking aim on Union troops on the cliffs across the river, 300.0 m away. (See Figure 3-12.) The muzzle speed of the cannonballs is 67.0 m/s. In order to hit the edge of the cliff, the cannonballs have to be fired at an angle of elevation of 31°. How high are the cliffs?

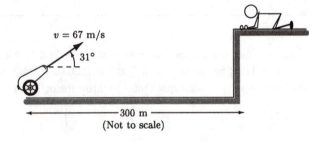

$v = 67$ m/s

31°

300 m
(Not to scale)

Figure 3-12

Picture the Problem. Use the x motion to calculate the time it takes for the cannonball to move the 300 m to the cliffs, then use this time in the y equation of motion to determine the cliff height. **Try it yourself.** Work the problem on your own, in the spaces provided, to get the final answer.

1. Assume the origin is at the cannon muzzle, which we assume to be ground height. Find the time required to travel the 300 m horizontally.	
2. Use this time in the equation for the vertical motion to find the height of the cliffs.	$y = 46.4\,\text{m}$

Example #9. In a game of American football, a quarterback throws a pass from his 20-yard line with an initial velocity of 22.0 m/s at an angle of 40° above the horizontal. The receiver starts running downfield from the 30-yard line 1.00 s before the pass is thrown. With what speed must he run in order to catch the pass? (Assume that the pass is caught at the same height above the ground at which it was thrown. If you need to convert units, 1 m = 1.09 yd.)

Picture the Problem. This is a classic "chase" problem, in which we are interested in the conditions required for two objects to be in the same place at the same time. For the receiver to catch the ball, the ball must return to its original height. We can use the projectile motion equations to determine the horizontal displacement of the ball during its flight. We then equate the final position of the ball to the position of the receiver to solve for the receiver's speed.

1. Use the vertical equation of motion of the ball to find out how long it is in the air.	$y = y_0 + v_{0y}t + \frac{1}{2}a_y t^2$ $0 = 0 + (22.0\,\text{m/s})(\sin 40°)t - \frac{1}{2}gt^2$ $t = 2.88\,\text{s}$
2. Use the horizontal equation of motion of the ball to determine its displacement from the quarterback.	$x = x_0 + v_{0x}t + \frac{1}{2}a_x t^2$ $x = 0 + (22.0\,\text{m/s})(\cos 40°)(2.88\,\text{s}) + 0$ $x = 48.5\,\text{m}\left(\dfrac{1.09\,\text{yd}}{1\,\text{m}}\right) = 52.9\,\text{yd}$
3. The receiver must cover 10 fewer yards in one extra second. This distance and time give us the speed the receiver must have. This speed is comparable to the average speed of a world champion sprinter.	$v_x = \dfrac{42.9\,\text{yd}}{3.88\,\text{s}} = 11.1\,\text{yd/s}$

Example #10—Interactive. In a baseball game, a batter hits a fly ball directly toward the center fielder. The bat strikes the ball at a point 1.10 m directly above home plate, and the ball leaves the bat at a speed of 29.5 m/s at an angle of 35° above the horizontal. The center fielder is standing 116 m from home plate. If the center fielder starts running at the instant the ball is hit and if he catches the ball 1.10 m above the ground, how fast must he run to catch the ball? Assume that air resistance can be neglected.

Picture the Problem. The center fielder must be in the same position as the ball to catch it. The projectile motion equations of the ball will give us the horizontal displacement of the ball. We will then know how far the center fielder has to move, and how much time he has to move, so we can find his required average speed. **Try it yourself.** Work the problem on your own, in the spaces provided, to get the final answer.

1. Use the vertical equation of motion of the ball to find out how long it is in the air.	

2. Use the horizontal equation of motion of the ball to determine its displacement from the batter.	
3. Now you can calculate how far the center fielder must run to catch the ball, and you know how much time he has, so you can calculate his speed.	$v_x = 9.46 \, \text{m/s}$

Chapter **4**

Newton's Laws

I. Key Ideas

Newton's laws of motion apply to all objects. Often the motion of an object can be represented by the motion of a single point, and then the object can be treated as a particle. Forces, which are pushes or pulls on an object, affect the motion of objects. Newton's Laws are as follows:

Newton's First Law. An object at rest stays at rest *unless* acted on by an external force. An object in motion continues to travel with constant velocity *unless* acted on by an external force.

Newton's Second Law. The direction of the acceleration of an object is in the direction of the next external force acting on it. The acceleration is proportional to the next external force, $\vec{F}_{net}$, and inversely proportional to its mass. The net force acting on an object, also called the resultant force, is the vector sum of all the forces acting on it: $\vec{F}_{net} = \Sigma \vec{F}$. Thus,

$$\vec{F}_{net} = \sum \vec{F} = m\vec{a} \qquad \qquad \text{Newton's Second Law}$$

Newton's Third Law. Forces always occur in equal and opposite pairs. If object A exerts a force $\vec{F}_{A,B}$ on object B, an equal but opposite force $\vec{F}_{B,A}$ is exerted by object B on object A. Thus,

$$\vec{F}_{A,B} = -\vec{F}_{B,A} \qquad \qquad \text{Newton's Third Law}$$

Section 4-1. Newton's First Law: The Law of Inertia. When the motion of a particle is observed and measured, it must be observed and measured with respect to a specific reference frame. According to Newton's First Law, an object moves with constant velocity when the net force acting on it is zero. Reference frames in which this occurs are called **inertial reference frames.** Newton's three laws are only valid in inertial reference frames.

Section 4-2. Force, Mass, and Newton's Second Law. A force is an external influence on an object that causes it to accelerate relative to an inertial reference frame. Forces are pushes or pulls, like the pull (or tug) of a rope pulling a sled. One of your greatest conceptual challenges to master in physics will be the proper identification of all the forces acting on an object. Any time one particle touches another, they exert equal and opposite **contact forces** on each other via physical contact. Some forces do not require that the particles physically touch. These **action-at-a-distance forces**, like the pull of gravity, are forces that are exerted via fields, like the gravitational field of the earth, or electromagnetic fields. Until Chapter 22, the only action-at-a-distance force we will consider in any detail is the gravitational force.

The SI unit of force is the newton (N), and the SI unit of mass is the kilogram (kg). The kilogram is, by definition, the mass of the standard body, which is a platinum cylinder kept at the International Bureau of Weights and Measures in Sevres, France. A net force of **one newton**, acting on the standard body, results in an acceleration of $1\,\text{m/s}^2$. Forces are defined by the acceleration they produce on the standard body.

If equal net forces act on two different particles with different masses, they will generally undergo different accelerations. Newton's Second Law tells us that if the net forces are equal, the products of the corresponding masses and accelerations will also be equal. Therefore, measuring the accelerations of two different particles when the net force applied to each is the same allows the masses of the particles to be compared:

$$m_1 a_1 = m_2 a_2 \quad \text{or} \quad \frac{m_1}{m_2} = \frac{a_2}{a_1} \qquad\qquad\qquad \text{Definition of mass}$$

Section 4-3. The Force Due to Gravity: Weight. The weight $\vec{w}$ of an object is the gravitational force exerted on it. To determine the weight of an object, we measure the acceleration of an object when the only force acting on it is the gravitational force (the weight). When objects near the surface of the earth fall under the influence of *only* gravity, that is, when air resistance is negligible, they all fall with the same acceleration $\vec{g}$, which is $9.8\,\text{m/s}^2$ downward. For these objects, the force acting on them is their weight $\vec{w}$. It follows that for such falling objects, Newton's Second Law reduces to the relation between the object's weight and the object's mass, that is

$$\vec{w} = m\vec{g} \qquad\qquad\qquad \text{Weight of an object}$$

The weight of an object does not depend on whether or not the object is falling. That is, $\vec{g}$ is not the acceleration of the object. Rather, it is the force per unit mass exerted by the earth on the object, the gravitational field, that happens to be equal to the free-fall acceleration of an object near the surface of the earth.

$$\vec{g} = 9.81\,\text{N/kg} \qquad\qquad\qquad \text{Magnitude of the gravitational field}$$

When standing on a scale "weighing yourself," you are not measuring the gravitational force acting on you. Instead, you are measuring the magnitude of the force exerted by the scale upward on your feet. When objects are "weighed" in this manner, the quantity being measured is not the true weight but the **apparent weight**. When your acceleration is zero, your apparent weight equals your actual weight. When a shuttle astronaut is in orbit about the earth, the earth still exerts a gravitational force on her and thus her true weight is not zero. However, a shuttle astronaut in orbit is accelerating downward at the same rate as the shuttle, under the influence of the gravitational force alone. As a consequence, when she "stands" on a scale, the magnitude of the force exerted by the scale upward on her feet is zero, so her apparent weight is zero—the state commonly referred to as being "weightless."

Section 4-4. Forces in Nature. All observed natural forces can be described in terms of four basic interactions that occur between elementary particles:

1. The gravitational force
2. The electromagnetic force

3. The strong nuclear force
4. The weak nuclear force

For the next several chapters, we will be working almost exclusively with gravitational forces. The **contact forces** between touching objects are electromagnetic in nature, but until we get to Chapter 22, where electric forces are introduced, the only **action-at-a-distance** force we will consider in any detail is the gravitational force.

When working with two *solid* objects that are touching, like a box that is sliding down a ramp, the contacting surfaces deform slightly and exert forces on each other, forces that are distributed over the contacting surfaces. The contact force acting on a single surface is the vector sum of the distributed contact forces that act on that surface. It is often useful to think of this contact force as the sum of two distinct forces, one normal to and one parallel to the surface. (The word normal means perpendicular.) These forces are referred to as the **normal force** and the **friction force**, respectively.

If we pull on a *string* it stretches slightly and pulls back on us with an equal but opposite force. Under tension some strings, like monofilament fishing line, stretch a fair amount, while other strings, like cotton thread, stretch a nearly imperceptible amount. In physics textbook problems you should assume that the stretching of any string is negligible (as is its mass), unless it is stated otherwise. The strings in physics textbooks act much more like cotton thread than monofilament fishing line. Cotton acts much like a very stiff spring, one with a very large force constant.

What is a spring? It is a convenient agent for exerting forces on objects. It takes a force (a pull) to stretch a spring, and to maintain the spring in a stretched condition requires that the force be maintained. The greater the force, the greater the extension of the spring. Consider a particular spring attached to a 1 kg block on a frictionless horizontal surface, as shown in Figure 4-1. When the spring is pulled to the right, the spring stretches. The stretched spring pulls on the string causing the block to accelerate. Up to a certain limit (depending on the spring), the greater the extension of the spring, the greater the acceleration. By noting the extension needed to produce a particular acceleration, and by using Newton's Second Law ($\vec{a} = \vec{F}/m$), we can calibrate the spring once we realize that the net force acting on the block is simply the force from the spring. An acceleration of $1\,\text{m/s}^2$ means that the spring is exerting a force of 1 N; an acceleration of $2\,\text{m/s}^2$ means that the spring is exerting a force of 2 N, and so on.

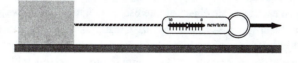

Figure 4-1

Using our 1-kg block, we can calibrate other springs in the same way. For common springs, the force exerted by the spring, F_x, is proportional to the extension Δx (for small extensions). This is known as Hooke's law.

$$F_x = -k\,\Delta x \hspace{6cm} \text{Hooke's law}$$

The larger the proportionality, or force constant k, the stronger a particular spring is. The negative sign in Hooke's law signifies that when the spring is stretched or compressed, the force the spring exerts is in the opposite direction.

Section 4-5. Problem Solving: Free-Body Diagrams. At this point, it is worth reminding yourself about the purpose of this Study Guide, and the problems presented in it. A typical physics problem describes a physical situation—such as a child swinging on a swing—and asks related questions. For example: If the speed of the child is 5.0 m/s at the bottom of her arc, what is the maximum height the child will reach? Solving such problems requires you to apply the concepts of physics to the physical situation, to generate mathematical relations, and to solve for the desired quantities. The problems presented here and in your textbook are exemplars; that is, they are examples that deserve imitation. When you master the methodology presented in the worked-out examples, you should be able to solve problems about a wide variety of physical situations.

To be successful in solving physics problems, study the techniques used in the worked-out example problems. A good way to test your understanding of a specific solution is to take a sheet of paper, and—without looking at the worked-out solution—reproduce it. If you get stuck and need to refer to the presented solution, do so. But then take a fresh sheet of paper, start from the beginning, and reproduce the entire solution. This may seem tedious at first, but it does pay off.

This is not to suggest that you reproduce solutions by rote memorization, but that you reproduce them by drawing on your understanding of the relations involved. By reproducing a solution in its entirety, you will verify for yourself that you have mastered a particular example problem. As you repeat this process with other examples, you will build your very own personal base of physics knowledge, a base of knowledge relating occurrences in the world around you— the physical universe—and the concepts of physics. The more complete the knowledge base that you build, the more success you will have in physics.

The only master equation that can be used to solve problems using Newton's laws is $\vec{a} = \vec{F}_{net}/m$. However, you will need to generate the elements of this equation from scratch for each and every problem. That is, you will have to identify the forces acting on each object and either calculate, or determine from information in the problem, the acceleration of each object in a problem.

Particular care must be given to properly identifying both the magnitude and direction of each of the forces acting on each separate object of interest. A diagram that shows schematically all the forces acting on an object is a ***free-body diagram.*** Creating such a diagram first requires you to isolate the object of interest. Once you have determined the object of interest, you can reduce the object to a point (particle) representation. Next draw the force vectors with their tails on the dot representing the object, pointing in the direction the force is applied. Label the forces appropriately. Determining the forces acting on an object is perhaps the most critical step of the problem-solving process. This chart will help you determine what forces might be influencing the motion of an object and in what direction they are acting.

Is the object in the earth's gravitational field?	Include the object's *weight* vector, $\vec{w}$, which always acts straight down, toward the center of the earth.
Is the object touching any other surfaces (table, wall, side of another block, etc.)?	For each surface the object is touching you should draw a *normal force* vector, $\vec{F}_n$, that is perpendicular to the surface. This force always tries to push the object away from the surface.
Do any of the surfaces above have friction?	For every surface that is not frictionless, you also need to draw a *frictional force* vector, $\vec{f}$, that is parallel to the surface, and in a direction that opposes the object's tendency to move. (You'll learn more about this in Chapter 5.)
Are there ropes or other pulling devices attached to the object?	You should have a *tension force* vector, $\vec{T}$, for each rope or pulling device. Remember that ropes always pull objects.
Is there some other applied force being delivered by a mechanism different from those above?	Include an *applied force* vector, $\vec{F}_{applied}$.

An excellent strategy for solving problems with Newton's laws is as follows.

1. Draw a neat pictorial diagram. List the known and unknown quantities.

2. Isolate the object of interest, and draw a free-body diagram showing each external force that acts on it. If there is more than one object of interest in the problem, draw a separate free-body diagram for each.

3. Choose a convenient coordinate system for each object and include it on that object's free-body diagram. If the direction of the acceleration is known, choose one coordinate axis to be parallel to the acceleration. For objects sliding along a surface, choose one coordinate axis parallel to the surface and the other perpendicular to it. If more than one object is accelerating, it is helpful to choose your coordinate systems so that both objects are accelerating in the positive direction at the same time. If the acceleration is zero, then select coordinate-axes directions that maximize the number of force vectors parallel to an axis.

4. Make a decision about the acceleration. At the very least, determine the line along which the object will accelerate.

5. Apply Newton's Second Law, $\Sigma \vec{F} = m\vec{a}$, in component form. This requires you to determine both the x- and y-components of each force, as well as the acceleration, and

solve two equations, one for the forces and acceleration in the *x*-direction and one in the *y*-direction.

6. For problems involving two or more objects, make use of Newton's Third Law, $\vec{F}_{A,B} = -\vec{F}_{B,A}$, and any constraints to simplify the equations arrived at in step 5.

7. Solve the resulting equations for the unknowns.

8. Check to see whether your results have the correct units and seem plausible. Substituting extreme values, like zero and very large positive and negative numbers, into your solution is a good way to check your work for errors.

Section 4-6. Newton's Third Law. Forces always come in pairs, which are referred to as **action-reaction pairs**. The two forces that make up an action-reaction pair always are equal in magnitude and oppositely directed, and they act on opposite objects. If you are standing at rest and want to start running into the ocean, Newton's Second Law dictates that you cannot accelerate toward the water unless a net external force pushes you in that direction. To start moving you must cause your surroundings to push you in the direction you want to go, and you do this by pushing your surroundings in a direction *opposite* to the direction you wish to go. That is, to start moving toward the water, you push the sand under your feet away from the water. That the sand will then push you toward the water follows from Newton's Third Law.

Section 4-7. Problems with Two or More Objects. In many mechanics problems, two or more objects that move are touching each other or are connected by a string. Such problems can be solved by treating each object separately (including the string). Draw a free-body diagram for each object, and then apply Newton's Second Law ($\Sigma \vec{F} = m\vec{a}$) to each. The resulting equations are then solved simultaneously for the unknown forces or accelerations. Two additional types of relations, one kinematic and one dynamic, are needed to solve multiple-object problems effectively. If two objects are connected by a string (or rope) that remains taut, the motions of the two objects will be related kinematically. For example, suppose I am dangling in a crevasse while hanging onto one end of a rope, and you are on top of a glacier—holding onto the other end and being dragged toward the edge of crevasse; our motions will be identical, except that your motion will be horizontal while mine will be vertical.

When one object is touching another object, like a rope pulling a wagon, the force exerted by the rope on the wagon affects the motion of the wagon, whereas the force exerted by the wagon on the rope affects the motion of the rope. These two forces are equal but opposite, a dynamic relation (Newton's Third Law).

II.   Physical Quantities and Key Equations

Physical Quantities

Acceleration due to gravity $\vec{g} = 9.81\,\text{m/s}^2$, downward; $g = |\vec{g}| = 9.81\,\text{m/s}^2$

Gravitational field $\vec{g} = 9.81\,\text{N/kg}$, downward

Newton $1 \text{ N} = 1 \text{ kg} \cdot \text{m/s}^2$

Key Equations

Newton's Second Law $\vec{a} = \dfrac{\vec{F}_{net}}{m}$ or $\vec{F}_{net} = m\vec{a}$

Newton's Third Law $\vec{F}_{AB} = -\vec{F}_{BA}$

Hooke's law $F_x = -k\,\Delta x$

Weight $\vec{w} = m\vec{g}$

III. Potential Pitfalls

Except for action-at-a-distance forces, like gravity, two objects must be touching in order to exert a force on each other.

Notice that the action-reaction force pairs referred to by Newton's Third Law never act on the same object. Thus they never cancel when the net force acting on an object is calculated.

Always choose a coordinate system that is fixed to an inertial reference frame. Newton's laws must be applied in inertial reference frames. For most of the problems considered in this chapter, the earth is considered an inertial reference frame and all coordinate systems will be fixed relative to the earth.

Be sure to put only those forces on a free-body diagram for which you can identify an external physical source. Acceleration is not a force, and neither is the product of mass and acceleration ($m\vec{a}$). Any force on your free-body diagram should be identified as either an action-at-a-distance force or a contact force. With rare exceptions, the only action-at-a-distance force considered during the first semester of a general physics course is the gravitational force. In addition to action-at-a-distance forces, the diagram should contain at least one force vector for each thing that touches the object.

Only those forces that act on an object go on that object's free-body diagram. Forces the object exerts on other things go on their free-body diagrams, not on its. The net force acting on an object is the name for the vector sum of all the forces that act on the object.

Be aware of the positive direction of the coordinate axes used in each problem. For example, the sign of the *x* component of a vector depends on the choice of the positive *x* direction (the direction of increasing *x*).

IV. True or False Questions and Responses

True or False

____ 1. If a particle moves with constant velocity, no forces are acting on it.

____ 2. Force is a vector quantity.

____ 3. The mass of an object may be determined in terms of the acceleration produced by a known force acting on it.

____ 4. If an object moves at constant speed, the net force acting on it must be zero.

____ 5. The weight of an object is the force that Earth's gravity exerts on it.

____ 6. The weight of an astronaut in a near-earth orbit is zero.

____ 7. When all the forces that act on a particle and the mass of the particle are known, Newton's laws provide a complete description of the particle's motion.

____ 8. An object rests on a table top. The upward force by the table surface on the object and the downward gravitational force exerted by the earth on the object form an action-reaction pair and thus must be equal in magnitude and oppositely directed.

____ 9. That the tension in a light string connecting two objects is the same throughout its length follows from Newton's third law (action equals reaction).

____ 10. Newton's laws of motion are not valid in noninertial reference frames.

____ 11. The forces that bind atoms together into molecules are electromagnetic in origin.

____ 12. Contact forces between macroscopic objects are electromagnetic in origin.

____ 13. Hooke's law states that the force exerted on a compressed or extended spring is directly proportional to the compression or extension of that spring.

____ 14. When a pony pulls a stationary cart and sets it into motion, the force with which the pony pulls the cart forward exceeds the force with which the cart pulls back on the pony.

____ 15. When a large, fully loaded eighteen-wheeler (large truck) runs head on into a Geo Metro (small car), the magnitude of the force exerted by the truck on the car is the same as that exerted by the car on the truck.

____ 16. When a large, fully loaded eighteen-wheeler (large truck) runs head on into a Geo Metro (small car), the magnitude of the car's acceleration is greater than that of the truck.

Responses to True or False

1. False. The vector sum of all the forces acting on it must be zero.

2. True   3.   True

4. False. For the acceleration—and thus the net force—to be zero, the body must move at constant speed in the same direction.

5. True

6. False. The astronaut's apparent weight is zero. The true weight—the gravitational force exerted on the astronaut by the earth—is the force that keeps the astronaut moving in a curved path around the earth.

7. False. Knowledge of all the forces and the mass specifies only acceleration. However, knowledge of the initial position and velocity is also needed to completely specify the motion.

8. False. These forces both act on the same object, so they can't possibly be an action-reaction pair. Together they constitute the net force acting on the object. Because the acceleration is zero, it follows from Newton's Second Law ($\vec{F}_{net} = m\vec{a}$) that the net force is zero. If the net force due to two forces is zero, the forces must be equal in magnitude and oppositely directed.

9. False. The tensions at two separate points in a string are not necessarily equal and are not an action-reaction pair. The net force on a segment of string equals the product of its mass and acceleration. To the degree that its mass equals zero, the net force equals zero.

10. True   11. True   12. True   13. True

14. False. The forces are an action-reaction pair, so they are always equal in magnitude and oppositely directed.

15. True   16. True

V. Questions and Answers

Questions

1. Suppose that only one force acts on an object. Can you tell in what direction the object is moving?

2. When you jump into the air, you have (for a short time) an upward acceleration. What external agent is exerting the upward force on you? What is the Newton's Third Law reaction force to the upward force exerted by this agent?

3. A car is being driven up a long straight hill at constant speed. What forces act on it and what is the net force?

4. Suppose a force $\vec{F}$ stretches a spring a distance Δx from its relaxed length, and another force $\vec{F}'$ stretches it by $2\Delta x$. How can you tell if $\vec{F}'$ is in fact equal to $2\vec{F}$?

5. Why do you seem to be thrown forward when a car in which you are riding stops abruptly?

6. Why do you seem to be thrown outward when a car in which you are riding makes a sharp turn?

7. An object of mass m is being weighed in an elevator that has an upward directed acceleration $\vec{a}$. What is the result if the weighing is done using (a) a spring scale and (b) a balance?

Answers

1. Knowing the direction of the force will give you only the direction of the acceleration, not the direction of the velocity.

2. The floor exerts an upward contact force, $\vec{F}_n$, on the soles of your feet. The reaction force is the downward push of the soles of your feet on the floor.

3. The forces acting on the car are the gravitational force of the earth, the contact force of the road, and the contact force of the air (air resistance). Because the magnitude and direction of the velocity are constant, the acceleration must be zero. Because the acceleration is zero, it follows from Newton's Second Law ($\vec{F}_{net} = m\vec{a}$) that the net force is also zero.

4. To tell whether or not $\vec{F}' = 2\vec{F}$, place a block on a frictionless table top and accelerate it with the spring extension first at Δx and then at $2\Delta x$. For each of these spring extensions, measure the acceleration of the block. If the second acceleration is twice the first, then $\vec{F}' = 2\vec{F}$.

5. You are not thrown forward. You and your seat are initially moving forward at the same speed. When the brakes are applied, the seat slows down but inertia keeps you moving forward.

6. You are not thrown outward. You and your seat are initially moving forward at the same speed. When the car makes a sharp turn—say to the left—your seat moves with it, and you are pulled to the left literally by the seat of your pants. The inertia of your head and upper torso cause them to lag behind, and it seems as if you are being thrown to the right.

7. (a) The spring scale reading is proportional to the apparent weight of the mass m. When a spring scale is used, the spring must exert an additional upward force to cause the mass to accelerate upward. To do this, the spring must stretch; thus the reading on the spring scale increases. (b) A balance compares the mass m with that of an object of known mass. When a balance is used, the arm of the balance supporting the mass must exert an additional upward force to cause it to accelerate upward. However, the other arm of the balance must also exert an additional upward force on the counterweight—also of mass m—to give it the required acceleration $\vec{a}$. Thus if the balance is balanced when there is no acceleration, the balance will remain balanced when the acceleration differs from zero.

VI. Problems, Solutions, and Answers

Example #1. A 2.00-kg block slides to the right at constant velocity along a frictionless table top, as shown in Figure 4-2. Determine the magnitude and direction of all forces acting on the block.

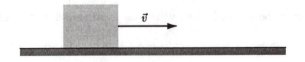

Figure 4-2

1. Draw a free-body diagram of the block. There is only one object touching the block, the table top. Thus there are only two forces acting on the block, the gravitational force of the earth (the weight) and the contact force of the table top. We don't necessarily know the direction of the force of the table, so we draw the vector representing that force in an arbitrary direction.	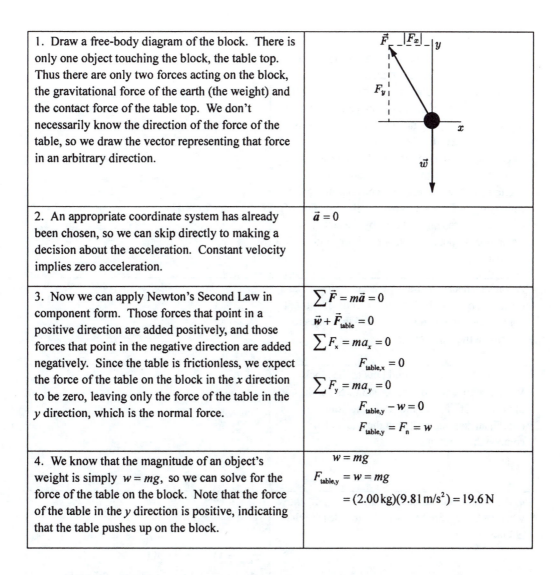
2. An appropriate coordinate system has already been chosen, so we can skip directly to making a decision about the acceleration. Constant velocity implies zero acceleration.	$\vec{a} = 0$
3. Now we can apply Newton's Second Law in component form. Those forces that point in a positive direction are added positively, and those forces that point in the negative direction are added negatively. Since the table is frictionless, we expect the force of the table on the block in the x direction to be zero, leaving only the force of the table in the y direction, which is the normal force.	$\sum \vec{F} = m\vec{a} = 0$ $\vec{w} + \vec{F}_{table} = 0$ $\sum F_x = ma_x = 0$ $\quad F_{table,x} = 0$ $\sum F_y = ma_y = 0$ $\quad F_{table,y} - w = 0$ $\quad F_{table,y} = F_n = w$
4. We know that the magnitude of an object's weight is simply $w = mg$, so we can solve for the force of the table on the block. Note that the force of the table in the y direction is positive, indicating that the table pushes up on the block.	$w = mg$ $F_{table,y} = w = mg$ $\quad = (2.00\,\text{kg})(9.81\,\text{m/s}^2) = 19.6\,\text{N}$

Example #2—Interactive. A 2.00-kg block on a horizontal, frictionless table is steadily pulled to the right by a string, as shown in Figure 4-3. In 5.00 s the speed of the block increases from zero to 10 m/s with constant acceleration. Determine the magnitude and direction of all forces acting on the block. **Try it yourself.** Work the problem on your own, in the spaces provided, to get the final answer.

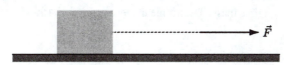

Figure 4-3

1. Draw a free-body diagram. There are two objects touching the block, the table top and the string. Thus there are three forces acting on it, the gravitational force due to the earth (the weight), the contact force of the table top, and the contact force of the string. If the direction of a force is not known, draw the vector representing that force in an arbitrary direction. Use a result from the previous problem—that the contact force exerted by a frictionless surface is normal to the contacting surfaces—to give you the direction of the force exerted on the block by the table top. Remember to put a coordinate axis on your diagram.	
2. Make a decision about the acceleration. The block clearly does not accelerate vertically, but it may accelerate horizontally. Use kinematics to determine the value of a_x.	
3. Apply Newton's Second Law to the block. Write out the equation for Newton's Second Law in vector form for the block. Using your coordinate system, write out the equivalent component equations.	
4. Use the acceleration calculated in step 2, and what we know about the weight, to calculate the unknown forces.	$F_{\text{rope}} = 4\,\text{N}$ to the right $F_{\text{table}} = w = 19.6\,\text{N}$, the force of the table is directed up, and the weight is down.

Example #3. As is shown in Figure 4-4, a 25.0-kg traffic light is suspended from two light strands of wire with negligible mass. Determine the tension in each strand.

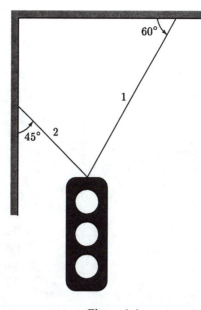

Figure 4-4

1. Draw a free-body diagram of the traffic light. There are two objects touching traffic light, the two strands. Thus there are three forces acting it, the gravitational force of the earth (the weight) and the tension forces of the two strands. The force exerted by one of the strands is in the direction of that strand. The weight is straight down. The object is not accelerating, so we choose a conventional coordinate system.	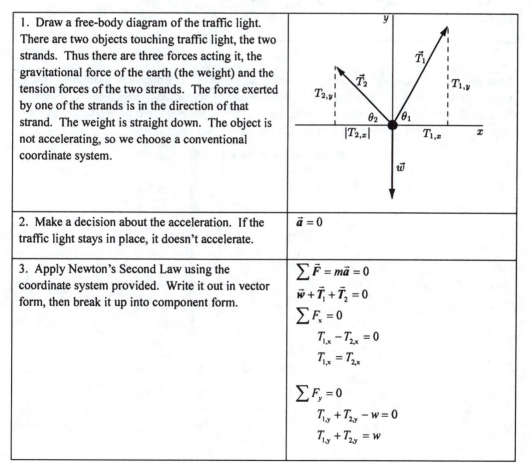
2. Make a decision about the acceleration. If the traffic light stays in place, it doesn't accelerate.	$\vec{a} = 0$
3. Apply Newton's Second Law using the coordinate system provided. Write it out in vector form, then break it up into component form.	$\sum \vec{F} = m\vec{a} = 0$ $\vec{w} + \vec{T}_1 + \vec{T}_2 = 0$ $\sum F_x = 0$ $T_{1,x} - T_{2,x} = 0$ $T_{1,x} = T_{2,x}$ $\sum F_y = 0$ $T_{1,y} + T_{2,y} - w = 0$ $T_{1,y} + T_{2,y} = w$

4. Using the geometry of the problem, we can rewrite the magnitude of the components of $\vec{T_1}$ and $\vec{T_2}$ in terms of their magnitudes and the angles given.	$T_{1,x} = T_1 \cos\theta_1$ $\qquad$ $\left\|T_{2,x}\right\| = T_2 \cos\theta_2$ $T_{1,y} = T_1 \sin\theta_1$ $\qquad$ $T_{2,y} = T_2 \sin\theta_2$
5. Now we can rewrite the x- and y-component equations from step 3.	$T_1 \cos\theta_1 = T_2 \cos\theta_2$ $T_1 \sin\theta_1 + T_2 \sin\theta_2 = w = mg$
6. Now that we have two equations and two unknowns (T_1 and T_2), we can arrive at a solution. Solve the first equation for T_1 and substitute that into the second equation to solve for T_2.	$T_1 = T_2 \dfrac{\cos\theta_2}{\cos\theta_1}$ $T_2 \dfrac{\cos\theta_2}{\cos\theta_1} \sin\theta_1 + T_2 \sin\theta_2 = mg$ $T_2 = \dfrac{mg}{\frac{\cos\theta_2}{\cos\theta_1}\sin\theta_1 + \sin\theta_2}$ $T_2 = \dfrac{(25.0\,\text{kg})(9.81\,\text{m/s}^2)}{\frac{\cos 45°}{\cos 60°}\sin 60° + \sin 45°} = 127\,\text{N}$
7. Finally, take this value for T_2 and use it to solve for T_1.	$T_1 = (127\,\text{N})\dfrac{\cos 45°}{\cos 60°} = 180\,\text{N}$

Example #4—Interactive. A 10.0-kg object is supported by a string that is attached to the ceiling. The object is pulled to the side by a second string connected to the object, as shown in Figure 4-5. Determine the tensions in the two strings. **Try it yourself.** Work the problem on your own, in the spaces provided, to get the final answer.

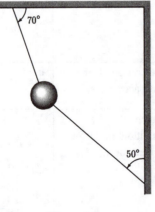

Figure 4-5

1. Draw a free-body diagram of the suspended object. Include an appropriate coordinate system.	
2. Make a decision about the acceleration.	
3. Apply Newton's Second Law for the object. Write out the equation for Newton's Second Law in vector form. Write out the equivalent component equations using the coordinate system chosen in step 2.	
4. Using the geometry of the problem, rewrite the components of $\vec{T}_1$ and $\vec{T}_2$ in terms of their magnitudes and the angles given.	
5. Rewrite the x- and y-component equations from step 3, using the components from step 4.	
6. Solve the two equations in step 5 for the two unknowns.	$T_1 = 150\,\text{N}$ $T_2 = 67\,\text{N}$

Why is T_1 larger than the weight of the 10.0-kg object?

Example #5. Two blocks, connected by a light string as shown in Figure 4-6, are being pulled across a frictionless horizontal table top by a second light string. Block A has twice the mass of block B, the blocks are gaining speed as they move toward the right, and the strings remain taut at all times. Find the ratio T_1 / T_2 of the two tensions.

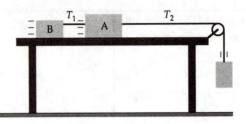

Figure 4-6

1. Draw a separate free-body diagram for each block. There are two objects touching the block B, the table top and a string, and three touching block A, the table top and the two strings. Thus, in addition to a weight force there are two forces acting on block B and three on block A. The table top is frictionless so the forces exerted by it on the blocks are normal to its surface.	**Block B**: y axis, $\vec{F}_{n,B}$ (up), $\vec{T}_{1,B}$ (right), $\vec{w}_B$ (down), x axis. **Block A**: y axis, $\vec{F}_{n,A}$ (up), $\vec{T}_{1,A}$ (left), $\vec{T}_{2,A}$ (right), $\vec{w}_A$ (down), x axis.
2. Make a decision about the acceleration. The blocks are speeding up to the right along the table. Assuming the strings remain taught, the blocks are constrained to move together, so both blocks must have the same acceleration.	$\vec{a}_A = a_{x,A}\hat{i}$ $\vec{a}_B = a_{x,B}\hat{i}$
3. Apply Newton's Second Law to block A. Write out the equation for Newton's Second Law in vector form for the block. Write out the equivalent component equations using the coordinate system provided.	$\sum \vec{F} = m\vec{a}_A$ $\vec{w}_A + \vec{F}_{n,A} + \vec{T}_{1,A} + \vec{T}_{2,A} = m_A \vec{a}_A$ $\sum F_x = m_A a_{x,A}$ $\quad T_{2,x,A} - T_{1,x,A} = m_A a_{x,A}$ $\sum F_y = 0$ $\quad F_{n,A} - w_A = 0$ $\quad F_{n,A} = w_A$
4. Do the same thing for block B.	$\sum \vec{F} = m\vec{a}_B$ $\vec{w}_B + \vec{F}_{n,B} + \vec{T}_{1,B} = m_B \vec{a}_B$ $\sum F_x = m_B a_{x,B}$ $\quad T_{1,x,B} = m_B a_{x,B}$ $\sum F_y = 0$ $\quad F_{n,B} - w_B = 0$ $\quad F_{n,B} = w_B$

5. Assuming the strings remain taut, but do not stretch, both blocks are constrained to accelerate together.	$\vec{a}_A = \vec{a}_B = a_x \hat{i}$ $a_{x,A} = a_{x,B} = a_x$
6. The magnitude of the tension in an ideal massless string is everywhere the same. See Remark 4 below.	$T_{1,A} = T_{1,B} = T_1$
7. Using the equations for the x-components of Newton's Second Law, we can now solve for the ratio of T_1 / T_2.	B: $T_1 = m_B a$ A: $T_2 - T_1 = m_A a$ $T_2 - m_B a = m_A a$ $T_2 = (m_A + m_B)a$ $\dfrac{T_1}{T_2} = \dfrac{m_B a}{(m_A + m_B)a} = \dfrac{m_B}{m_A + m_B} = \dfrac{m}{3m} = \dfrac{1}{3}$

Remark 1: In applying Newton's Second Law to the two blocks only the x-component equations were utilized. The y-component equations are valid but do not contribute to solving for the tensions as long as friction is negligible.

Remark 2: Tension T_1 is the only force accelerating block B. To give block A the same acceleration requires twice the force since block A has twice the mass. Block A has a force T_1 holding it back. For the net force on block A to equal $2T_1$ in the forward direction, it follows that T_2 must equal $3T_1$.

Remark 3: Applying Newton's Second Law to block A resulted in an equation where each term had a subscript A. Likewise, in repeating this process for block B the each term had a subscript B. This subscripting was done so that the symbol for each force or acceleration term be unique. It is advisable to follow this practice in order to avoid mix-ups.

Remark 4: In Step 6, we set $T_{1,A} = T_{1,B}$. These tensions represent the tension in rope 1 at the right and left ends, respectively. The tensions in these two ends are only approximately equal. By drawing a free-body diagram (see Figure 4-7) of string 1 and by applying Newton's Second Law, we now show that the tensions at opposite ends are equal only in the limit that the string's mass approaches zero.

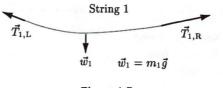

Figure 4-7

Applying Newton's Second Law to the string gives:

$$\vec{T}_{1R} + \vec{T}_{1L} + m_1\vec{g} = m_1\vec{a}_1$$

where m_1 is the mass of string 1. Taking the limit of both sides as $m_1 \to 0$ we obtain

$$\vec{T}_{1R} + \vec{T}_{1L} = 0$$

which means the tension forces acting at opposite ends of the string are *equal in magnitude and oppositely directed.*   This is true even though the force of gravity pulls downward on the string and even though it is being accelerated to the right. The gravitational force is proportional to the string's mass, so a negligible mass results in a negligible gravitational force and, as a result, a negligible sag. Since the mass of the string being accelerated is negligible, to cause it to accelerate to the right requires that the tension force toward the right exceeds the tension force toward the left by only a negligible amount.

Example #6—Interactive.   Two blocks, connected by a light string as shown in Figure 4-8, are sliding toward the left across a frictionless horizontal table top. A second light string attached to block B pulls it to the right and as a result the blocks lose speed. Block A has two times the mass of block B and the strings remain taut at all times. Find the ratio T_1/T_2 of the two tensions. **Try it yourself.** Work the problem on your own, in the spaces provided, to get the final answer.

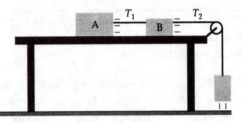

Figure 4-8

1. Draw a separate free-body diagram for each block. There are three objects touching block B, the table top and two strings, and two touching block A, the table top and one string. Thus, in addition to a weight force there are three forces acting on block B and two on block A. Include a reasonable coordinate system.	
2. Make a decision about the acceleration.	
3. Apply Newton's Second Law to block A. Write out the equation for Newton's Second Law in vector form for the block. Write out the equivalent component equations using your coordinate system.	

4. Repeat this process for block B.	
5. Apply constraints for the accelerations and tensions.	
6. Solve for the ratio.	$\dfrac{T_1}{T_2} = \dfrac{2}{3}$

Why is this result different from that in Example 5?

Chapter **5**

Applications of Newton's Laws

I. Key Ideas

Recommendations for problem solving relevant to this chapter are discussed at some length in Section 4-5 of both the main text and this Study Guide. These recommendations should be reread and learned well. It is strongly recommended that you develop the habit of approaching a problem by first making a drawing of the situation. Then, for each object, draw a free-body diagram illustrating the forces acting on it.

When working with objects that are in contact, like a box sliding down a ramp, the contacting surfaces deform slightly and exert forces on each other. These forces are distributed over the contacting surfaces. The net contact force acting on a single surface is the vector sum of the distributed contact forces that act throughout that surface. It is often useful to think of this contact force as the sum of two distinct contact forces, one normal to and one parallel to the surface. (The word normal means perpendicular.) These two forces are referred to as the **normal force** and the **friction force,** respectively.

Section 5-1. Friction. The friction force is called the force of *static friction,* $\vec{f}_s$, when the surfaces in contact *do not* slide on each other. Static frictional forces oppose any tendencies for the contacting surfaces to slide. For a given pair of contacting surfaces, the magnitude of the static frictional force f_s ranges anywhere from zero to a maximum value $f_{s,max}$ that is proportional to the magnitude of the normal force F_n. This empirically established relation is expressed as

$$f_s \le \mu_s F_n \quad \text{or} \quad f_{s,max} = \mu_s F_n \qquad\qquad \text{Static friction}$$

where μ_s is the **coefficient of static friction**.

When skidding occurs, that is, when the surfaces in contact slide across each other, the frictional force is called the force of *kinetic friction,* $\vec{f}_k$. The magnitude of the force of kinetic friction is given by the relation

$$f_k = \mu_k F_n \qquad\qquad \text{Kinetic friction}$$

where μ_k is the coefficient of kinetic friction. The coefficients of static and kinetic friction are small for slippery surfaces and larger for surfaces that do not easily slide across one another. Values for μ_s and μ_k for a sampling of materials are listed in Table 5-1 on page 120 of the textbook. The two coefficients depend upon the nature of the two surfaces in contact.

When an ideal, rigid wheel rolls *at constant speed* along an ideal, rigid horizontal road without slipping, no frictional force slows its motion. However, because real tires and roads continually deform and because the tread and road are continually peeled apart, the road exerts a force of **rolling friction,** $\vec{f_r}$, that opposes the motion. The **coefficient of rolling friction,** μ_r, is the ratio of the magnitudes of the rolling frictional force f_r and the normal force F_n:

$$f_r = \mu_r F_n \qquad\qquad \text{Rolling friction}$$

Guidelines for Solving Problems Involving Friction

1. Select a coordinate system with the x axis in the direction of motion and the y axis in the normal direction. With this choice the normal force F_n appears only in the y component equation and the frictional force f_s or f_k appears only in the x component equation.

2. With the above choice of axes first solve for F_n using the y component equation, and then solve for f_k using the relation $f_k = \mu_k F_n$. Substitute this result for f_k into the x component equation and solve for the desired unknown. (If the friction is static and a maximum, then substitute $f_{s,max}$ for f_k and μ_s for μ_k.)

Section 5-2. Motion Along a Curved Path. The direction of the acceleration vector is always in the direction of the change in the velocity vector $\Delta\vec{v}$ during a sufficiently short time interval. As discussed in Section 3-5, the acceleration of a particle moving along a circular path has both a tangential component and a centripetal component. As a result, a particle moving along a curved path might experience two components of force. A *centripetal* component of the force directed toward the center of the circle is responsible for making the particle move in a circle, and the *tangential* component of the force will cause the particle to speed up or slow down.

It is important to realize that the centripetal and tangential forces are not new forces to be added to a free-body diagram. The centripetal and tangential forces are the two components of the resultant force acting on the particle. This resultant force is due to all or some of the forces you have already learned about: tension, normal force, friction force, etc. We only use the words centripetal and tangential as descriptors for those forces that happen to be involved in circular motion for a specific situation.

Section 5-3. Drag Forces. When an object moves through a fluid such as air or water, the fluid exerts a **drag force** that opposes the motion of the object through the fluid. Unlike the force of kinetic friction, this force increases with the speed of the object. This drag force increases with the first power of the speed for slow-moving objects and with the square of the speed for faster-moving objects. The magnitude of the drag force is related to the speed by the expression

$$F_{drag} = bv^n \qquad\qquad \text{Fluid drag}$$

Here b depends on the size, shape, and roughness of the object along with the density and the viscosity of the fluid; and n varies from 1 to 2 as the speed increases and the flow becomes more turbulent. For low speeds, where turbulence is negligible, n equals 1, and for higher speeds, where turbulence dominates, n equals 2. When an object is falling straight down through a fluid, the drag force increases as the speed of the object increases until it reaches terminal speed v_t. At terminal speed the acceleration is zero, so the drag force equals the object's weight. It follows that

$$v_t = \left(\frac{mg}{b}\right)^{1/n}$$
\hfill Terminal speed

For a sky diver, terminal speed is about 60 m/s (134 mi/hr).

Section 5-4. Numerical Methods: Euler's Method. In many circumstances it is possible to integrate the equation gotten by applying Newton's 2nd Law, $\vec{F}_{net} = m\vec{a}$, to an object to determine its position in terms of a simple algebraic function of time. However, in some circumstances this is either very inconvenient or actually impossible to do. In such situations we resort to numerical methods. Numerical methods require knowledge of the dependence of the acceleration on position, velocity, and time.

The basic idea of Euler's method of numerical integration is to divide the time interval into a large number of short intervals and to then use the equations for motion with constant acceleration to solve for the changes in velocity and position of an object during the first short time interval. These new values of velocity and position are then used to update the value of the acceleration that is used to calculate the changes in the velocity and position during the second short time interval, and so forth.

II. Physical Quantities and Key Equations

Physical Quantities

There are no new physical quantities for this chapter.

Key Equations

Static friction $\qquad\qquad$ $f_{s,max} = \mu_s F_n$ $\qquad$ or $\qquad$ $f_s \leq \mu_s F_n$

Kinetic friction $\qquad\qquad$ $f_k = \mu_k F_n$

Rolling friction $\qquad\qquad$ $f_r = \mu_r F_n$

Fluid drag $\qquad\qquad$ $F_{drag} = bv^n$

Terminal speed $\qquad\qquad$ $v_t = \left(\frac{mg}{b}\right)^{1/n}$

III. Potential Pitfalls

The value of the force of static friction f_s is not always equal to $\mu_s F_n$. The *maximum* value that f_s can have is $\mu_s F_n$. Remember that while static friction is always directed so as to oppose relative motion between the contacting surfaces, static friction does not necessarily oppose motion. For example, when you start walking on a horizontal surface, like the floor of a classroom, it is the force of static friction exerted by the surface of the floor on the soles of your shoes that pushes you in the direction that you start to move. This force, which is in the forward direction, opposes relative motion between your foot and the floor by preventing your foot from sliding backward.

Do not treat the net force as a separate force. The net force is just a name for the vector sum of all the forces acting on an object. The centripetal force is not a separate force either. It is the name given to the centripetal component of the net force. For motion along a general curved path, the centripetal direction is the direction toward the center of curvature of the path; thus for a circular path it is toward the center of the circle.

When two surfaces come into contact, both a normal force component and a frictional force component of the contact force act on each surface.

You should not assume that the normal force on an object is equal to its weight. It is equal to the weight only for specific situations. The normal force is best determined by applying Newton's 2nd Law, $\vec{F} = m\vec{a}$, in component form to the object. In doing this, choose one coordinate axis to be parallel to the normal force and the other parallel to the frictional force. For an object sliding along a flat surface the normal component of the acceleration is zero.

When you use numerical methods to do integration, do not set the time intervals too small or the round-off errors will be too large. The trick is to set them small enough to determine the value of the final velocity or position to the desired accuracy, but not so small that the round-off errors become excessive. One way to tell if you have set the time intervals small enough is to reduce them by a factor of four and then recalculate the results. If the results do not appreciably change, the size of the time interval is probably satisfactory.

IV. True or False Questions and Responses

True or False

_____ 1. The force of static friction is the force that must be exerted to start two surfaces sliding against each other.

_____ 2. It takes more force to keep a given pair of surfaces sliding against one another than it does to get them started.

_____ 3. The speed of an object falling through a fluid under the influence of gravity approaches a limiting value.

_____ 4. The coefficient of kinetic friction is directly proportional to the normal force.

_____ 5. The magnitude of the kinetic frictional force is directly proportional to the magnitude of the normal force as long as the kinetic coefficient of friction remains constant.

_____ 6. The magnitude of the static frictional force is directly proportional to the magnitude of the normal force as long as the static coefficient of friction remains constant.

Responses to True or False

1. False. The force of static friction is the component of the contact force that opposes the sliding of the surfaces against each other. Another force creates the tendency to slide.

2. False. It takes more force to start the two surfaces sliding against each other than it does to keep them sliding. Anyone who has had to slide a heavy object like a piano should know this.

3. True.

4. False. While the force of kinetic friction is directly proportional to the normal force, the coefficient of kinetic friction only depends upon the nature of the surfaces.

5. True.

6. False. It is only the maximum static frictional force that is proportional to the normal force.

V. Questions and Answers

Questions

1. Why are curved roads banked?

2. Do the coefficients of either static or kinetic friction ever exceed the number one?

3. When you are standing still and then start walking, what outside force acts upon you to cause your acceleration?

4. When a car travels around an unbanked curve, friction between the tires and the road provides the needed centripetal force. Is this friction kinetic or static?

5. You are driving a car with rear-wheel drive. On an icy road the car starts to skid off the road so you turn the front wheels toward the direction of the skid. Why?

6. Certain racing cars have an airfoil over the top of the car. The airfoil deflects air upward. What advantage is gained by deflecting the air upward when the car turns corners?

Answers

1. Curved roads are banked so the normal force of the road on the tires has a centripetal component. This component supplements the frictional force; that way friction alone does not have to provide the necessary centripetal force, and there is less danger of skidding off the curve.

2. Yes, they sometimes exceed one. There is nothing special about a coefficient of friction of unity. The coefficient of static friction between a dragster (a car used in drag races) tire and the track is typically about three.

3. The floor applies a frictional force to the soles of your shoes in response to your shoes pushing the floor.

4. Hopefully, it is static friction. If it were kinetic, the tires would skid on the surface.

5. You turn into the skid to reestablish static friction between the front tires and the road. Static friction can occur only when the wheels are rolling in the direction of the car's motion.

6. According to Newton's 3rd Law, when the foil deflects the air upward the air will push the foil downward. This increases the normal force of the car on the road, which is desirable because the maximum frictional force is proportional to the normal force, and it is this frictional force that provides the necessary centripetal acceleration.

VI. Problems, Solutions, and Answers

Example #1. A 2.00-kg block on a horizontal table is pulled to the right by a constant force of 16.0 N, as shown in Figure 5-1. The acceleration of the block is $3.00\,\text{m/s}^2$. Determine the coefficient of kinetic friction between the block and the table top.

Figure 5-1

Picture the Problem. Draw a free-body diagram of the block, and use Newton's 2nd Law to determine the frictional force and the coefficient of kinetic friction.

1. Draw a free-body diagram of the block.	
2. Write Newton's 2nd Law in its vector form for the block. This is really separate equations for each of the x and y components.	$\sum \vec{F} = \vec{F}_{\text{app}} + \vec{f}_k + \vec{F}_n + \vec{w} = m\vec{a}$
3. Sum the forces in the vertical direction and use Newton's 2nd Law to determine the normal force. The acceleration in this direction is zero.	$\sum F_y = F_n - w = 0$ $F_n = w = (2.00\,\text{kg})(9.81\,\text{m/s}^2) = 19.6\,\text{N}$
4. Sum the forces in the horizontal direction and use Newton's 2nd Law and the given acceleration to find the frictional force.	$\sum F_x = F_{\text{app}} - f_k = ma$ $f_k = F_{\text{app}} - ma = 16.0\,\text{N} - (2.00\,\text{kg})(3.0\,\text{m/s}^2)$ $= 10.0\,\text{N}$

5. Use the relation between the normal force and the frictional force to calculate the coefficient of kinetic friction.	$f_k = \mu_k F_n$ $\mu_k = f_k / F_n = (10.0\,\text{N})/(19.6\,\text{N}) = 0.510$

Example #2—Interactive   A 2.00-kg block on a horizontal table is pulled by a constant force of 16.0 N at an angle of 20° above the horizontal, as shown in Figure 5-2. The kinetic coefficient of friction is 0.510. Determine the acceleration of the block.

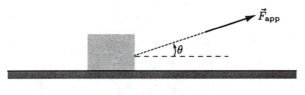

Figure 5-2

Picture the Problem. Draw a free body diagram of the block, and use Newton's 2nd Law in the vertical direction to determine the normal force, and in the horizontal direction to determine the acceleration. **Try it yourself.** Work the problem on your own, in the spaces provided, to get the final answer.

1. Draw a free-body diagram of the block.	
2. Write Newton's 2nd Law in its vector form for the block.	
3. Sum the forces in the vertical direction and use Newton's 2nd Law to determine the normal force. The acceleration in this direction is zero.	
4. Use the relation between the normal force and the frictional force to calculate the coefficient of kinetic friction.	
5. Sum the forces in the horizontal direction and use Newton's 2nd Law to find the acceleration.	$a = 3.9\,\text{m/s}^2$

Example #3. A 0.100-kg steel ball is suspended from the ceiling by a light string 0.400 m long, as shown in Figure 5-3. A physics instructor moves the ball to the side and then flicks it so that the ball passes through the lowest point in the arc with a speed of 0.800 m/s. Determine the tension in the string when the ball is passing through the lowest point in its arc.

Picture the Problem. At the bottom of the arc, the ball will have a centripetal acceleration directed upward. This acceleration, and the mass of the ball, can be used to find the tension in the string using Newton's 2nd law.

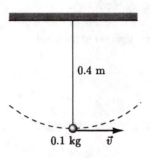

0.4 m

0.1 kg $\vec{v}$

Figure 5-3

1. Draw a free-body diagram of the ball at the lowest point in its arc. The tension from the string is the only force acting on the ball other than its weight.	$\vec{T}$ $\qquad$ $\hat{j}$ $\qquad$ $\hat{i}$ $\qquad$ $\vec{w}$
2. Write Newton's 2nd Law in its vector form for the ball.	$\sum \vec{T} + \vec{w} = m\vec{a}$
3. Apply Newton's 2nd Law in the vertical direction. Because the ball is at the bottom of its circular arc, we know that the centripetal component of the acceleration is directed in the positive $\hat{j}$ direction.	$\sum F_y = T - w = ma_y$ $w = mg$ $a_y = a_c = v^2 / R$ $T = w + ma_y = mg + m\dfrac{v^2}{R}$ $T = \left(0.100\,\text{kg}\right)\left(\left(9.81\,\text{m/s}^2\right) + \dfrac{\left(0.800\,\text{m/s}\right)^2}{0.400\,\text{m}}\right)$ $T = 1.14\,\text{N}$
4. Apply Newton's 2nd Law in the horizontal direction. There can be no acceleration in the $\hat{i}$ direction because there are no forces along that direction.	$\sum F_x = T_x - w_x = ma_x = 0$

Remark. The tension in the string exceeds the weight of the ball by 0.16 N. You may not have expected the tension to exceed the weight of the ball. The ball is accelerating upward so the net force acting on it must also be upward. For this to occur the tension must exceed the weight.

Example #4—Interactive. A boy is swinging a 0.100-kg yo-yo in a vertical circle of radius 0.400 m by a light string, as shown in Figure 5-4. When the string is horizontal and the yo-yo is moving straight upward, the speed of the ball is 2.00 m/s. At that instant, determine the tension in the string and the rate of change of the yo-yo's speed.

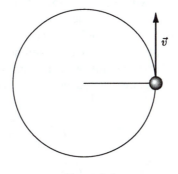

Figure 5-4

Picture the Problem. At the point of interest, the ball will have a centripetal acceleration directed to the left, toward the center of the circle. This acceleration can be used to find the tension in the string from Newton's 2nd Law applied to the horizontal direction. However, because the ball's weight acts tangential to the motion, in the downward direction, the ball will also have a changing speed, which can be found by applying Newton's 2nd Law to the vertical forces and acceleration. **Try it yourself.** Work the problem on your own, in the spaces provided, to get the final answer.

1. Draw a free-body diagram of the yo-yo.	
2. Write Newton's 2nd Law in vector form for the yo-yo.	
3. Apply Newton's 2nd Law in the horizontal direction to find the tension in the string.	$T = 1.00\,\text{N}$
4. Apply Newton's 2nd Law in the vertical direction to find the rate at which the ball's speed is changing.	$a_y = -9.81\,\text{m/s}^2$

Example #5. A car is moving at 33.0 m/s down a hill that makes an angle of 20° with the horizontal. If the coefficient of static friction between the tires and the road is 0.580, what is the shortest distance in which the car can stop?

Picture the Problem. Draw both a sketch (Figure 5-5) and free-body diagram of the car. The latter will include the weight of the car, as well as a normal force and frictional force. Use a rotated coordinate system so that the x axis is parallel to the plane. This choice means the normal force is entirely in the y direction, and the acceleration will be only along the x direction. Use Newton's 2nd Law to find the acceleration of the car, and kinematics to find the stopping distance. The shortest stopping distance will require the maximum static friction force between the tires and the road.

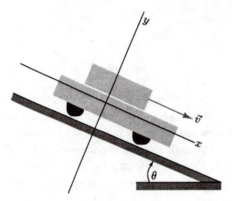

Figure 5-5

1. Draw a free-body diagram.	
2. Write Newton's 2nd Law in vector form, which is really separate equations for each of the x and y components.	$\sum \vec{F} = \vec{F}_n + \vec{f}_s + \vec{w} = m\vec{a}$ $\sum F_x = f_{s,x} + F_{n,x} + w_x = ma_x$ $\sum F_y = f_{s,y} + F_{n,y} + w_y = ma_y$
3. We will first consider the equation for the y components. The car does not accelerate perpendicular to the plane, so this component of the acceleration is zero. We also need the y components of all the forces on the free-body diagram. We can then solve for F_n.	$a_y = 0 \qquad\qquad f_{s,y} = 0$ $w_y = -mg\cos\theta \qquad F_{n,y} = F_n$ $\sum F_y = 0 + F_n - mg\cos\theta = m(0)$ $F_n = mg\cos\theta$

4. Before considering the equation for the x components, we need the x components of each force. This enables us to solve for the acceleration of the car.	$a_x \neq 0$ $\qquad\qquad\qquad$ $f_s = -\mu_s F_n$ $w_x = mg\sin\theta$ $\qquad\qquad$ $F_{n,x} = 0$ $\sum F_x = -\mu_s mg\cos\theta + 0 + mg\sin\theta = ma_x$ $a_x = \mu_s g\cos\theta - g\sin\theta$
5. Now that we know the acceleration, we can find the distance traveled by the car using the time-independent kinematic equation.	$v^2 = v_0^2 + 2a_x\,\Delta x$ $\Delta x = \dfrac{v^2 - v_0^2}{2a_x}$ $\Delta x = \dfrac{0 - (33.0\,\text{m/s})^2}{-2g(\mu_s\cos\theta - \sin\theta)}$ $\Delta x = \dfrac{1089\,\text{m}^2/\text{s}^2}{2(9.81\,\text{m/s}^2)(0.580\cos 20^\circ - \sin 20^\circ)} = 273\,\text{m}$

Remark. For a real challenge, try solving this problem using a coordinate system with a horizontal x axis and a vertical y axis. The algebra is more complex, but it can certainly be done.

Example #6—Interactive. A block is projected at an initial speed of 3.50 m/s straight up a ramp which makes a 26° angle with the horizontal. The coefficient of kinetic friction between the block and the incline is 0.300. How far up the incline will it slide before stopping?

Picture the Problem. Draw both a sketch and a free-body diagram of the car. The latter will include the weight of the car, as well as a normal force and frictional force. Use a rotated coordinate system so that the x axis is parallel to the plane. Use Newton's 2nd Law to find the acceleration of the car and kinematics to find the stopping distance. **Try it yourself.** Work the problem on your own, in the spaces provided, to get the final answer.

1. Draw a sketch and a free-body diagram.	
2. Write Newton's 2nd Law in vector form for the block.	
3. Consider the y components first and find an expression for the normal force.	

4. Use the *x* components to find the acceleration.	
5. Use the time-independent kinematic equation to find the distance the block travels.	$x = 0.883\,\text{m}$

Example #7.   In Figure 5-6, a block of wood of unknown mass m_1 rests on a 5.00-kg block, which in turn rests on a table top. The blocks are connected by a light string that passes over a frictionless peg. The coefficient of kinetic friction at both surfaces is $\mu_k = 0.330$. The upper block is pulled to the left by a 60.0-N force, and the lower block is pulled to the right by the string that passes around the peg. The blocks are moving at a constant speed. Determine the mass m_1 of the upper block.

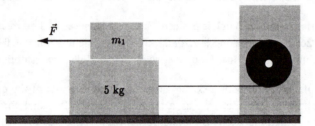

Figure 5-6

Picture the Problem. Draw free-body diagrams of both blocks. Remember that by Newton's 3rd Law, the friction and normal forces of the top block on the bottom block are equal and opposite to those same forces of the bottom block on the top block. Apply Newton's 2nd Law to both blocks. The acceleration of both blocks should be zero.

1. Draw a free-body diagram of each block. The top block experiences a normal force $\vec{F}_{n1}$ from the bottom block. Because the top block will tend to slide to the left with respect to the bottom block, the frictional force $\vec{f}_{k1}$ from the bottom block will oppose this relative motion, and so be directed to the right. The top block also experiences the applied force, tension from the string, and weight. The bottom block experiences both a friction $\vec{f}_{k2}$ and a normal $\vec{F}_{n2}$ force from the top block, which are Newton's 3rd Law pairs to the forces from the bottom block on the top	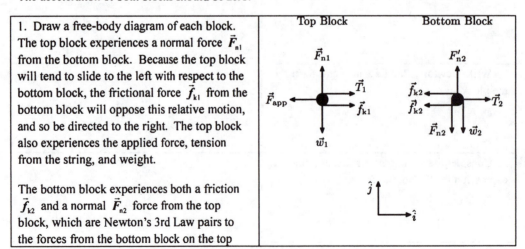

block. The lower block will also experience friction $\vec{f}'_{k2}$ and normal $\vec{F}'_{n2}$ forces from the table top. The applied force is not exerted on the bottom block, so the only other forces it feels are its weight and the tension.	
2. Apply Newton's 2nd law to the top block.	$\vec{F}_{app} + \vec{w}_1 + \vec{T}_1 + \vec{f}_{k1} + \vec{F}_{n1} = m_1\vec{a}_1 = 0$
3. Work with the equation for the y components to find the normal force.	$F_{app,y} + w_{1y} + T_{1y} + f_{k1y} + F_{n1y} = 0$ $0 - m_1 g + 0 + 0 + F_{n1} = 0$ $F_{n1} = m_1 g$
4. Now use the x equation, making sure to relate the kinetic friction force to the normal force we just solved for.	$F_{app,x} + w_{1x} + T_{1x} + f_{k1x} + F_{n1x} = 0$ $-F_{app} + 0 + T_1 + f_{k1} + 0 = 0$ $-F_{app} + T_1 + \mu_k m_1 g = 0$
5. Apply Newton's 2nd Law to the bottom block.	$\vec{w}_2 + \vec{T}_2 + \vec{f}_{k2} + \vec{F}_{n2} + \vec{f}'_{k2} + \vec{F}'_{n2} = m_2\vec{a}_2 = 0$
6. Work with the equation for the y components to find the normal force $\vec{F}'_{n2}$. Remember that $\vec{F}_{n1}$ and $\vec{F}_{n2}$ are 3rd Law pairs.	$w_{2y} + T_{2y} + f_{k2y} + F_{n2y} + f'_{k2y} + F'_{n2y} = 0$ $-w_2 + 0 + 0 - F_{n2} + 0 + F'_{n2} = 0$ $F'_{n2} = w_2 + F_{n2} = m_2 g + F_{n1} = (m_1 + m_2)g$
7. Work with the equation for the x components to solve for the tension. Remember to relate the normal force we just found with the friction force.	$w_{2x} + T_{2x} + f_{k2x} + F_{n2x} + f'_{k2x} + F'_{n2x} = 0$ $0 + T_2 - f_{k2} + 0 - f'_{k2} + 0 = 0$ $T_2 = f_{k2} + f'_{k2} = f_{k2} + \mu_k(m_1 + m_2)g$
8. Because $\vec{f}_{k2}$ and $\vec{f}_{k1}$ are 3rd Law pairs, their magnitudes must be equal. Substituting f_{k1} in for f_{k2}, we can solve for the tension. Because we can neglect the mass of the string, the tension in the rope is everywhere the same. That is, $T_1 = T_2 = T$.	$T = \mu_k m_1 g + \mu_k(m_1 + m_2)g = \mu_k(2m_1 + m_2)g$
9. Substituting the expression from step 8 into the result from step 4, we can solve for m_1.	$-F_{app} + \mu_k(2m_1 + m_2)g + \mu_k m_1 g = 0$ $3\mu_k m_1 g = F_{app} - \mu_k m_2 g$ $m_1 = \dfrac{1}{3}\left(\dfrac{F_{app}}{\mu_k g} - m_2\right)$ $= \dfrac{1}{3}\left(\dfrac{60.0\,\text{N}}{0.330\left(9.81\,\text{m/s}^2\right)} - \left(5.00\,\text{kg}\right)\right) = 4.51\,\text{kg}$

Example #8—Interactive. Assume the mass of the upper block in the previous problem is 4.53 kg. If the force pulling the upper block to the left is equal to 75.0 N, determine the acceleration of each block.

Picture the Problem. Except that the accelerations are no longer zero, follow the same procedure as the previous problem. Because of the physical constraint of the string, the motion of the upper block to the left is identical to the motion of the lower block to the right. To simplify the mathematics, it is often easier to use a different coordinate system for each block, so that both blocks move in the "positive" direction at the same time. **Try it yourself.** Work the problem on your own, in the spaces provided, to get the final answer.

1. Draw a free-body diagram of each block. Use a different coordinate system for each so that both blocks move in the "positive" direction at the same time.	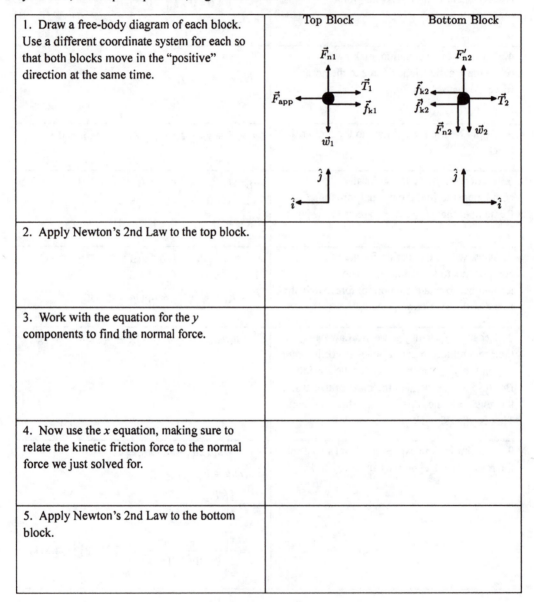
2. Apply Newton's 2nd Law to the top block.	
3. Work with the equation for the y components to find the normal force.	
4. Now use the x equation, making sure to relate the kinetic friction force to the normal force we just solved for.	
5. Apply Newton's 2nd Law to the bottom block.	

6. Work with the equation for the y components to find the normal force $\vec{F}'_{n2}$.	
7. Work with the equation for the x components to solve for the tension.	
8. Substituting the expression from step 7 into the result from step 4, we can solve for a.	$a = 1.55\,\text{m/s}^2$

Chapter **6**

Work and Energy

I.   Key Ideas

Under certain conditions, energy is transferred between objects through the forces they exert on each other. This kind of energy transfer is called work. Work—like force, mass, and numerous other terms that have both a scientific and an everyday usage—has a precise scientific meaning which, like kinetic energy, arises from the work-kinetic energy theorem. Kinetic energy is energy associated with motion. Another form of energy is potential energy, energy associated with the configuration of a system. Potential energy changes are related to the work done by forces that are independent of the path traveled.

Section 6-1. Work and Kinetic Energy. In one dimension, the **work** done by a constant force on an object is the product of the force and the displacement of the point of application of the force. If the force and the displacement are in different directions (as shown in Figure 6-1), then only the component of the force in the direction of the displacement is multiplied by the displacement.

$$W = F_x \, \Delta x = F \cos \theta \, \Delta x$$

Work in one dimension by a constant force

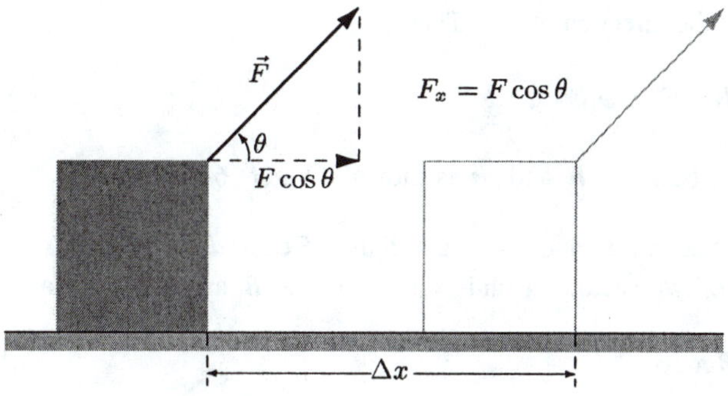

Figure 6-1

Work (which can be either positive or negative) is a scalar quantity and thus has no direction. When the force acts at right angles to the displacement, the work done by the force is zero.

When an object undergoes a displacement, each force acting on it may, in principle, do work. The **total work** W_{total} done on an object is the algebraic sum of the work done by each of the forces acting on it. An extended object can be modeled as a particle only if all of its parts undergo *identical* displacements during any and all time intervals. That is, an object is a particle if it is perfectly rigid and moves without rotating. The following discussion applies only to objects that can be modeled as particles. It follows that for a particle, the total work done on it equals the product of its displacement and the net force acting on it.

$$W_{\text{total}} = F_{\text{net},x}\Delta x \qquad\qquad \text{Total work}$$

The **work-kinetic energy theorem**, which follows from Newton's 2nd Law ($\Sigma \vec{F} = m\vec{a}$), states that the total work W_{total} done on a particle equals its change in kinetic energy K.

$$W_{\text{total}} = \Delta K = \tfrac{1}{2}mv_f^2 - \tfrac{1}{2}mv_i^2 \qquad\qquad \text{Work-kinetic energy theorem}$$

where the kinetic energy of a particle is defined to be

$$K = \tfrac{1}{2}mv^2 \qquad\qquad \text{Kinetic energy}$$

When a variable force does work, we can determine the work done by dividing the net displacement into a large number of very short displacements. We can estimate the work done in each short displacement by multiplying that displacement and the value of the force at the beginning of that displacement. The overall work done by a force acting over a displacement equals the sum of the work done by the force acting over a sequence of infinitesimal displacements that make up the displacement. That is,

$$W = \int_{x_1}^{x_2} F_x\,dx = \text{area under the } F_x\text{-versus-}x \text{ curve} \qquad\qquad \text{Work done by a variable force}$$

Section 6-2. The Dot Product. What if a particle does not move in a straight line? How does one calculate the work done by a force on a particle that moves along a curved path? The differential amount of work done by the force $\vec{F}$ acting over a small displacement $d\vec{s}$ is equal to ds times the component of $\vec{F}$ in the direction of $d\vec{s}$. That is,

$$dW = F_s\,ds = F\cos\phi\,ds = \vec{F}\cdot d\vec{s} \qquad\qquad \text{Work}$$

where ϕ is the angle between $\vec{F}$ and $d\vec{s}$ as shown in Figure 6-2.

Multiplying the vectors $\vec{F}$ and $d\vec{s}$ to get $F_s\,ds = F\cos\theta\,ds$ gives their **scalar or dot product**. The scalar product of two vectors $\vec{A}$ and $\vec{B}$ is written $\vec{A}\cdot\vec{B}$ and is defined as

$$\vec{A}\cdot\vec{B} = AB\cos\phi \qquad\qquad \text{Dot product}$$

where ϕ is the angle between them. This product is related to the x, y, and z components of the two vectors by

$$\vec{A} \cdot \vec{B} = A_x B_x + A_y B_y + A_z B_z \qquad\qquad \text{Dot product}$$

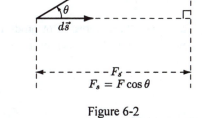

Figure 6-2

The dot product of a unit vector and a vector $\vec{A}$ gives the component of $\vec{A}$ in the direction of the unit vector. For example,

$$\hat{i} \cdot \vec{A} = A \cos \phi = A_x \qquad\qquad \text{Projecting } \vec{A} \text{ onto the } x \text{ axis}$$

$A \cos \phi$ is the projection of $\vec{A}$ onto the x axis, where ϕ is the angle between $\vec{A}$ and $\hat{i}$. Consequently, taking the dot product of a unit vector and a vector can be thought of as projecting the vector onto a line in the direction of the unit vector. This may not seem particularly significant now, but thinking of the dot product of a unit vector and a vector in this way will prove useful in later chapters.

Power is the rate at which a force does work, or the work done per unit time. The instantaneous power P associated with the instantaneous force $\vec{F}$ acting on an object is

$$P = \frac{dW}{dt} = \frac{\vec{F} \cdot d\vec{s}}{dt} = \vec{F} \cdot \frac{d\vec{s}}{dt} = \vec{F} \cdot \vec{v} \qquad\qquad \text{Power}$$

where $\vec{v}$ is the velocity of the point of application of the force. The SI unit of power, one joule per second, is called the watt (W).

$$1 \text{ W} = 1 \text{ J/s}$$

More generally, the term power refers not just to the rate of doing work but to any rate of energy transfer.

Since power is the rate of change of work, and the work done on a particle equals the change in kinetic energy, it follows that the total power associated with the net force on a particle equals the rate of change of the particle's kinetic energy. That is,

$$P_{total} = \vec{F}_{net} \cdot \vec{v} = \frac{dW_{total}}{dt} = \frac{dK}{dt}$$

Rearranging this relation, we find that

$$dK = P_{total}\, dt \quad \text{or} \quad \Delta K = P_{total}\, \Delta t \text{ (constant power)} \qquad \text{Kinetic energy and power}$$

if the total power is constant.

While a complex system like a car cannot be modeled as a particle, the incremental relation can be applied to a car if P_{total} includes the power delivered to the car from the engine, all external forces, and internal and external frictional forces.

Section 6-3. Work and Energy in Three Dimensions. Dot product notation provides for the general definition of the work W done by a force $\vec{F}$ on a particle moving from point 1 to point 2:

$$W = \int_{\vec{s}_1}^{\vec{s}_2} \vec{F} \cdot d\vec{s} \qquad\qquad \text{Work (general definition)}$$

and the total work done on a particle becomes

$$W_{total} = \int_{\vec{s}_1}^{\vec{s}_2} \vec{F}_{net} \cdot d\vec{s} = K_2 - K_1 = \Delta K \qquad \text{Work-kinetic energy equation in three dimensions}$$

Section 6-4. Potential Energy. For most forces, the work done by the force depends on the path of the particle it acts on. For example, if you push your chair from point A to point B, the amount of work you do depends on the path over which you push the chair. Other things being equal, the longer the path, the more work you do. Yet, even when the path lengths are equal, the work done is not necessarily equal. For example, if one path is over a slick newly waxed floor while the other path is over a thick carpet, more work is required to push the chair from A to B over the thick carpet than over the waxed floor. That is, pushing harder (exerting a greater force) over the same distance means you do more work.

There are situations, however, when the work done by a force depends only on the initial and final positions and not on the path taken. Gravitational forces are an example of this. The work done by the force of gravity on a particle of mass m as it moves from a height y_i to a height y_f is $-mg(y_f - y_i)$. For example, when you lift a book of mass m off your desk and place it on a shelf, the work done by the gravitational force on the book is $-mg(y_s - y_d)$, where y_s and y_d are the heights of the shelf and of the desk top. Whether you pick up the book and place it directly on the shelf or carry it around with you all day before placing it on the shelf, the work done on the book *by gravity* remains $-mg(y_s - y_d)$. **Conservative forces**, such as the gravitational force, are forces where the work done depends only on the initial and final positions and not on the path taken. If a force is conservative, then the work done by that force on a particle that returns to its initial position must be zero. To be conservative a force must be a function of position only.

The change in **potential energy** U associated with a conservative force is the negative of the work done by that force. That is

$$\Delta U = U_f - U_i = -W = -\int_{\vec{s}_1}^{\vec{s}_2} \vec{F} \cdot d\vec{s}$$ Potential energy

Consequently, near the surface of the earth the change in gravitational potential energy is related to the change in position by the relationship $U - U_0 = +mg(y - y_0)$, where y_0 and y are the initial and final heights of the particle. U, a function of y, is called a potential energy function. If we define U_0 as mgy_0, this may be expressed as

$$U = U_0 + mgy$$ Gravitational potential energy

Your physics book and the earth attract each other via gravitational forces. When objects attract each other, the farther apart they are, the greater potential energy they have. Where is this potential energy? Potential energy is an abstract concept, so that is a challenging question. You should think of the potential energy as an inherent part of the earth-book system. Unfortunately, physicists sometimes talk as if this potential energy is associated only with the book and its position; but actually the potential energy is always shared between the book and the earth. The potential energy of the book-earth system depends on the relative position of the book with respect to the earth, and not just on the position of the book.

The potential energy of the earth-book system increases when you raise the book. To raise the book you must do positive work on it. However, while you are doing this positive work the gravitational force on the book is doing negative work. It is the negative of the work done by the gravitational force that equals the increase in the system's gravitational potential energy.

Another example of a conservative force is the force exerted by a spring that obeys Hooke's law $(F_x = -kx)$. The potential energy function associated with such a spring is

$$U = \tfrac{1}{2}kx^2$$ Potential energy of a spring

where $x = 0$ is the equilibrium position of the spring.

In one dimension, the change in potential energy associated with a conservative force $\vec{F} = F_x \hat{i}$ is $dU = -F_x dx$. Thus the force can be expressed as the negative of the derivative of the potential energy function:

$$F_x = -\frac{dU}{dx}$$ Force and potential energy

It follows that on a graph of the potential energy function U versus the position x, the conservative force F_x is the negative of the slope of the curve. On a U versus x graph, a positive slope indicates a negatively directed force, and a negative slope indicates a positively directed force. The important thing to note is that *the force is in the direction such as to accelerate the particle toward lower potential energy*. If the conservative force F_x is the only force acting on a particle, then the particle is in **equilibrium** when the slope of the potential energy versus position curve is zero, that is, when F_x is zero.

Figure 6-3 demonstrates the different types of equilibrium. A particle is in **stable equilibrium** if small displacements away from equilibrium result in a restoring force that pushes the particle back toward its equilibrium position (point a). Consequently, when a particle is in stable equilibrium it is at a potential energy minimum. For example, a ball resting at the lowest point in a valley is in stable equilibrium. A particle is in **unstable equilibrium** if small displacements from equilibrium result in a force that pushes the particle even further away from equilibrium (point b). Thus when a particle is in unstable equilibrium, it is at a potential energy maximum, like a ball poised at the crest of a hill. A particle is in **neutral equilibrium** if the force remains zero following any small displacement from an equilibrium position (point c).

A conservative force always tends to accelerate a particle toward a position of lower potential energy.

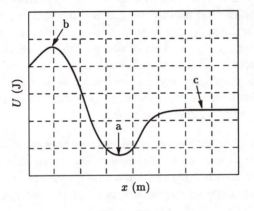

Figure 6-3

II. Physical Quantities and Key Equations

Physical Quantities

Energy $1\,J = 1\,N{\cdot}m = 1\,kg{\cdot}m^2/s^2 = 0.738\,ft{\cdot}lb$

                    $1\,kW{\cdot}hr = 3.6\,MJ$

Power $1\,W = 1\,J/s$

                    $1\,hp = 550\,ft{\cdot}lb/s = 746\,W$

Key Equations

Work in one dimension by a constant force $W = F_x\,\Delta x = F\cos\theta\,\Delta x$

Work in one dimension by a variable force $W = \int_{x_1}^{x_2} F_x\,dx$

Kinetic energy $K = \frac{1}{2}mv^2$

Dot product	$\vec{A} \cdot \vec{B} = AB\cos\phi = A_x B_x + A_y B_y + A_z B_z$
Projecting $\vec{A}$ onto the x axis	$\hat{i} \cdot \vec{A} = A\cos\phi = A_x$
Work in three dimensions by a variable force	$W = \int_{s_1}^{s_2} \vec{F} \cdot d\vec{s}$
Work-kinetic energy theorem	$W_{total} = \Delta K = \frac{1}{2}mv_f^2 - \frac{1}{2}mv_i^2$
Increment of work	$dW = F_s\,ds = F\cos\phi\,ds = \vec{F} \cdot d\vec{s}$
Instantaneous power	$P = \dfrac{dW}{dt} = \vec{F} \cdot \vec{v}$
Kinetic energy and power	$\Delta K = P_{total}\,\Delta t$ (constant power)
Conservative force defined	If $\int_{s_1}^{s_2} \vec{F} \cdot d\vec{s}$ is path independent, then $\vec{F}$ is conservative
Potential energy	$\Delta U = U_f - U_i = -W = -\int_{s_1}^{s_2} \vec{F} \cdot d\vec{s}$ (where $\vec{F}$ is conservative)
Force and potential energy	$F_x = -\dfrac{dU}{dx}$ (where $\vec{F} = F_x\hat{i}$ is conservative)
Gravitational potential energy	$U = U_0 + mgy$
Potential energy of a spring	$U = \frac{1}{2}kx^2$

III. Potential Pitfalls

This chapter contains many words whose technical meanings differ from their meaning in everyday usage. These include *work*, *kinetic energy*, *potential energy*, and *power*. Be sure to understand the technical definitions.

Do not think the work has to be nonzero just because the force and the displacement are each nonzero. The work done by a force is equal to the dot product of the force and the displacement of the point of application of the force. This work is zero when the displacement of the point of application of the force is either zero or is directed perpendicular to the force vector.

Do not think of work as having a direction just because the force and the displacement do. Work is a scalar quantity, and thus it has no direction. The sign of a work term represents the sense of an energy change. When the work is positive, energy is transferred to the object on which the force acts.

Do not think that the work-kinetic energy theorem can always be applied to a block sliding along a flat surface. If there is kinetic friction between the block and the surface then the block cannot be modeled as a particle. It follows that the theorem does not apply. Only if the surface is frictionless can the block be modeled as a particle.

Only *changes* in potential energy have physical meaning. The choice of the reference point for potential energy is always arbitrary, so choose a reference point that is convenient for a particular situation. Although kinetic energy can never be negative, potential energy can be either positive or negative.

Do not associate potential energy with a single particle. Potential energy depends on the system configuration and is associated with the entire system. If only one particle in a system moves, it is common parlance to say that the particle's potential energy has changed, but in reality it's the potential energy of a larger system that has changed.

Be aware of the sign in the definition of potential energy. When the displacement is in the direction of a conservative force, the change in potential energy is negative. For example, when an object moves downward, in the direction of the gravitational force, the gravitational potential energy decreases.

IV. True or False Questions and Responses

True or False

_____ 1. Like kinetic energy, work is necessarily a positive quantity.

_____ 2. The work-kinetic energy theorem is best applied to problems where the force on a particle is known explicitly as a function of time.

_____ 3. When several forces act on a *particle*, the total work done by all of them is always equal to $\int_{\vec{s}_1}^{\vec{s}_2} \vec{F}_{net} \cdot d\vec{s}$ where $\vec{F}_{net}$ is the net force acting on the particle.

_____ 4. When several forces act on an *object*, the total work done by all of them is always equal to $\int_{\vec{s}_1}^{\vec{s}_2} \vec{F}_{net} \cdot d\vec{s}$ where $\vec{F}_{net}$ is the net force acting on the object.

_____ 5. The scalar or dot product of any two perpendicular vectors is always zero.

_____ 6. When a particle moves in the direction of a conservative force the potential energy of the particle increases.

_____ 7. Kinetic energy and work have the same dimensions.

_____ 8. The reference point for gravitational potential energy must be mean sea level.

_____ 9. An athlete throws a ball. To throw the ball again so that it moves with twice the speed requires that the athlete do twice as much work on the ball.

_____ 10. A particle is in stable equilibrium at a point where its potential energy is a maximum.

_____ 11. Particles that attract via conservative forces have more potential energy when they are close together than when they are far apart.

_____ 12. Particles that repel via conservative forces have more potential energy when they are close together than when they are far apart.

Responses to True or False

1. False. Work is negative when the force and displacement are oppositely directed.

2. False. Work is computed by evaluating an integral where the integration variable is an increment of position, not time. Thus the work-kinetic energy theorem is best applied to problems where the force on a particle is known explicitly as a function of position.

3. True

4. False. The statement is true only if the points of application of each force undergo equal displacements. Although this condition must hold for a particle, it does not necessarily hold for an object. For example, if you compress a spring by pushing on opposite ends with your hands the two ends do not move through identical displacements. (Even if they are identical in magnitude they will not be identical in direction.)

5. True

6. False. If a particle moves in the direction of a conservative force its potential energy decreases.

7. True. The dimensions are $[M][L]^2/[T]^2$.

8. False. The reference point is arbitrary.

9. False. For the ball to have twice its original speed, it must have four times its original kinetic energy. Because the total work done on the ball equals the change in its kinetic energy, the athlete must do four times as much work.

10. False. When a particle is at a potential energy maximum, small displacements of the particle from the equilibrium position will result in a force that pushes the particle farther from the equilibrium position. Thus, the particle is in unstable equilibrium.

11. False. If one of the particles moves in the direction of the conservative force the potential energy will, as always, decrease. They will also become closer together. Thus they have less potential energy when they are closer together.

12. True

V. Questions and Answers

Questions

1. You pick a book up off your desk and place it on a bookshelf on the wall above the desk. What is the total work done by all forces acting on the book? Explain.

2. A pendulum consists of a 1-inch-diameter steel ball of mass m suspended from the ceiling by a string. As the ball swings from its lowest position (where it has speed v_0) up to its highest position (where it reverses direction), what is the total work done on the ball by all the forces acting on it? How much work is done by each force acting on it? Explain.

3. A block of mass m, released from rest, slides down a frictionless incline and reaches the bottom with speed v_f. What is the total work done on the block? How much work is done by each force acting on it? Explain.

4. When you get up from your chair and start walking toward the door, your kinetic energy increases. What force does work on you to cause this increase in your kinetic energy? Where does this kinetic energy come from?

5. A car of mass m travels up a long, straight hill of height h at constant speed. Assume air drag is negligible. How much work is done by each force acting on the car? What is the total work done on it by external forces? Explain.

6. Your roommate solves an assigned physics problem and gets a negative value for the change in potential energy of the pendulum bob in Question 2. Is this possible? Explain.

Answers

1. The work-kinetic energy theorem relates the total work done on a particle with the change in the particle's kinetic energy. The total work done on the book equals the change in kinetic energy of the book. The initial speed of the book is zero and the final speed of the book is zero, thus the change in kinetic energy is zero minus zero which equals zero. The total work done on the book is zero. (The gravitational force does negative work on the book, but your hand does positive work. The work-kinetic energy theorem tells us that the sum of the two works must equal zero.)

2. The total work done on the ball equals its change in kinetic energy. Its final kinetic energy is zero and its initial kinetic energy is $\frac{1}{2}mv_0^2$, where m is its mass and v_0 is its velocity at its lowest point. Thus the total work done on the ball is $-\frac{1}{2}mv_0^2$. Two forces act on the ball: the gravitational force and the tension force exerted by the string. The work done by the tension force is zero because the ball's velocity, and therefore the infinitesimal increments of its displacement, is always perpendicular to this force. The power delivered by the tension force $\vec{T}$ is given by $P = \vec{T} \cdot \vec{v} = Tv\cos 90° = 0$. If the direction of a force and the velocity of its point of application are perpendicular, the rate at which that force is doing work is zero. It follows that the total work done on the ball equals the work done by only the gravitational force: $W_g = -\frac{1}{2}mv_0^2$.

3. The total work done on the block equals its change in kinetic energy. Its initial kinetic energy is zero and its final kinetic energy is $\frac{1}{2}mv_f^2$, where m is its mass and v_f is its velocity as it reaches the bottom of the incline. Thus the total work is $\frac{1}{2}mv_f^2$. Because the normal force is perpendicular to the displacement, the work done by it is zero. The only other force acting on the block is the gravitational force. The work W_g done by this force is equal to the total work done on the block; that is, $W_g = \frac{1}{2}mv_f^2$.

4. Two forces act on you when you start walking: the force of gravity and the contact force of the floor on your feet. Assuming that you are walking on a horizontal floor, the force of gravity does no work because it acts at right angles to your displacement. The contact force of the floor only acts on a foot when that foot is in contact with the floor. Thus, whenever this force acts on it, the foot is at rest so the displacement of the point of application of the contact force is zero. Therefore the contact force does no work. Because neither force acting on you does any work, the total work done on you is zero. Your increased kinetic energy comes from the chemical potential energy that is stored in your muscles. These chemical changes constitute changes in your internal structure. An object whose structure changes *cannot* be modeled as a particle, but the work-kinetic energy theorem only holds for particles. For this reason this theorem does not apply to your motion. This type of problem is analyzed using another theorem, the work-energy theorem, which is presented in Chapter 7.

5. Because the structure of the car does not remain fixed as the car moves, the car cannot be modeled as a particle. This means the work-kinetic energy theorem does not apply. Neglecting air resistance, there are two forces acting on the car, the force of gravity and the contact force of the road on the tires. The force of gravity does work $W_g = -mgh$, where m is the mass of the car and h is the altitude gained by the car. The work done by the contact force of the road on the car is zero because the displacement of the point of application of this contact force is zero. (Unless the car is skidding, the part of the tire in contact with the road is not moving.) Thus the total work done on the car equals $-mgh$. If air drag were not negligible it would do additional negative work on the car. The car's speed, and hence kinetic energy, does not change because the energy stored in the fuel of the car gets converted to propel the car forward. This is a change in the internal structure of the car, so once again, the work-kinetic energy theorem does not apply.

6. It is possible, but only if your roommate makes a mistake. The pendulum bob moves upward and the conservative force acting on it, its weight, is downward. The potential energy must increase if the bob moves either directly or obliquely upward. The value of the potential energy depends on the reference point chosen. If the chosen reference point is on the ceiling, then the potential energy will be negative as long as the ball remains below the ceiling. The point is, any *change* in potential energy is unaffected by the choice of reference point.

VI. Problems, Solutions, and Answers

Example #1. A 2.00-kg particle is subjected to a single force F_x that varies with position as shown in Figure 6-4. From the graph, determine the work done by the force when the particle moves from $x = 0$ to (*a*) $x = 4\,\text{m}$ and (*b*) $x = -3\,\text{m}$. (*c*) If the particle is projected from the origin with a speed of 1 m/s, determine its speed when it is at $x = 4\,\text{m}$.

Picture the Problem. The work done by a force is equal to the area under the F_x-versus-x curve. When calculating the area, you must remember to include both the sign of the force and the displacement.

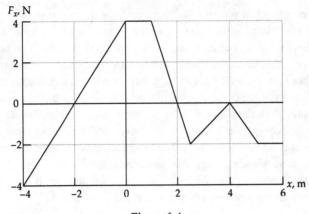

Figure 6-4

1. To find the area, first divide the region between the curve and the x axis into rectangles and triangles.	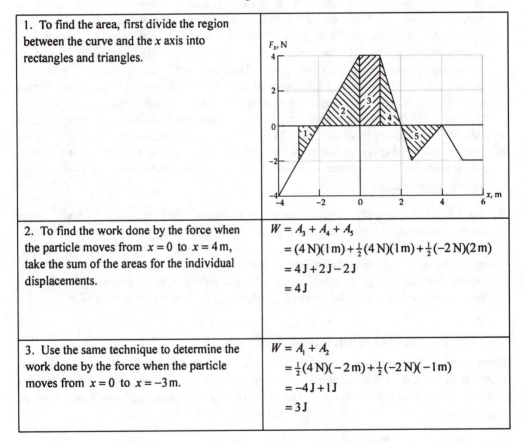
2. To find the work done by the force when the particle moves from $x = 0$ to $x = 4\,\text{m}$, take the sum of the areas for the individual displacements.	$W = A_3 + A_4 + A_5$ $= (4\,\text{N})(1\,\text{m}) + \frac{1}{2}(4\,\text{N})(1\,\text{m}) + \frac{1}{2}(-2\,\text{N})(2\,\text{m})$ $= 4\,\text{J} + 2\,\text{J} - 2\,\text{J}$ $= 4\,\text{J}$
3. Use the same technique to determine the work done by the force when the particle moves from $x = 0$ to $x = -3\,\text{m}$.	$W = A_1 + A_2$ $= \frac{1}{2}(4\,\text{N})(-2\,\text{m}) + \frac{1}{2}(-2\,\text{N})(-1\,\text{m})$ $= -4\,\text{J} + 1\,\text{J}$ $= 3\,\text{J}$

4. To find the speed of the particle at $x = 4$ m, use the work-kinetic energy theorem to relate the work done on the particle to the change in its kinetic energy. Solve the resulting expression for the final speed.	$W = \Delta K = K_f - K_i = \frac{1}{2}mv_f^2 = \frac{1}{2}mv_i^2$ $v_f = \sqrt{\dfrac{2W}{m} + v_i^2}$ $= \sqrt{\dfrac{2(4\,\text{J})}{2\,\text{kg}} + (1\,\text{m/s})^2}$ $= 2.24\,\text{m/s}$

Remark. Because work is related to kinetic energy, we can use the work-kinetic energy theorem to find initial or final velocities, without knowing, or having to calculate, the acceleration.

Example #2—Interactive. A 2.00-kg particle is subjected to a single force F_x that varies with position as shown in Figure 6-5. From the graph, determine the work done by the force when the particle moves from $x = 0.00$ to (*a*) $x = 6.00$ m and (*b*) $x = -4.00$ m. (*c*) If the particle is projected from $x = -4.00$ m with a velocity of 5.00 m/s in the $+x$ direction, determine its speed when it is at $x = 6.00$ m.

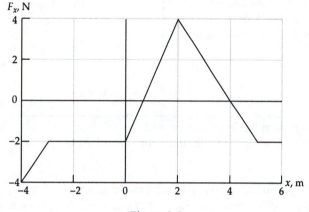

Figure 6-5

Picture the Problem. The work done by a force is equal to the area under the F_x-versus-x curve. When calculating the area, you must remember to include both the sign of the force and the displacement. **Try it yourself.** Work the problem on your own, in the spaces provided, to get the final answer.

1. To find the area, first divide the region between the curve and the x axis into rectangles and triangles.	

2. To find the work done by the force when the particle moves from $x = 0$ to $x = 6$ m, take the sum of the areas for the individual displacements.	$W = 3.00\,\text{J}$
3. Use the same technique to determine the work done by the force when the particle moves from $x = 0$ to $x = -4$ m.	$W = 9.00\,\text{J}$
4. To find the speed of the particle at $x = 6\,\text{m}$, use the work-kinetic energy theorem to relate the work done on the particle to the change in its kinetic energy.	$v_f = 4.36\,\text{m/s}$

Remark. The work done as the particle moves from $x = -4\,\text{m}$ to $x = 0\,\text{m}$ is $-9.00\,\text{J}$, the opposite of the work done as the particle moves from $x = 0\,\text{m}$ to $x = -4\,\text{m}$, because the displacement is in the opposite direction.

Example #3. Starting from rest, a 2.00-kg block is pulled up a frictionless 37.0° incline by a force $\vec{F}$ of 15.0 N directed up the incline, as shown in Figure 6-6. (*a*) Determine the speed of the block after it has traveled a distance *d*. (*b*) Determine the power being delivered by the force when *d* equals 0.500 m.

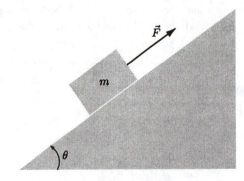

Figure 6-6

Picture the Problem. There are three forces acting on the block: the gravitational force, the normal force, and the applied force $\vec{F}$. (*a*) By the work-kinetic energy theorem the speed of the block is related to the total work done on the block by the sum of these forces. (*b*) To find the power, all we need is the force, the speed calculated in (*a*), and the angle between the applied force and the velocity.

1. Draw a free-body diagram of the particle with all the forces labeled and the direction of motion indicated.	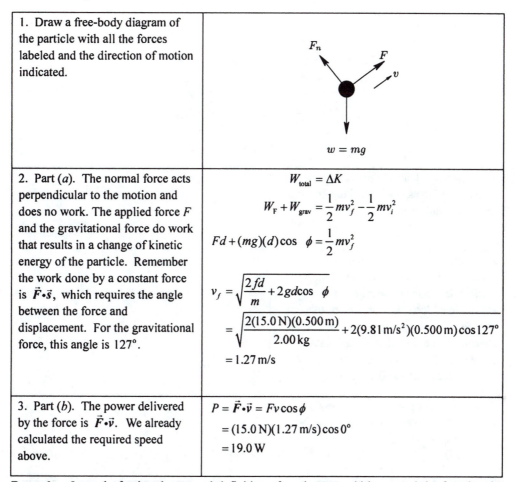
2. Part (*a*). The normal force acts perpendicular to the motion and does no work. The applied force F and the gravitational force do work that results in a change of kinetic energy of the particle. Remember the work done by a constant force is $\vec{F}\cdot\vec{s}$, which requires the angle between the force and displacement. For the gravitational force, this angle is 127°.	$$W_{\text{total}} = \Delta K$$ $$W_F + W_{\text{grav}} = \frac{1}{2}mv_f^2 - \frac{1}{2}mv_i^2$$ $$Fd + (mg)(d)\cos\ \phi = \frac{1}{2}mv_f^2$$ $$v_f = \sqrt{\frac{2fd}{m} + 2gd\cos\ \phi}$$ $$= \sqrt{\frac{2(15.0\,\text{N})(0.500\,\text{m})}{2.00\,\text{kg}} + 2(9.81\,\text{m/s}^2)(0.500\,\text{m})\cos 127°}$$ $$= 1.27\,\text{m/s}$$
3. Part (*b*). The power delivered by the force is $\vec{F}\cdot\vec{v}$. We already calculated the required speed above.	$$P = \vec{F}\cdot\vec{v} = Fv\cos\phi$$ $$= (15.0\,\text{N})(1.27\,\text{m/s})\cos 0°$$ $$= 19.0\,\text{W}$$

Remarks. Instead of using the general definition of work, we could have used the fact that the work done by gravity is equal to the negative change in the gravitational potential energy of the system.

Example #4—Interactive. A 5.00-kg block rests on a frictionless 53.0° incline, as shown in Figure 6-7. The block is pushed by a constant horizontal force $\vec{F}$ of 96.0 N. (*a*) Determine the speed of the block after it has traveled a distance d of 0.500 m. (*b*) Determine the power being delivered by the force when d equals 0.500 m.

Picture the Problem. There are three forces acting on the block: the gravitational force, the normal force, and the applied force $\vec{F}$. (a) By the work-kinetic energy theorem the speed of the block is related to the total work done on the block by the sum of these forces. (b) To find the power, all we need is the force, the speed calculated in (a), and the angle between the applied force and the velocity. **Try it yourself.** Work the problem on your own, in the spaces provided, to get the final answer.

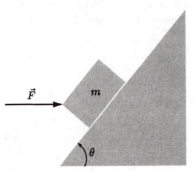

Figure 6-7

1. Draw a free-body diagram of the particle with all the forces labeled and the direction of motion indicated.	
2. Part (*a*). The normal force acts perpendicular to the motion and does no work. The applied force *F* and the gravitational force do work that results in a change of kinetic energy of the particle. Remember the work done by a constant force is $\vec{F} \cdot \vec{s}$, which requires the angle between the force and displacement.	$v_f = 1.93 \, \text{m/s}$
3. Part (*b*). The power delivered by the force is $\vec{F} \cdot \vec{v}$. We already calculated the required speed above.	$P = 112 \, \text{W}$

Example #5. The potential-energy function for a conservative force $\vec{F} = F_x \hat{i}$ acting on a 2.00-kg particle is shown in Figure 6-8. For what values of *x* is the force (*a*) zero; (*b*) directed leftward; (*c*) directed rightward? (*d*) What values of *x* are equilibrium positions? (*e*) For each equilibrium position state whether it is a position of stable, neutral, or unstable equilibrium. (*f*) For what value(s) of *x* is the magnitude of the force greatest?

Picture the Problem. For a conservative force, $F_x = -dU/dx$, so we need to calculate the slope of the potential-energy function to answer questions about the force. Equilibrium conditions depend on whether or not the potential-energy function is at a maximum, minimum, or constant.

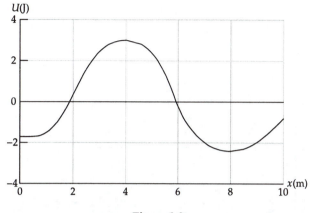

Figure 6-8

(a) The force is zero where $dU/dx = 0$.	$0 \le x \le 0.7\,\mathrm{m}$, $x = 4\,\mathrm{m}$, $x = 8\,\mathrm{m}$
(b) The force is directed leftward, in the $-x$ direction, where $F_x < 0$, or $dU/dx > 0$.	$0.7\,\mathrm{m} < x < 4\,\mathrm{m}$, $8\,\mathrm{m} < x < 10\,\mathrm{m}$
(c) The force is directed rightward, in the $+x$ direction, where $F_x > 0$, or $dU/dx < 0$.	$4\,\mathrm{m} < x < 8\,\mathrm{m}$
(d) Equilibrium positions are positions at which the force is zero.	$0 \le x \le 0.7\,\mathrm{m}$, $x = 4\,\mathrm{m}$, $x = 8\,\mathrm{m}$
(e) A stable equilibrium occurs at a potential-energy minimum, an unstable equilibrium occurs at a potential energy maximum, and a neutral equilibrium occurs where the potential-energy function is constant.	$0 \le x \le 0.7\,\mathrm{m}$, neutral equilibrium $x = 4\,\mathrm{m}$, unstable equilibrium $x = 8\,\mathrm{m}$, stable equilibrium
(f) The magnitude of the force is greatest where the slope of the potential-energy function is the steepest.	$x \approx 2\,\mathrm{m}$, $x \approx 5.8\,\mathrm{m}$

Example #6—Interactive. The potential-energy function for a conservative force $\vec{F} = F_x \hat{i}$ acting on a 2.00-kg particle is shown in Figure 6-9. For what values of x is the force (a) zero; (b) directed leftward; (c) directed rightward? (d) What values of x are equilibrium positions? (e) For each equilibrium position state whether it is a position of stable, neutral, or unstable equilibrium. (f) For what value(s) of x is the magnitude of the force greatest?

Picture the Problem. For a conservative force, $F_x = -dU/dx$, so we need to calculate the slope of the potential-energy function to answer questions about the force. Equilibrium conditions depend on whether or not the potential-energy function is at a maximum, minimum, or constant. **Try it yourself.** Work the problem on your own, in the spaces provided, to get the final answer.

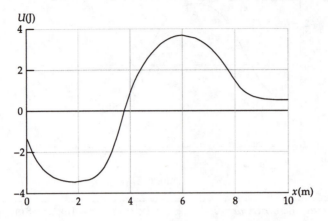

Figure 6-9

(a) The force is zero where $dU/dx = 0$.	$x = 2\,\text{m}, \; x = 6\,\text{m}, \; \approx 9\,\text{m} \leq x \leq 10\,\text{m}$
(b) The force is directed leftward, in the $-x$ direction, where $F_x < 0$, or $dU/dx > 0$.	$2\,\text{m} < x < 6\,\text{m}$
(c) The force is directed rightward, in the $+x$ direction, where $F_x > 0$, , or $dU/dx < 0$.	$0\,\text{m} < x < 2\,\text{m}, \; 6\,\text{m} < x < \approx 9\,\text{m}$
(d) Equilibrium positions are positions at which the force is zero	$x = 2\,\text{m}, \; x = 6\,\text{m}, \; \approx 9\,\text{m} \leq x \leq 10\,\text{m}$
(e) A stable equilibrium occurs at a potential-energy minimum, an unstable equilibrium occurs at a potential energy maximum, and a neutral equilibrium occurs where the potential-energy function is constant.	$x = 2\,\text{m}$, stable equilibrium $x = 6\,\text{m}$, , unstable equilibrium $\approx 9\,\text{m} \leq x \leq 10\,\text{m}$, neutral equilibrium
(f) The magnitude of the force is greatest where the slope of the potential-energy function is the steepest.	$x \approx 3.5\,\text{m}$

Example #7. Olympic weightlifter Tara Nott, who is in the under-45 kg class, can lift 102.5 kg in the clean-and-jerk event. In the "clean" stage, she lifts the barbell from the ground to her chest, approximately 1.25 m. If this single movement takes 0.5 s, how much power does Tara deliver to the weights during this portion of the lift? Assume the barbell is raised at a constant rate.

Picture the Problem. Power is $\vec{F} \cdot \vec{v}$, so both force and velocity must be calculated to find the power. Newton's laws can be used to determine the force, and kinematics will be used to find the speed.

1. If the barbell is lifted at a constant rate, it experiences no acceleration, so from Newton's 2nd Law, there is no net force on the barbell. Tara's force is exerted upward, and the gravitational force is exerted downward.	$\sum F_y = F_{\text{Tara}} - mg = 0$ $\begin{aligned} F_{\text{Tara}} &= mg \\ &= (102.5\,\text{kg})(9.81\,\text{m/s}^2) \\ &= 1000\,\text{N} \end{aligned}$
2. Use kinematics to find the speed at which the barbell is raised	$x = x_0 + v_0 t + \frac{1}{2}at^2$ $1.25\,\text{m} = v_0 (0.5\,\text{s})$ $v_0 = \dfrac{1.25\,\text{m}}{0.5\,\text{s}} = 2.5\,\text{m/s}$
3. The power can now be calculated. Since Tara exerts an upward force on the barbell, and the barbell is lifted upward, both the velocity and force are in the same direction.	$P = \vec{F} \cdot \vec{v}$ $= (1000\,\text{N})(2.5\,\text{m/s})\cos 0°$ $= 2500\,\text{W}$

Remarks. That's a lot of power from a relatively small woman. If that mechanical power was converted to electrical power, she could fully light twenty-five 100 W light bulbs during that 0.5 s burst. However, the maximum sustained output power of a human being is about 450 W.

Chapter **7**

Conservation of Energy

I. Key Ideas

Energy is a central concept in all of science. As far as anyone can tell, energy is never created or destroyed; rather, it is transformed from one form to another and transferred from one location to another.   Forms of energy include kinetic energy (energy associated with motion), potential energy (energy associated with the relative positions of interacting objects), thermal energy (energy associated with random molecular motions and configurations), and others. Under certain conditions, energy is transferred between objects through the forces they exert on each other. This kind of energy transfer is called work. Work, like force, mass, and numerous other terms that have both a scientific and an everyday usage, has a precise scientific definition.

Section 7-1.   The Conservation of Mechanical Energy. Kinetic energy and potential energy together constitute what is called **mechanical energy**. The total kinetic energy K of a system of particles is the sum of the kinetic energies of the individual particles. Potential energy is energy associated with the configuration of a system. For any particular configuration the actual value of the potential energy is arbitrary since only changes in potential energy are physically meaningful. For a specific configuration, called the reference configuration, the potential energy is set to zero. The forces exerted on each particle in the system by the other particles in the system are called internal forces. If a system's configuration changes, the accompanying change in potential energy during the process is the negative of the total work done by all internal conservative forces. Thus, for a given configuration the system's potential energy U is its change in potential energy as it moves from its reference configuration to its present one.

For example, consider a system consisting of only a crate and the earth, and the initial configuration has the crate resting on the floor. We specify this as the reference configuration where the system's potential energy is zero. You now lift the crate some distance above the floor. (You are not part of this system.) The gravitational forces of the earth on the crate and the crate on the earth are both conservative internal forces. As the crate rises, the gravitational force on it does negative work (the force and the displacement are oppositely directed) and the gravitational force of the crate on the earth does no work (the earth's displacement is essentially zero). Thus, during the lift the change in the earth-crate system's potential energy equals the negative of the work done on the crate by the earth's gravitational pull. This potential energy change equals the system's new potential energy.

The **total mechanical energy** E of a system is equal to its total kinetic energy plus its potential energy. That is,

$$E_{mech} = K_{sys} + U_{sys} = K + U_g + U_s + \ldots \qquad \text{Total mechanical energy}$$

where U_s is the potential energy associated with elastic deformations, such as that associated with a stretched spring, and U_g is the gravitational potential energy, such as that associated with the water behind a dam. Other forms of potential energy will be added in later chapters.

If, while a system changes, no external forces do work on it and if the internal forces doing work are all conservative (the work done by a conservative force doesn't depend on the path of the system), its total mechanical energy remains constant. That is,

$$E_{mech} = \text{constant} \qquad \text{or} \qquad K_i + U_i = K_f + U_f \qquad \text{Conservation of mechanical energy}$$

Section 7-2. The Conservation of Energy. There are two ways that the total energy of a system can change. One way is if one or more external forces do work on the system and the other is if heat is transferred to or from the system. A discussion of energy transfer via heat is explored in Chapter 18. This chapter will consider only energy transferred by work. In this context the relation between work and energy can be expressed

$$W_{ext} = \Delta E_{mech} + \Delta E_{therm} + \Delta E_{chem} + \ldots \qquad \text{Work-energy theorem}$$

where W_{ext} is the total work done on the system by external forces; E_{mech} is the system's mechanical energy; E_{chem} is its chemical energy, such as the energy stored in your muscles that powers your physical actions; and E_{therm} is its thermal energy, the energy associated with the random motion of molecules. The equation is not complete as there are other forms of energy, like nuclear energy that is stored the nucleus of atoms.

Thermal energy is increased if two surfaces slide on each other. It can be shown that if both surfaces are within a system,

$$\Delta E_{therm} = f_k d \qquad \text{Frictional dissipation of mechanical energy}$$

where ΔE_{therm} is the system's change in thermal energy, f_k is the frictional force on one of the surfaces, and d is the distance that one surface slides relative to the other surface.

Consider a system consisting of planet Earth, a massless spring, a ramp, and a block of mass m as shown in Figure 7-1. Initially the block is at rest and the spring is neither stretched nor compressed.

A student is pulling on the block with a force $\vec{F}$. The block gains height h as it slides a distance L up the ramp. Applying the work-energy theorem to the system we have:

$$\begin{aligned} W_{ext} &= \Delta E_{mech} + \Delta E_{therm} + \Delta E_{chem} \\ &= \Delta K + \Delta U_s + \Delta U_g + \Delta E_{therm} + \Delta E_{chem} \end{aligned}$$

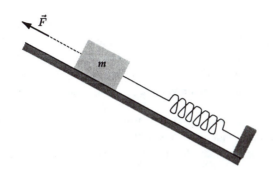

Figure 7-1

It is useful for bookkeeping purposes to make a table of the work done by the external forces and a second table for the various energy terms.

External forces	Work done
F	FL

	K	U_g	U_s	E_{therm}	E_{chem}
initial	0	0	0	n.a.*	n.a.
final	$\frac{1}{2}mv^2$	Mgh	$\frac{1}{2}kL^2$	n.a.	n.a.
change (Δ)	$\frac{1}{2}mv^2$	Mgh	$\frac{1}{2}kL^2$	$f_k L$	0

* Not applicable

We don't know the initial thermal energy, but we do know the final thermal energy exceeds the initial by $f_k L$. The change in chemical energy is zero. Substituting these values into the work-energy equation gives

$$FL = \frac{1}{2}mv^2 + \frac{1}{2}kL^2 + mgh + f_k L$$

Had the system included the student, the change in chemical energy would not be zero because chemical energy is transformed into mechanical and thermal energy by the student's muscles. The decrease in the student's chemical energy has to be at least as large as the work FL done by her on the system. However, the student was not included in the system, so this change of chemical energy does not appear in this application of the work-energy theorem.

Section 7-3. Mass and Energy. Fundamental subatomic particles come in pairs. For each particle there is a corresponding antiparticle. If a particle-antiparticle pair combine they are transformed from material particles into photons (electromagnetic energy), and the energy of the photons can be accounted for only if we accept the fact that each material particle has an intrinsic energy called its **rest energy** E_0. A particle's rest energy is given by the famous equation

$$E_0 = mc^2 \qquad\qquad \text{Rest energy}$$

where m is its mass and c is the speed of light in a vacuum.

Fundamental particles combine to form atomic nuclei, so we can treat an atomic nucleus as a system of these particles. Since the particles that are bound together in a nucleus attract each other, energy must be supplied to separate them. The minimum energy required to separate a nucleus into its constituent particles is called its **binding energy**. Atomic and nuclear energies are usually measured in units called **electron volts (eV)**, where $1.00\,\text{eV} = 1.602 \times 10^{-19}\,\text{J}$.

Energy is equivalent with mass. Since the energy of a bound nucleus is less than the energy when the same particles are widely separated, the mass of the system in the nucleus, with its particles bound together, is less than the sum of the masses of the widely separated constituent particles.

Section 7-4. Quantization of Energy. For a bound system, like an atomic nucleus, an atom, or a molecule, the total energy of the system is found to be quantized. That is, the smallest increases or decreases of a bound system's total energy occur in finite amounts, so the system's energy levels are discrete (quantized). It is common practice to label each energy level of a system with one or more integers called quantum numbers. Because these energy changes are small in comparison to the changes that typically occur in our everyday life, the finite size of the smallest increments in energy were not noticed until about a century ago.

Electromagnetic energy (radio waves, microwaves, light waves, X-rays, and gamma rays) is always absorbed or emitted by an amount directly proportional to the frequency of the radiation. That is,

$$E = hf \qquad\qquad \text{Energy quantum for electromagnetic radiation}$$

where the proportionality constant h, called Planck's constant, equals $6.626 \times 10^{-34}\,\text{J·s}$ and f represents the frequency of the radiation.

II. Physical Quantities and Key Equations

Physical Quantities

Electron volt	$1.00\,\text{eV} = 1.602 \times 10^{-19}\,\text{J}$
Planck's constant	$h = 6.626 \times 10^{-34}\,\text{J·s}$

Key Equations

Total mechanical energy	$E_{\text{mech}} = K_{\text{sys}} + U_{\text{sys}} = K + U_{\text{g}} + U_{\text{s}} + \ldots$
Conservation of mechanical energy	$E_{\text{mech}} = \text{constant} \quad\text{or}\quad K_{\text{i}} + U_{\text{i}} = K_{\text{f}} + U_{\text{f}}$
Work-energy theorem	$W_{\text{ext}} = \Delta E_{\text{mech}} + \Delta E_{\text{therm}} + \Delta E_{\text{chem}} + \ldots$
Frictional dissipation of mechanical energy	$\Delta E_{\text{therm}} = f_{\text{k}} d$
Rest energy	$E_0 = mc^2$
Energy quantum for electromagnetic radiation	$E = hf$

III. Potential Pitfalls

Don't confuse a system's total mechanical energy with its total energy. The total energy includes the total mechanical energy and any additional forms of energy.

Only changes in potential energy have physical meaning. The choice of the reference point for potential energy is always arbitrary, so choose a reference configuration that is convenient for a particular situation. Although kinetic energy can never be negative, potential energy can be either positive or negative.

Keep in mind that the amount of mechanical energy that is transformed into thermal energy when two surfaces slide across each other is $f_k d$, where f_k is the kinetic frictional force and d is the distance one surface moves relative to the other surface. Don't mistake $f_k d$ for the work done by the kinetic frictional force. It isn't. The work done by f_k cannot be directly calculated because the displacements of the points where this force is applied are not directly observable.

Don't think that the mass of a system of particles is always equal to the sum of the masses of constituent fundamental particles. If the system is bound then the energy is less than it would be if the particles are widely separated. Since the mass of a system is proportional to its energy, a bound system has less mass than a system of the same particles if they are widely separated. The difference is called the binding energy. These distinctions are always valid, but they are most significant for subatomic systems.

IV. True or False Questions and Responses

True or False

_____ 1. The work done by a kinetic frictional force f_k equals c where d is the distance one surface slides relative to the other surface.

_____ 2. The mechanical energy dissipated by a kinetic frictional force f_k equals $f_k d$, where d is the distance one surface slides relative to the other surface.

_____ 3. A block sliding on a horizontal floor is brought to rest by a kinetic frictional force. All of the dissipated kinetic energy appears as the thermal energy of the block.

_____ 4. The mass of a nucleus is exactly equal to the sum of the masses of its constituent protons and neutrons.

_____ 5. The initial and final energies of a system are the sums of all types of energy (kinetic, potential, chemical, thermal, . . .) at two specific times, the initial and final times.

_____ 6. External work refers to a process in which the system is displaced from an initial to a final configuration as a result of external forces that act on the system.

_____ 7. A ball rolling on a horizontal surface will eventually come to rest when it completely runs out of energy.

_____ 8. A wooden ball that is not moving has no energy.

_____ 9.   Energy is a force.

Responses to True or False

1. False. The displacement of the point of application of the frictional force is not equal to the bulk displacement of the object that the force acts on.

2. True.

3. False. Some of the dissipated kinetic energy appears as the thermal energy of the block, almost all the rest as the thermal energy of the floor.

4. False. For a bound system like an atomic nucleus, energy has to be added to separate it into widely separated protons and neutrons. Energy and mass are proportional so the mass of the nucleus is less than the sum of the masses of the constituent protons and neutrons.

5. True.

6. True.

7. False. It will eventually come to rest but it doesn't run out of energy. When the ball finally stops rolling, all of its initial kinetic energy has been transformed into thermal energy, some of which resides in the ball, some in the surface, and the rest in the air.

8. False. It is not rolling when it has no kinetic energy, but it still has other forms of energy including chemical and thermal.

9. False. Energy is not a force.

V. Questions and Answers

Questions

1. A block of mass m, released from rest, slides down a frictionless incline and reaches the bottom with speed v_f. What is the change in the total mechanical energy of the *block-slide-earth* system? Explain.

2. When you step on the accelerator (gas pedal) your car's kinetic energy increases. What external force, if any, does work on the car to cause this increase in its kinetic energy? Where does this kinetic energy come from?

3. A boy pulls a wagon up a long, straight hill at constant speed. What is the change in total energy of the *boy-wagon-hill-earth* system? Explain.

4. A boy pulls a wagon up a long, straight hill at constant speed. Does the total energy of the *wagon-hill-earth* system change? Explain.

5. A blob of putty falls on the floor (plop). Neglecting air resistance, does the total energy of the *putty-floor-earth* system change? Explain.

Answers

1. Zero. There are no external forces acting on the block-slide-earth system, so the total energy of the system remains constant. The slide is frictionless so no mechanical energy is dissipated by friction. The only force doing work on the block is the conservative force of gravity. Therefore, for this system mechanical energy is conserved. The decrease in gravitational potential energy equals the increase in kinetic energy.

2. No external force does work on the car. The static frictional force on the tires by the pavement does no work because that part of each tire in contact with the pavement is at rest. (We are assuming the tires do not slip on the pavement.) The increase in the car's kinetic energy equals the decrease in the chemical energy stored in the gasoline and the oxygen in the atmosphere. The kinetic energy comes from the gasoline-oxygen mixture.

3. The total energy of this system does not change. There are no external forces acting on this system, so its total energy remains unchanged.

4. The total energy of this system increases. As the boy pulls the wagon up the hill, the force he exerts on the wagon does positive work. This is the only external work done on the system, so the total external work done on the system is positive. This means the system's total energy increases, in accord with the work-energy theorem. Where does this energy come from? The chemical energy stored in the boy's muscles.

5. No. The total energy of the putty-floor-earth system does not change because there are no external forces on the system, so no work is done on it by external forces. During the fall the system's gravitational potential energy is transformed into the kinetic energy of the blob and during the plop this kinetic energy is dissipated via friction within the putty and between the putty and the floor.

VI. Problems, Solutions, and Answers

Example #1. A pendulum consists of a compact 0.500-kg particle suspended from a 2.00-m long string that is attached to a mount on the ceiling. If the pendulum is released from rest with the string taut and horizontal, how fast will the particle be moving when the string makes an angle of 30° with the vertical while the particle is on its downward arc? Air drag is negligible.

Picture the Problem. If you choose your system to consist of the pendulum and the earth, there are no external forces acting, and the total mechanical energy of the system will remain constant. You can find the speed of the particle at the final height by equating the initial and final mechanical energies.

1. Draw a sketch of the situation, labeling the points of interest. 1 refers to the initial position, and 2 refers to the final position.	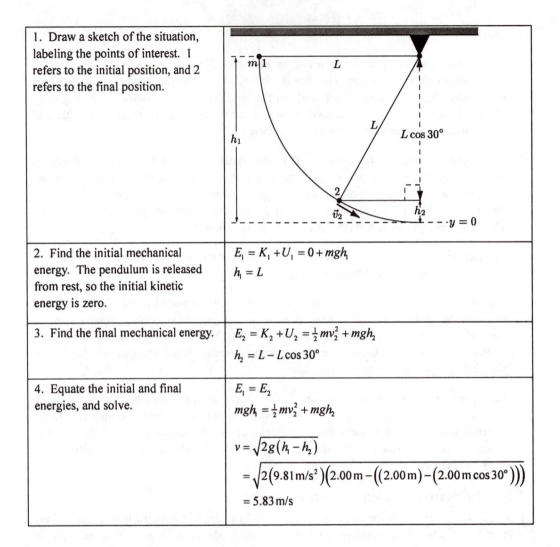
2. Find the initial mechanical energy. The pendulum is released from rest, so the initial kinetic energy is zero.	$E_1 = K_1 + U_1 = 0 + mgh_1$ $h_1 = L$
3. Find the final mechanical energy.	$E_2 = K_2 + U_2 = \frac{1}{2}mv_2^2 + mgh_2$ $h_2 = L - L\cos 30°$
4. Equate the initial and final energies, and solve.	$E_1 = E_2$ $mgh_1 = \frac{1}{2}mv_2^2 + mgh_2$ $v = \sqrt{2g(h_1 - h_2)}$ $\quad = \sqrt{2(9.81\,\text{m/s}^2)(2.00\,\text{m} - ((2.00\,\text{m}) - (2.00\,\text{m}\cos 30°)))}$ $\quad = 5.83\,\text{m/s}$

Example #2—Interactive. A 3.00-kg block sliding along a horizontal surface at 2.00 m/s begins to slide up a 20.0-cm high hill. Assuming friction is negligible on all surfaces, what is the speed of the block after it reaches the top of the hill?

Picture the Problem. If you choose your system to consist of the block, the hill, and the earth, there are no external forces acting, and the total mechanical energy of the system will remain constant. You can find the speed of the block at the final height by equating the initial and final mechanical energies. **Try it yourself.** Work the problem on your own, in the spaces provided, to get the final answer.

1. Draw a sketch of the situation, labeling the points of interest.	

2. Find the initial mechanical energy.	
3. Find the final mechanical energy.	
4. Equate the initial and final energies, and solve.	$v = 0.28\,\text{m/s}$

Example #3. A pendulum consists of a compact particle of mass m suspended from a string of length L attached to a mount on the ceiling, as shown in Figure 7-2. The particle is released from rest with the string horizontal. As the particle passes through the lowest point in its path, the string strikes a skinny peg a distance bL above the lowest point. As the particle passes through point P, a distance bL directly above the peg, determine an expression for the tension in the string in terms of m, g, L, and b. Find the value of b in the limit that the tension in the string approaches zero. Assume air drag is negligible. For certain values of b, the equation yields a negative value for the tension. Is that physically possible? Explain.

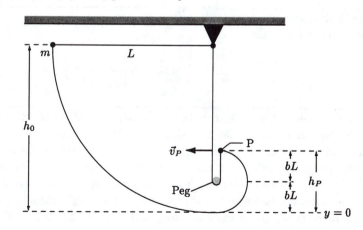

Figure 7-2

Picture the Problem. Choose the system to be the string, particle, ceiling, string mount, peg, and earth. Because no external forces act on this system, we can use conservation of energy to find the speed of the particle at point P. We can then relate this speed to the centripetal acceleration of the particle at that point, and solve for the tension using Newton's 2nd Law.

1. Find the energy at the initial position.	$E_i = K_1 + U_1 = 0 + mgh_0$
2. Find the energy at point P.	$E_P = K_P + U_P = \frac{1}{2}mv_P^2 + mgh_P$
3. Equate the initial energy with the energy at point P to find the speed of the particle at P.	$E_i = E_P$ $mgh_0 = \frac{1}{2}mv_P^2 + mgh_P$ $gL = \frac{1}{2}v_P^2 + g(2bL)$ $v_P = \sqrt{2gL(1-2b)}$
4. Using the speed at P, find the centripetal acceleration of the particle at that point.	$a_c = \dfrac{v^2}{R} = \dfrac{2gL(1-2b)}{bL} = \dfrac{2g(1-2b)}{b}$
5. Draw a free-body diagram of the particle at point P to help apply Newton's 2nd Law.	
6. Apply Newton's 2nd Law to solve for the tension in the rope.	$\vec{F} = m\vec{a}$ $w + T = ma_c$ $mg + T = m\dfrac{2g(1-2b)}{b}$ $T = mg\left(\dfrac{2(1-2b)}{b} - 1\right) = mg\left(\dfrac{2}{b} - 5\right)$
7. Rearrange the above equation to solve for b. Then take the limit of that expression as the tension goes to zero.	$b = \dfrac{2}{\frac{T}{mg} + 5}$ $\lim\limits_{T \to 0} b = \lim\limits_{T \to 0}\left(\dfrac{2}{\frac{T}{mg} + 5}\right) = \dfrac{2}{5} = 0.4$

If the peg is placed so that $b > 0.4$, our solution in step 6 yields negative values for the tension. In reality, this does not occur. Our analysis to determine an expression for T assumes the particle moves in the circular arcs shown in Figure 7-2. This assumption holds as long as the string remains taut, that is, as long as the tension remains greater than zero. An exception to this occurs if $b = 0.4$ In this case the tension remains greater than zero except for the single instant that the particle passes through the topmost point of the arc.   If the peg is placed so $b > 0.4$, the tension still never becomes negative because the string will become slack and the particle never reaches point P.

Example #4—Interactive. A small block of mass m slides down a frictionless incline and around a circular loop-the-loop of radius R, as shown in Figure 7-3. In terms of R, determine the height h above the bottom of the loop such that as the block passes through the highest point on the loop-the-loop the force of the track on the block equals half the weight of the block.

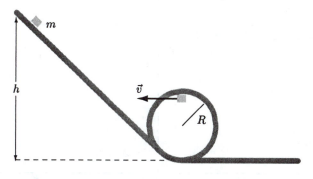

Figure 7-3

Picture the Problem. Choose the system to be the ramp, the block, and the earth. Applying Newton's 2nd Law, find the centripetal acceleration of the block at the top of the loop given that the normal force equals half the weight of the block. The speed of the block at the top of the loop can be found from the centripetal acceleration. Finally, conservation of energy can be used to determine the initial height of the block. **Try it yourself.** Work the problem on your own, in the spaces provided, to get the final answer.

1. Draw a free-body diagram of the particle at the top of the loop to help apply Newton's 2nd Law.	
2. Apply Newton's 2nd Law to solve for the centripetal acceleration at the top of the loop.	
3. Use the centripetal acceleration at the top of the loop to determine the speed of the block at that point.	

4. Find the energy of the system when the block is at the top of the loop.	
5. Find the energy of the system when the block is at its initial height h.	
6. Equate the initial energy with the energy at the top of the loop to find the starting height of the block.	$h = 2.75R$

Example #5. Blocks of mass m_1 and m_2 ($m_1 > m_2$) are hung over a massless, frictionless pulley using a string of negligible mass, as shown in Figure 7-4. When the system is released from rest, block 2 is in contact with the floor. Following release, block 1 falls a distance h to the floor as block 2 is pulled upward through the same distance. To what maximum height does block 2 rise above its starting position if $m_1 = 6.00\,\text{kg}$, $m_2 = 5.00\,\text{kg}$, and $h = 1.50\,\text{m}$?

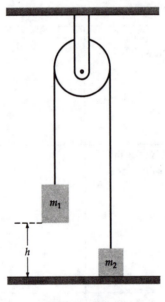

Figure 7-4

Picture the Problem. You might initially think that block 2 will simply rise a distance h. However, this is not the case, because when block 2 reaches this height, it will have some speed that will continue to carry it upward an additional distance. This problem consists of one segment while block 1 is falling, and a second segment during which block 1 remains on the floor and block 2 continues to rise. For segment one, choose the system to be the two blocks, the string, the pulley, the pulley mount, the ceiling, and the earth, so that mechanical energy of the system is conserved. For segment two, consider the system to consist of only block 2 and the earth, again allowing the conservation of mechanical energy.

1. Determine the initial energy of the system for segment one. It is easiest if we assume the floor is at $y = 0$.	$E_i = K_{1i} + U_{1i} + K_{2i} + U_{2i}$ $= 0 + m_1 gh + 0 + 0$
2. Determine the final energy of the system for segment one.	$E_f = K_{1f} + U_{1f} + K_{2f} + U_{2f}$ $= \frac{1}{2} m_1 v_{1f}^2 + 0 + \frac{1}{2} m_2 v_{2f}^2 + m_2 gh$
3. Set the initial and final energies equal to each other and solve for the final speed of block 2. Because of the coupling between block 1 and block 2, their final speeds are equal. This speed will be the initial speed of block 2 for the second segment.	$m_1 gh = \frac{1}{2} m_1 v_{1f}^2 + 0 + \frac{1}{2} m_2 v_{2f}^2 + m_2 gh$ $gh(m_1 - m_2) = \frac{1}{2} v_{2f}^2 (m_1 + m_2)$ $v_{2f} = \sqrt{\dfrac{2gh(m_1 - m_2)}{(m_1 + m_2)}}$
4. Find the initial energy of the much smaller system for segment two.	$E_i = K_i + U_i$ $= \frac{1}{2} m_2 v_{2f}^2 + m_2 gh$
5. Find the final energy of the system for segment two.	$E_f = K_f + U_f$ $= 0 + m_2 gh_2$
6. Set the initial and final energies equal to each other, and solve for h_2.	$\frac{1}{2} m_2 v_{2f}^2 + m_2 gh = m_2 gh_2$ $h_2 = \dfrac{v_{2f}^2}{2g} + h = \dfrac{\dfrac{2gh(m_1 - m_2)}{(m_1 + m_2)}}{2g} + h$ $h_2 = \dfrac{h(m_1 - m_2)}{(m_1 + m_2)} + h$ $h_2 = \dfrac{(1.5\,\mathrm{m})\big((6\,\mathrm{kg}) - (5\,\mathrm{kg})\big)}{(6\,\mathrm{kg} + 5\,\mathrm{kg})} + 1.5\,\mathrm{m} = 1.64\,\mathrm{m}$

Example #6—Interactive. Blocks of mass m_1 and m_2 are hung over a massless, frictionless pulley by a light string, as shown in Figure 7-5. There is no friction between the block and the incline. When the system is released from rest in the position shown, block 1 descends a distance h to the ground as the string pulls block 2 along the incline through the same distance. What is the maximum total displacement along the incline that block 2 undergoes when $m_1 = m_2 = 5.00\,\mathrm{kg}$, $h = 1.00\,\mathrm{m}$, and $\theta = 30°$?

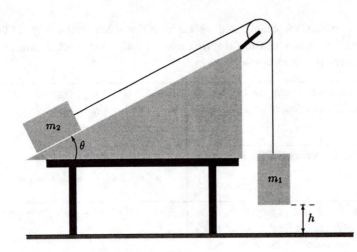

Figure 7-5

Picture the Problem. This problem consists of one segment while block 1 is falling and a second segment during which block 1 remains on the floor and block 2 continues to rise up the ramp. For segment one, choose the system to be the two blocks, the string, the pulley, the pulley mount, the ramp, and the earth, so that mechanical energy of the system is conserved. For segment two, consider the system to consist of only block 2 and the earth, again allowing the conservation of mechanical energy. **Try it yourself.** Work the problem on your own, in the spaces provided, to get the final answer.

1. Determine the initial energy of the system for segment one. It is easiest if we assume the floor is at $y = 0$.	
2. Determine the final energy of the system for segment one.	
3. Set the initial and final energies equal to each other and solve for the final speed of block 2. Because of the coupling between block 1 and block 2, their final speeds are equal. This speed will be the initial speed of block 2 for the second segment.	
4. Find the initial energy of the much smaller system for segment two.	
5. Find the final energy of the system for segment two.	

6. Set the initial and final energies equal to each other, and solve for the final height of the block.	
7. The displacement of block 2 two can be found geometrically from the height block 2 rises and the angle.	$d = 1.5\,\text{m}$

Example #7.   You are inside a crate that is released from rest at the top of a ramp inclined $30°$ with the horizontal. After sliding 4.00 m down the ramp the crate runs into a spring bumper which it compresses as it slows. You and the crate have a mass of 80.0 kg, the spring is massless and has a spring constant of 500 N/m, and the coefficient of kinetic friction between the block and the ramp is 0.300. What is the maximum distance that the spring is compressed?

Picture the problem.   A sketch of the system is shown in Figure 7-6. The points of interest are when you are initially at the top of the ramp and at the bottom of the ramp when you have fully compressed the spring. Choosing the system as you, the crate, the ramp, the spring, and the earth, the initial potential energy of the system will be converted into thermal energy through friction, and into the elastic potential energy stored in the compressed spring. There are no external forces acting on the system, so the total energy of the system remains unchanged.

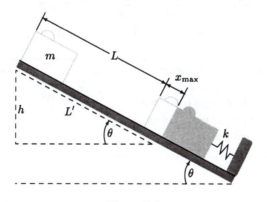

Figure 7-6

1. Write out the work-energy theorem as a guide for solving the problem.	$W_{\text{ext}} = \Delta K + \Delta U_g + \Delta U_s + \Delta E_{\text{therm}}$

2. Determine the work done by external forces. Since there are no external forces, this is zero.	$W_{ext} = \int \vec{F}_{ext} \cdot d\vec{s} = 0$
3. Determine the initial, final, and change in the kinetic energy. You and the crate are at rest both initially and at the maximum compression of the spring.	$K_i = 0$ $K_f = 0$ $\Delta K = 0$
4. Determine the initial, final, and change in the gravitational potential energy. You and the crate start out at height h and end at a height referenced as zero.	$U_{gi} = mgh$ $U_{gf} = 0$ $\Delta U_g = -mgh$
5. Determine the initial, final, and change in the elastic potential energy of the spring. The spring is initially uncompressed and at the end has a maximum compression.	$U_{si} = 0$ $U_{sf} = \frac{1}{2}kx^2_{max}$ $\Delta U_s = \frac{1}{2}kx^2_{max}$
6. Determine the amount of mechanical energy converted to heat.	$\Delta E_{therm} = f_k \left(L + x_{max} \right)$
7. Draw a free-body diagram of you and the crate combined, and use Newton's 2nd Law to find the normal force.	 $\sum \vec{F}_y = F_n - w_y = 0$ $F_n = w\cos\theta = mg\cos\theta$
8. Substitute everything into the work-energy theorem.	$W_{ext} = \Delta K + \Delta U_g + \Delta U_s + \Delta E_{therm}$ $0 = 0 - mgh + \frac{1}{2}kx^2_{max} + f_k \left(L + x_{max} \right)$
9. From the figure, we can see that the height h is related geometrically to the distance $L + x_{max}$ that the crate slides. We also know that the kinetic friction force is related to the normal force. Making these substitutions, we find:	$0 = -mg\left(L + x_{max} \right)\sin\theta + \frac{1}{2}kx^2_{max}$ $\quad + \mu_k mg\cos\theta\left(L + x_{max} \right)$

10. This is a quadratic equation in x_{max}. Solve it by rearranging it into standard form.	$0 = \frac{1}{2}kx_{max}^2 + mgx_{max}\left(\mu_k \cos\theta - \sin\theta\right)$ $\quad + mgL\left(\mu_k \cos\theta - \sin\theta\right)$ $= ax_{max}^2 + bx_{max} + c$ $a = \dfrac{k}{2} = \dfrac{500\,\text{N/m}}{2} = 250\,\text{N/m}$ $b = mg\left(\mu_k \cos\theta - \sin\theta\right)$ $\quad = (80\,\text{kg})(9.81\,\text{m/s}^2)(0.3\cos30° - \sin30°)$ $\quad = -189\,\text{N}$ $c = mgL\left(\mu_k \cos\theta - \sin\theta\right) = bL$ $\quad = (-189\,\text{N})(4.00\,\text{m}) = 756\,\text{N}\cdot\text{m}$
11. Substituting the values for a, b, and c into the quadratic equation finally yields the maximum spring compression. Because the *distance* the block moves while the spring is compressed is never negative, we have $x_{max} = 2.16\,\text{m}$.	$x_{max} = \dfrac{-b \pm \sqrt{b^2 - 4ac}}{2a}$ $\quad = 2.16\,\text{m}, \quad -1.40\,\text{m}$

Example #8—Interactive. Starting from rest, Buck the sled dog drags a 45.0-kg sled up a 5.00-m long ramp that is inclined 35° with the horizontal. The kinetic coefficient of friction between the ramp and the sled is 0.350. Using the work-energy theorem, determine how much work Buck must do on the sled just to drag it up the ramp. Assume Buck's paws do not slip on the ramp, and the initial and final velocities of the sled are zero.

Picture the Problem. The points of interest are when the sled is initially at rest at the bottom of the ramp and when the sled has reached the top of the ramp. Choosing the system as the sled, the ramp, and the earth, the force of Buck's pulling will be an external force that does work on the system. The work done by Buck will be converted into gravitational potential energy, kinetic energy, and thermal energy through friction. **Try it yourself.** Work the problem on your own, in the spaces provided, to get the final answer.

1. Draw a sketch of the problem, identifying the important points.	

2. Write out the work-energy theorem as a guide for solving the problem.	
3. Determine the work done by external forces, in this case, the work done by Buck.	
4. Determine the initial, final, and change in the kinetic energy.	
5. Determine the initial, final, and change in the gravitational potential energy.	
6. Determine the amount of mechanical energy converted to heat.	
7. Draw a free-body diagram of the sled, and use Newton's 2nd Law to find the normal force.	
8. Substitute everything into the work-energy theorem.	
9. Solve for the work done by Buck.	$W = 1.90 \times 10^3 \text{ J}$

Chapter 8

Systems of Particles and Conservation of Linear Momentum

I. Key Ideas

In our study of Newton's laws, we have treated extended objects—objects like blocks, balls, and automobiles—as particles. The justification of these treatments is presented in this chapter. We will see that the total force exerted on a system of particles equals the product of the total mass of the system and the acceleration of a point called the center of mass. The motion of a system of particles will be divided in two terms: the motion of the center of mass and the motion of various parts of the system relative to the center of mass.

We will introduce the concept of momentum, which is the product of the mass of a particle and its velocity. Momentum is considered a fundamental concept in physics because the total momentum of an isolated system always remains constant.

Section 8-1. The Center of Mass. The position of the **center of mass** $\vec{R}_{cm}$ of a system is related to the positions of the particles that make up the system by the equation

$$M\vec{R}_{cm} = m_1\vec{r}_1 + m_2\vec{r}_2 + \ldots = \sum_i m_i\vec{r}_i \qquad \text{Center of mass}$$

where m_i is the mass of the ith particle, $M = \Sigma m_i$ is the total mass of the system, and $\vec{r}_i$ and $\vec{R}_{cm}$ are the position vectors of the ith particle and the center of mass, respectively.

For a continuous object, the summation in the above expression is replaced by the integral

$$M\vec{R}_{cm} = \int \vec{r}\, dm \qquad \text{Center of mass for continuous object}$$

where dm is the mass of an arbitrary infinitesimal piece of the object and $\vec{r}$ is the position vector of that piece.

The vector $\vec{r}_i$ can be expressed in terms of its components as $\vec{r}_i = x_i\hat{i} + y_i\hat{j} + z_i\hat{k}$, and the vector $\vec{R}_{cm}$ can be expressed in terms of its components as $\vec{R}_{cm} = X_{cm}\hat{i} + Y_{cm}\hat{j} + Z_{cm}\hat{k}$.

The center-of-mass equation can be expressed in component form as the three equations

$$MX_{cm} = \sum_i m_i x_i$$

$$MY_{cm} = \sum_i m_i y_i$$

$$MZ_{cm} = \sum_i m_i z_i$$

Center of mass (component form)

It can be shown that the center of mass of a system consisting of a number of objects, like a collection of bricks, is determined by the expression

$$M\vec{R}_{cm} = m_a \vec{r}_{cm,a} + m_b \vec{r}_{cm,b} + \ldots$$

Center of mass for a composite system

where a, b, etc., refer to the individual objects (bricks).

The gravitational potential energy of a system of particles is expressed as $U = \sum m_i g y_i$. Factoring out the g from this sum leaves us with the expression

$$U = \sum m_i g y_i = g \sum m_i y_i = MgY_{cm}$$

Gravitational potential energy

That is, near the surface of the earth, the gravitational potential energy of a complex system depends only on the height of the center of mass of the system.

Section 8-2. Finding the Center of Mass by Integration.
Sometimes the mass distribution of a system can be treated as a continuous smear of mass, as opposed to a discrete collection of point particles. In these cases, the center of mass is calculated by doing an integral, which is nothing more than adding up the contributions of infinitely many *infinitesimal* pieces, just as you would do for a discrete mass distribution. If the mass is distributed along a line, then each infinitesimal length dl has a differential mass $dm = \lambda\, dl$, where λ is the mass per unit length. If the mass is distributed on a surface, then each infinitesimal piece of the surface has an area dA and mass $dm = \sigma\, dA$, where σ is the mass per unit area. Lastly, if the mass is distributed throughout a volume then each infiitesimal piece has volume dV and mass $dm = \rho\, dV$, where ρ is the mass per unit volume.

Section 8-3. Motion of the Center of Mass.
As an object moves, the bulk motion of the object is often of particular importance. For example, when a car moves along the highway, the complex motion of the pistons, valves, and other parts may be of interest, but it is the bulk motion of the car that is of foremost interest. After all, a car is primarily a transportation vehicle. It is the motion of the center of mass that best represents the bulk motion.

The **velocity of the center of mass** $\vec{V}_{cm}$ equals the rate of change of the position of the center of mass. To find the velocity of the center of mass we differentiate both sides of the center-of-mass equation with respect to time. This gives

$$M\vec{V}_{cm} = m_1 \vec{v}_1 + m_2 \vec{v}_2 + \ldots = \sum_i m_i \vec{v}_i$$

Velocity of center of mass

By differentiating again we get

$$M\vec{A}_{cm} = m_1\vec{a}_1 + m_2\vec{a}_2 + \ldots = \sum_i m_i\vec{a}_i$$ Acceleration of center of mass

where $\vec{A}_{cm}$ is the **acceleration of the center of mass**.

Newton's 2nd Law, $\vec{F}_{net} = m\vec{a}$, holds for individual particles. To consider Newton's 2nd Law for a system of particles, we must look at the forces on each individual particle. If one particle of the system exerts a force on a second particle of the system, then by Newton's 3rd Law, we know that the second particle must exert a force of equal magnitude, but in the opposite direction, on particle one. That is, these *internal* forces sum to zero. They will have no net effect on the center-of-mass motion of the system. Because of this, only external forces can affect the motion of the center of mass of a system.

$$\vec{F}_{net,ext} = \sum_i \vec{F}_{i,ext} = M\vec{A}_{cm}$$

where the net external force on the system, $\vec{F}_{net,ext}$, is the sum of all the external forces acting on the system. The implication of the equation is that the center of mass of a system moves like a particle of mass $M = \sum m_i$ under the influence of the net external force acting on the system.

Section 8-4. Conservation of Linear Momentum. The **linear momentum** of a particle is defined as the product of its mass and its velocity:

$$\vec{p} = m\vec{v}$$ Linear momentum of a particle

It is useful to write Newton's 2nd Law in term of momentum. This can be done by realizing that the derivative of the momentum with respect to time equals the product of the mass and the acceleration, that is,

$$\frac{d\vec{p}}{dt} = \frac{d(m\vec{v})}{dt} = m\frac{d\vec{v}}{dt} = m\vec{a}$$

Thus we can express Newton's 2nd Law as

$$\vec{F}_{net} = \frac{d\vec{p}}{dt}$$ Newton's 2nd Law for a particle

The concept of momentum is important because if the net external force acting on a system is zero, the total momentum of the system $\vec{P} = \Sigma\,\vec{p}_i$ remains constant. The velocity of the center of mass is related to the momentum by the equation

$$\vec{P} = \sum_i \vec{p}_i = \sum_i m_i\vec{v}_i = M\vec{V}_{cm}$$ Linear momentum for systems

Differentiating both sides of this equation with respect to time results in an equation relating the acceleration of the center of mass of the system $\vec{A}_{cm}$ and its total momentum $\vec{P}$. Substituting from this equation into Newton's 2nd Law for particles results in

$$\vec{F}_{net,ext} = \frac{d\vec{P}}{dt}$$ Newton's 2nd Law for systems

As long as the net external force on a system is zero, the total momentum of the system remains constant, as does the velocity of the center of mass. This result is known as the **law of conservation of momentum.**

If $\vec{F}_{net,ext} = 0$, then $\vec{P}$ = constant   Conservation of momentum

Section 8-5. Kinetic Energy of a System. The kinetic energy of a system can be expressed as the sum of two terms: the kinetic energy associated with the center-of-mass motion and the kinetic energy associated with the motions relative to the center of mass:

$$K = \tfrac{1}{2}MV_{cm}^2 + K_{rel}$$ Kinetic energy for systems

where the total kinetic energy is still the sum of the kinetic energy of each particle: $K = \Sigma\tfrac{1}{2}m_i v_i^2$. The relative kinetic energy is $K_{rel} = \Sigma\tfrac{1}{2}m_i u_i^2$, where u_i is the speed of the ith particle relative to the center of mass.

Consider two identical bicycle wheels, one in your hand and another on the unicycle you are riding. The velocities of the centers of mass of these wheels are the same since they must both move along with the unicycle. However, the unicycle wheel is spinning because it is rolling and the wheel in your hand is not. Consequently, the spinning wheel has additional kinetic energy due to its spinning motion. This additional kinetic energy is K_{rel}, the kinetic energy due to motion relative to the center of mass.

Section 8-6. Collisions. In a collision two objects approach and interact strongly for a short time. During the brief time of the collision, any external forces are much smaller than the forces of interaction between objects. Thus, during the collision the only significant forces acting on the two-object system are the forces of interaction between them. These forces are equal and opposite, so the total momentum of the system remains unchanged. Also, the collision time is usually so small that any displacement of the objects during the collision can be neglected.

Impulse and Average Force. Forces transfer momentum to the objects they act on. The measure of the momentum transferred by a force is the **impulse** $\vec{I}$. The impulse imparted by the force $\vec{F}$ during the time interval from t_i to t_f is defined by the equation

$$\vec{I} = \int_{t_i}^{t_f} \vec{F}\, dt$$ Impulse

Integrating Newton's 2nd Law, $\vec{F}_{net} = d\vec{p}/dt$, and applying the definition of impulse gives

$$\vec{I}_{net} = \int_{t_i}^{t_f} \vec{F}_{net}\, dt = \vec{p}_f - \vec{p}_i = \Delta\vec{p}$$ Impulse-momentum theorem

where $\vec{I}_{net}$ is the impulse associated with the net force $\vec{F}_{net}$ acting on the particle.

The **average force** is defined as

$$\vec{F}_{av} = \frac{1}{\Delta t} \int_{t_i}^{t_f} \vec{F}\, dt$$ Time average of a force

where Δt is the elapsed time $t_f - t_i$. Using the definition of impulse it follows that

$$\vec{I} = \vec{F}_{av}\, \Delta t$$ Impulse and average force

Collisions in One Dimension. When two objects collide the momentum of the two-object system remains constant. In one dimension, directions are indicated by signs, so the explicit use of vectors can be avoided. Thus we can write the conservation of momentum equation for two objects as

$$m_1 v_{1i} + m_2 v_{2i} = m_1 v_{1f} + m_2 v_{2f}$$ Conservation of momentum in one dimension

For **perfectly inelastic collisions,** where the objects stick together, both objects have the same final velocity—the velocity of their center of mass. Combining this condition with the previous equation gives

$$m_1 v_{1i} + m_2 v_{2i} = (m_1 + m_2) V_{cm}$$ Perfectly inelastic collision

For **elastic collisions**, the initial and final kinetic energies are the same, thus

$$\frac{1}{2} m_1 v_{1i}^2 + \frac{1}{2} m_2 v_{2i}^2 = \frac{1}{2} m_1 v_{1f}^2 + \frac{1}{2} m_2 v_{2f}^2$$ Elastic collision

Using this equation combined with the conservation of momentum equation, it can be shown that, for an elastic collision, the magnitude of the relative velocity of approach equals the magnitude of the relative velocity of departure. That is,

$$v_{2f} - v_{1f} = -(v_{2i} - v_{1i})$$ Relative velocities in an elastic collision

Most collisions fall somewhere between being elastic and perfectly inelastic. The degree of elasticity varies with the **coefficient of restitution**, e, which is defined as the ratio of the relative speed of recession to the relative speed of approach:

$$e = \frac{v_{2f} - v_{1f}}{-(v_{2i} - v_{1i})}$$ Coefficient of restitution

For an elastic collision, $e = 1$; for a perfectly inelastic collision, $e = 0$.

Collisions in Three Dimensions. For two and three dimensions, the vector nature of momentum is important. Perfectly inelastic collisions can be analyzed in a straightforward manner using the equation

$$m_1 \vec{v}_{1i} + m_2 \vec{v}_{2i} = (m_1 + m_2) \vec{V}_{cm}$$ Perfectly inelastic collision

For all collisions, the total momentum is conserved in each dimension. However, given the masses and initial velocities, conservation of momentum alone is not sufficient to determine the final velocities. The final velocities will not be the same for elastic collisions as for inelastic collisions. For elastic collisions, in addition to the conservation of kinetic energy, you also will need additional information such as the angles at which the objects depart after the collision.

Section 8-7. The Center-of-Mass Reference Frame. Any reference frame in which the center of mass of a system remains stationary is called a **center-of-mass reference frame** for that system. Relative to this frame both the velocity and acceleration of the center of mass are always zero, and measurements are usually made from a coordinate system attached to this frame and with the origin at the center of mass. In this frame the total momentum of the system, which equals the product of the total mass of the system and the velocity of the center of mass, also equals zero. Consequently, a center-of-mass reference frame is also a **zero-momentum reference frame**.

In the analysis of problems it is frequently useful to transform back and forth between a given frame of reference and the center-of-mass reference frame. To do this we make use of the relative velocity formula $\vec{v}_{AC} = \vec{v}_{AB} + \vec{v}_{BC}$ developed in Chapter 3. In words this states: the velocity of particle A relative to reference frame C equals the velocity of A relative to frame B plus the velocity of frame B relative to frame C. If A is the ith particle of a system, B is a center-of-mass reference frame, and C is an arbitrary reference frame then this formula can be written $\vec{v}_i = \vec{u}_i + \vec{V}_{cm}$, where $\vec{v}_i$ is the velocity of particle i relative to an arbitrary frame, $\vec{u}_i$ is the velocity of the same particle relative to a center-of-mass frame, and $\vec{V}_{cm}$ is the velocity of the center-of-mass frame relative to the arbitrary frame. It is often useful to rearrange the terms and write the formula as

$$\vec{u}_i = \vec{v}_i - \vec{V}_{cm} \qquad \text{Velocity relative to center of mass}$$

This transformation can be used to show that for an elastic collision, the velocity of each object relative to the center of mass is merely reversed:

$$u_{1i} = -u_{1f} \quad \text{and} \quad u_{2i} = -u_{2f} \qquad \text{1D elastic collision in the center-of-mass frame}$$

Section 8-8. Systems with Continuously Varying Mass: Rocket Propulsion. Applying conservation of momentum to systems with variable mass is a little more challenging than what we have done so far. Why would we want to do this? Consider an interesting form of propulsion that occurs when a system ejects a fluid in one direction and the rest of the system accelerates in the opposite direction. This kind of propulsion is known as **rocket propulsion**. When a balloon is inflated, and then released, the balloon propels itself through the air in this manner.

When a rocket engine is firing, exhaust gasses are ejected out the back of the engine. The engine exerts a force on these gasses causing them to eject out of the back. It is the reaction force, the force of the gasses on the engine, that propels the engine forward. If a rocket engine, initially at rest in empty space where no external forces act on it, begins a burn, the center of mass of the fuel-engine system will remain at rest as the engine and the exhaust gasses move off in opposite directions.

Applying conservation of momentum and Newton's 2nd Law to a rocket results in an equation known as the *rocket equation*:

$$M \frac{d\vec{v}}{dt} = \vec{u}_{ex} \left| \frac{dM}{dt} \right| + F_{net,\,ext}$$ Rocket equation

where M is the instantaneous mass of the rocket and the remaining fuel; u_{ex} is the speed of the exhaust relative to the rocket; and $F_{net,\,ext}$ is the net external force acting on the rocket. The mass M decreases as the burned fuel is exhausted; thus dM/dt, the rate of change of the mass of the rocket and remaining fuel, is negative. The quantity $\vec{u}_{ex} |dM/dt|$ is called the **thrust**. It is the contact force exerted by the combustion products on the rocket.

II. Physical Quantities and Key Equations

Physical Quantities

There are no new physical quantities for this chapter.

Key Equations

Center of mass $$M\vec{R}_{cm} = m_1\vec{r}_1 + m_2\vec{r}_2 + \ldots = \sum_i m_i\vec{r}_i$$

Center of mass (component form)

$$MX_{cm} = \sum_i m_i x_i$$

$$MY_{cm} = \sum_i m_i y_i$$

$$MZ_{cm} = \sum_i m_i z_i$$

Velocity of center of mass $$M\vec{V}_{cm} = m_1\vec{v}_1 + m_2\vec{v}_2 + \ldots = \sum_i m_i\vec{v}_i$$

Acceleration of center of mass $$M\vec{A}_{cm} = m_1\vec{a}_1 + m_2\vec{a}_2 + \ldots = \sum_i m_i\vec{a}_i$$

Center of mass for continuous object $$M\vec{R}_{cm} = \int \vec{r}\, dm$$

Gravitational potential energy $$U = \sum m_i g y_i = g \sum m_i y_i = MgY_{cm}$$

Newton's 2nd Law for systems $$\vec{F}_{net,ext} = \sum_i \vec{F}_{i,ext} = M\vec{A}_{cm}$$

$$\vec{F}_{net,ext} = \frac{d\vec{P}}{dt}$$

Linear momentum of a particle $$\vec{p} = m\vec{v}$$

Newton's 2nd Law for particles	$\vec{F}_{net} = \dfrac{d\vec{p}}{dt}$
Linear momentum for systems	$\vec{P} = \sum_i \vec{p}_i = \sum_i m_i \vec{v}_i = M\vec{V}_{cm}$
Kinetic energy for systems	$K = \tfrac{1}{2}MV_{cm}^2 + K_{rel}$
Impulse-momentum theorem	$\vec{I}_{net} = \int_{t_i}^{t_f} \vec{F}_{net}\, dt = \vec{p}_f - \vec{p}_i = \Delta\vec{p}$
Time average of a force	$\vec{F}_{av} = \dfrac{1}{\Delta t}\int_{t_i}^{t_f} \vec{F}\, dt$
Perfectly inelastic collision	$m_1\vec{v}_{1i} + m_2\vec{v}_{2i} = (m_1 + m_2)\vec{V}_{cm}$
Relative velocities in an elastic collision	$v_{2f} - v_{1f} = -(v_{2i} - v_{1i})$
Elastic collision	$\dfrac{1}{2}m_1 v_{1i}^2 + \dfrac{1}{2}m_2 v_{2i}^2 = \dfrac{1}{2}m_1 v_{1f}^2 + \dfrac{1}{2}m_2 v_{2f}^2$
Coefficient of restitution	$e = \dfrac{v_{2f} - v_{1f}}{-(v_{2i} - v_{1i})}$
Velocity relative to center of mass	$\vec{u}_i = \vec{v}_i - V_{cm}$
Rocket equation	$M\dfrac{d\vec{v}}{dt} = \vec{u}_{ex}\left|\dfrac{dM}{dt}\right| + F_{net,\,ext}$

III. Potential Pitfalls

The center of mass of a system does not necessarily coincide with the position of any of the particles or objects that make up a system.

Don't think that if momentum isn't conserved in one direction it isn't conserved in any direction. The law of conservation of momentum is a vector statement. If the component of the net external force on a system in a given direction is zero, the component of the system's momentum in that direction remains constant.

It is often the case that during collisions, impulses associated with external forces acting on the system are negligible compared to impulses associated with the internal forces between the colliding objects. Thus even for systems that are not isolated, it is often a good approximation to equate the system's pre- and post-collision momenta if the collision occurs in a sufficiently short time. During a collision, the momentum of the system remains constant to the degree that impulses associated with external forces are negligible.

When a perfectly inelastic one-dimensional collision occurs, the kinetic energy of the center-of-mass motion remains constant while the kinetic energy relative to the center of mass becomes zero. Thus the final kinetic energy of the system cannot be zero unless the velocity of the center of mass is zero.

Remember that if a collision is completely elastic, momentum conservation is not sufficient to solve the problem. You will also have to use conservation of kinetic energy.

IV. True or False Questions and Responses

True or False

_____ 1. The total momentum of a system is the product of its total mass and the velocity of its center of mass.

_____ 2. If the sum of the internal forces in a system remains zero, the momentum of the system necessarily remains constant.

_____ 3. If the x component of the total external force on a system remains zero, the x component of the momentum of the system necessarily remains constant.

_____ 4. During a collision, the kinetic energy relative to the center of mass remains constant.

_____ 5. Following a head-on, perfectly inelastic collision, the kinetic energy of the colliding particles necessarily equals zero.

_____ 6. When a head-on collision is perfectly elastic, the coefficient of restitution is zero.

_____ 7. The impulse associated with a force is a measure of the mechanical energy transferred.

_____ 8. It is possible to survive if the center of mass of a hand grenade passes through the center of your head. (Not to be tried at home!)

_____ 9. In a one-dimensional elastic collision between two objects, the velocity of each particle relative to the center of mass is reversed.

Responses to True or False

1. True.

2. False. Internal forces necessarily come in action-reaction pairs, so they always sum to zero. The rate of change of the momentum of the system equals the total external force on the system.

3. True.

4. Not necessarily. For perfectly inelastic collisions the kinetic energy relative to the center of mass is zero.

5. False. The velocity of the center of mass, along with the associated kinetic energy, is the same after the collision as before. If it is nonzero before the collision it is nonzero following the collision.

6. False. In a one-dimensional perfectly elastic collision the coefficient of restitution is one.

7. False. The impulse associated with a force is the measure of the momentum transferred.

8. True. If a thrown grenade explodes in flight, after the explosion no fragments remain near its center of mass. Consequently, it is possible for an object to go unscathed even though the center of mass passes through it.

9. True.

V. Questions and Answers

Questions

1. When a large truck collides head on with a small car, the force exerted by the car on the truck is equal in magnitude and oppositely directed to the force exerted by the truck on the car. Would the momentum of the car-truck system be conserved if this were not so?

2. When a spaceship in empty gravity-free space fires a rocket, does the momentum of the ship-fuel system change? Does the momentum of the ship change? Does the momentum of the fuel change?

3. The average force required to change the momentum of a system by a specific amount is less the longer the time interval in which the change takes place. Does a longer time interval also mean that a smaller maximum force is required?

4. Can a system have zero kinetic energy and nonzero momentum? Can it have zero momentum and nonzero kinetic energy?

Answers

1. If these forces were not equal in magnitude and oppositely directed, the changes in momentum of the car and the truck would not be equal in magnitude and oppositely directed, and thus the momentum of the car-truck system would not remain constant.

2. The momentum of the ship-fuel system remains constant. Any changes in the momentum of the ship are balanced by oppositely directed changes in the momentum of the exhausted fuel.

3. The impulse required to cause a specified change in momentum is fixed, thus the area under the force-versus-time curve is fixed. The longer the time interval, the smaller the maximum force required. The force must be at least as large as the average force at some time in the interval. Longer times mean a smaller average force, which means a smaller maximum force is possible.

4. When a system has zero kinetic energy, nothing is moving. (There are no negative terms in the kinetic energy sum. Thus the momentum must be zero. When a system has zero momentum, the center of mass is at rest. The system can still have kinetic energy due to the motion of parts of the system relative to the center of mass. A spinning ball necessarily has kinetic energy, but may have zero momentum.

VI. Problems, Solutions, and Answers

Example #1. A 2-kg particle is located at the origin and a 1-kg particle is located on the *x* axis at $x = 3\,\text{m}$, as shown in Figure 8-1. Determine the location of the center of mass of this two-particle system.

Figure 8-1

1. Use the formula for the *x* component of the center of mass to determine X_{cm}.	$$X_{cm} = \dfrac{\displaystyle\sum_i m_i x_i}{M}$$ $$= \dfrac{(2\,\text{kg})(0\,\text{m}) + (1\,\text{kg})(3\,\text{m})}{2\,\text{kg} + 1\,\text{kg}} = 1\,\text{m}$$

Example #2—Interactive. As shown in Figure 8-2, a 3-kg particle is located at the origin, a 1-kg particle is located on the *x* axis at $x = 2\,\text{m}$, and a 2-kg particle is located in the *xy* plane at the point (2 m, 3 m). Determine the location of the center of mass of this three-particle system. **Try it yourself.** Work the problem on your own, in the spaces provided, to get the final answer.

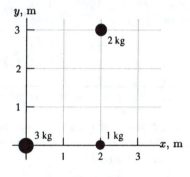

Figure 8-2

1. Use the formula for the *x* component of the center of mass to determine X_{cm}.	$X_{cm} = 1\,\text{m}$
2. Use the formula for the *y* component of the center of mass to determine Y_{cm}.	$Y_{cm} = 1\,\text{m}$

Example #3.   Find the location of the center of mass of a uniform thin rod with one end at the origin and the other end on the x axis at $x = L$.

Picture the Problem. Since the rod is uniform, the linear mass density of the rod is simply the total mass divided by the length: $\lambda = M / L$. Figure 8-3 shows the geometry of the problem needed to use the integral form for the center of mass calculation.

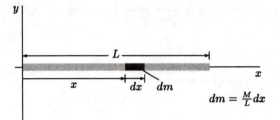

$$dm = \tfrac{M}{L} dx$$

Figure 8-3

1. Write the formula for center of mass as a guide.	$X_{cm} = \dfrac{1}{M} \displaystyle\int x\, dm$
2. We can assume the mass of the rod is M, but we still need to find the differential mass, dm.	$dm = \lambda\, dx = (M / L)\, dx$
3. The rod rests from $x = 0$ to $x = L$, which provides the limits of integration, so we can now find the center of mass.	$X_{cm} = \dfrac{1}{M}\displaystyle\int_0^L x (M / L)\, dx = \dfrac{1}{L}\displaystyle\int_0^L x\, dx$ $= \dfrac{1}{L}\dfrac{x^2}{2}\Big|_0^L = \dfrac{L}{2}$

As you should expect, the center of mass of the uniform rod is right in the middle.

Example #4—Interactive. Find the center of mass of a thin rod of length L whose density increases linearly with the distance from one end according to the formula $\lambda = (2\,\text{kg/m}^2)x$, where x is the distance in meters from one end of the rod.

Picture the Problem. Use the integral form of the center-of-mass formula. The rod is as shown in Figure 8-3, *except* that you cannot use the equation shown for the differential mass. **Try it yourself.** Work the problem on your own, in the spaces provided, to get the final answer.

1. Write out the center-of-mass formula as a guide.	$X_{cm} = \dfrac{1}{M}\displaystyle\int x\, dm$
2. Find an expression for the differential mass using the varying mass density provided.	

3. Because we are not given the mass of the rod, it must be calculated in terms of the given parameters. Integrate the differential mass to find the total mass.	
4. Now the center of mass calculation can be completed.	$X_{cm} = \dfrac{2}{3} L$

Example #5. As shown in Figure 8-4, a uniform 0.300-kg meter stick, free to rotate about a frictionless peg positioned at its 25-cm mark, is released from rest from a horizontal position. Determine the kinetic energy of the stick as it rotates through the vertical position, with the 100-cm mark directly below the rotation axis.

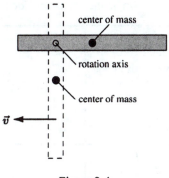

Figure 8-4

Picture the Problem. This is a conservation of energy problem. If we consider the peg, the stick, and the earth as the system, the change in potential energy of the center of mass of the stick will result in an equal but opposite change in the kinetic energy of the stick.

1. Write out the conservation of energy expression.	$K_f + U_f = K_i + U_i$
2. Set the $y = 0$ point so that the initial kinetic and final potential energies are zero, and solve.	$K_f = U_i = mgY_{cm}$ $= (0.300\,\text{kg})(9.81\,\text{m/s}^2)(0.25\,\text{m})$ $= 0.75\,\text{J}$

Example #6—Interactive. As shown in Figure 8-5, a uniform 0.300-kg meter stick, free to rotate about a frictionless peg located at its 20.0-cm mark, has a compact 0.200-kg weight fastened to the stick at the 80.0-cm mark. Determine (*a*) the location of the center of mass of the stick-weight system and (*b*) the change in the gravitational potential energy of the system when the stick rotates from a horizontal position to a vertical position with the weight directly above the peg.

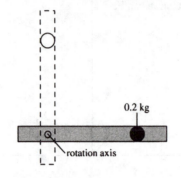

0.2 kg

rotation axis

Figure 8-5

Picture the Problem. Use the formula for the center of mass of a composite system to locate the center of mass of the stick-weight system. The change in the gravitational potential energy will be a result of the change in the height of the center of mass of the stick-weight system. **Try it yourself.** Work the problem on your own, in the spaces provided, to get the final answer.

1. Find the center of mass of the stick-weight system. Choose the origin to be at the zero-cm mark of the stick, with the positive x axis along the stick.	$X_{cm} = 62\,\text{cm}$
2. Find the change in the potential energy of the stick-weight system using the change in height of the system's center of mass.	$\Delta U = 2.06\,\text{J}$

Example #7. A uniform 3.00-kg meter stick is rotating freely about a fixed peg through its 25.0-cm mark. At the instant the stick's center of mass passes directly below the rotation axis as shown in Figure 8-6, it has a speed of 1.00 m/s. Determine the force exerted on the stick by the peg at this instant.

Picture the Problem. At the instant shown, the center of mass of the system is experiencing circular motion and a centripetal acceleration. Draw a free-body diagram of the stick, and use Newton's 2nd Law for systems to solve for the desired force.

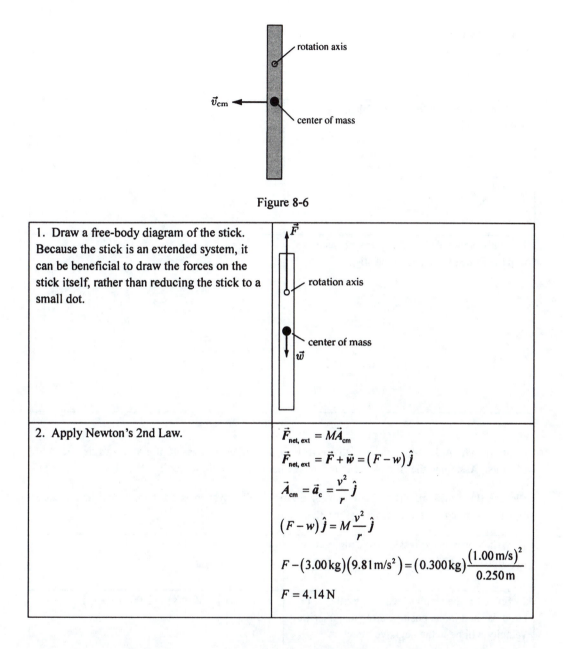

Figure 8-6

1. Draw a free-body diagram of the stick. Because the stick is an extended system, it can be beneficial to draw the forces on the stick itself, rather than reducing the stick to a small dot.	$\vec{F}$ rotation axis center of mass $\vec{w}$
2. Apply Newton's 2nd Law.	$$\vec{F}_{\text{net, ext}} = M\vec{A}_{\text{cm}}$$ $$\vec{F}_{\text{net, ext}} = \vec{F} + \vec{w} = (F - w)\hat{j}$$ $$\vec{A}_{\text{cm}} = \vec{a}_{\text{c}} = \frac{v^2}{r}\hat{j}$$ $$(F - w)\hat{j} = M\frac{v^2}{r}\hat{j}$$ $$F - (3.00\,\text{kg})(9.81\,\text{m/s}^2) = (0.300\,\text{kg})\frac{(1.00\,\text{m/s})^2}{0.250\,\text{m}}$$ $$F = 4.14\,\text{N}$$

Example #8—Interactive. As the meter stick referred to in Example #6 rotates through the vertical position, with the 100-cm mark directly below the rotation axis, the stick-weight system's center of mass has a speed of 1.3 m/s. Determine the force exerted on the stick by the peg at this instant. Compare this force with the weight of the stick.

Picture the problem. At the bottom of its motion the center of mass of the stick-weight system is experiencing circular motion and a centripetal acceleration. By drawing a free-body diagram of the stick-weight system and applying Newton's 2nd Law, the force of the peg on the stick can be found. **Try it yourself.** Work the problem on your own, in the spaces provided, to get the final answer.

1. Draw a free-body diagram of the stick-weight system.	
2. Apply Newton's 2nd Law for systems to find the force of the peg on the stick.	$F = 6.91\,\text{N}$

Example #9. A 3.00-kg rifle, initially at rest, fires a bullet of mass 10.0 g at a muzzle speed of 650 m/s. Assuming the rifle is free to move, at what speed would it recoil?

Picture the Problem. The momentum of the rifle-bullet system is conserved, and in this case equal to zero before and after firing the rifle.

1. Write down an expression for conservation of momentum in this case.	$\vec{P}_i = \vec{P}_f$ $0 = m_b \vec{v}_{b,f} + m_r \vec{v}_{r,f}$
2. Solve for the recoil speed of the rifle. If the bullet is fired in the positive direction, the rifle velocity will be in the negative direction.	$0 = (10.0\,\text{g})(650\,\text{m/s}) + (3.00\,\text{kg})(-v_r)$ $v_r = 2.17\,\text{m/s}$

Example #10—Interactive. A 900-kg car traveling north at 60.0 km/hr collides with a 1200-kg light truck traveling west at 50.0 km/hr. The vehicles stick together following the collision, as shown in Figure 8-7. Determine the velocity (magnitude and direction) of the wreck immediately following the collision. Neglect friction between the vehicles and the ground.

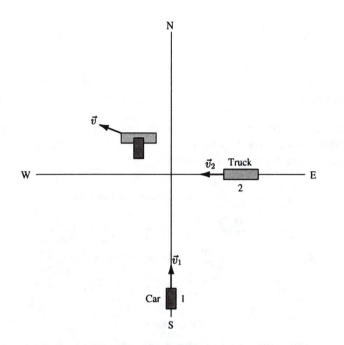

Figure 8-7

Picture the Problem. The momentum of the car-truck system before and after the collision must be the same. This is a two-dimensional problem, so momentum must be conserved in both the x and y directions. Let the positive x direction be East and the positive y direction be North. **Try it yourself.** Work the problem on your own, in the spaces provided, to get the final answer.

1. Write the conservation of momentum equation as a guide.	
2. Write the equation for conservation of momentum in the x direction.	
3. Write the equation for conservation of momentum in the y direction.	
4. Solve the equations from steps 2 and 3 to find the final speed and direction.	$v = 38.4\,\text{km/hr at } 42°\text{ north of west}$

Example #11. Consider a head-on elastic collision between a cue ball and the eight ball. Assuming the cue ball has an initial velocity $\vec{v}_0$, the eight ball is initially rest, and the balls have equal masses m, determine the velocities of each ball immediately following impact. To determine these velocities, transfer to the center-of-mass system, determine the post-collision velocities in that frame of reference, and then transfer back to the initial frame of reference.

Picture the Problem. Determine the velocity of the center of mass. Remember that in a one-dimensional elastic collision, like this one with two spheres, the velocities relative to the center of mass get reversed. Figure 8-8 helps to illustrate this.

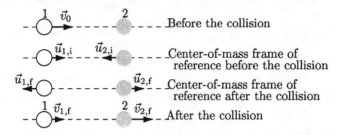

Figure 8-8

1. Find the velocity of the system's center of mass.	$$\vec{V}_{cm} = \frac{1}{M}\sum_i m_i \vec{v}_i = \frac{m_1\vec{v}_1 + m_2\vec{v}_2}{m_1 + m_2}$$ $$= \frac{m\vec{v}_0 + 0}{m+m} = \frac{\vec{v}_0}{2}$$
2. Find the initial velocities of the two balls relative to the center of mass velocity.	$$u_{1,i} = v_{1,i} - V_{cm}$$ $$= v_0 - \frac{v_0}{2} = \frac{v_0}{2}$$ $$u_{2,i} = v_{2,i} - V_{cm}$$ $$= 0 - \frac{v_0}{2} = -\frac{v_0}{2}$$
3. For elastic, 1D collisions, the speeds relative to the center of mass get reversed, and the speed of the center of mass remains constant. We can use this information to find the speeds of the balls after the collision.	$$u_{2,f} = u_{1,i} = \frac{v_0}{2} = v_{2,f} - V_{cm}$$ $$\frac{v_0}{2} = v_{2,f} - \frac{v_0}{2}$$ $$v_{2,f} = v_0$$ $$u_{1,f} = u_{2,i} = -\frac{v_0}{2} = v_{1,f} - V_{cm}$$ $$-\frac{v_0}{2} = v_{1,f} - \frac{v_0}{2}$$ $$v_{1,f} = 0\,\text{m/s}$$

Example #12—Interactive. Consider a head-on elastic collision between ball A and ball B. Ball A has mass m_A and velocity $\vec{v}_0$, ball B has mass m_B and is initially at rest, and the ratio of the masses is $m_A / m_B = \alpha$. Following the collision, determine the velocities of each ball in terms of α and $\vec{v}_0$. To determine these velocities, transfer to the center-of-mass system, determine the post-collision velocities in that frame of reference, and then transfer back to the initial frame of reference. Check your results for $\alpha = 1$, and when α approaches zero and infinity.

Picture the Problem. Determine the velocity of the center of mass. Remember that in a one-dimensional elastic collision, like this one with two spheres, the velocities relative to the center of mass get reversed. **Try it yourself.** Work the problem on your own, in the spaces provided, to get the final answer.

1. Find the velocity of the system's center of mass.	
2. Find the initial velocities of the two balls relative to the center of mass velocity.	
3. For elastic, 1D collisions, the speeds relative to the center of mass get reversed, and the speed of the center of mass remains constant. We can use this information to find the speeds of the balls after the collision.	$v_{A,f} = \left(\dfrac{1-\alpha}{1+\alpha}\right)$ and $v_{B,f} = \dfrac{2v_0}{1+\alpha}$

When α equals one, the two masses are equal, and you should get the same result as in Example #11. When α approaches infinity, the collision is analogous to that of a ping-pong ball colliding with a bowling ball. The velocity of the center of mass of this system is essentially zero. During the collision, the velocity of ball A (the ping-pong ball) is reversed and the velocity of ball B (the bowling ball) remains essentially zero.

When α approaches zero, the collision is analogous to that of a bowling ball colliding with a ping-pong ball. The velocity of the center of mass of this system equals the initial velocity of ball A (the bowling ball). From the perspective of the center-of-mass reference frame, the velocity of ball A (the bowling ball) remains zero while the velocity of ball B (the ping-pong ball) changes from $-\vec{v}_0$ to $\vec{v}_0$. Thus, following the collision ball B moves away from ball A with a velocity of $\vec{v}_0$. In the original frame of reference, ball A (the bowling ball) continues to move with velocity $\vec{v}_0$, while ball B (the ping-pong ball) moves with a velocity of $2\vec{v}_0$.

Example #13. A 2.00-kg sphere with a velocity of $8.00\,\text{m/s}\,\hat{i}$ runs into a stationary 8.00-kg sphere. Following the collision the 2.00-kg sphere moves with a velocity of $-6.00\,\text{m/s}\,\hat{j}$. Determine the magnitude and direction of the velocity of the 8-kg sphere following the collision. Is this collision elastic?

Picture the Problem. The momentum of the system is conserved in both the x and y directions.

1. Draw a sketch of the situation, showing the velocities of both spheres both before and after the collision.	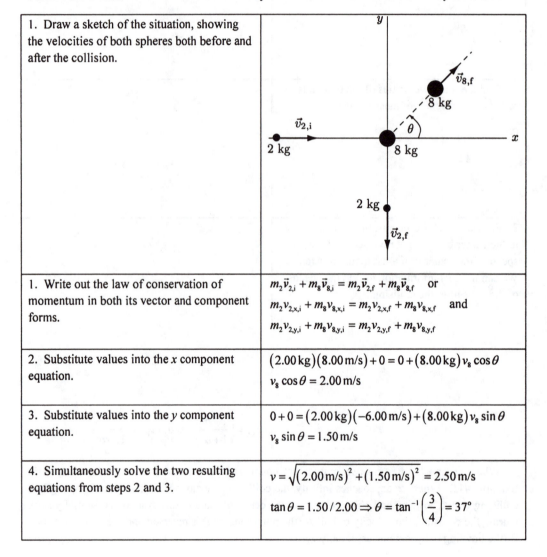
1. Write out the law of conservation of momentum in both its vector and component forms.	$m_2\vec{v}_{2,i} + m_8\vec{v}_{8,i} = m_2\vec{v}_{2,f} + m_8\vec{v}_{8,f}$ or $m_2 v_{2,x,i} + m_8 v_{8,x,i} = m_2 v_{2,x,f} + m_8 v_{8,x,f}$ and $m_2 v_{2,y,i} + m_8 v_{8,y,i} = m_2 v_{2,y,f} + m_8 v_{8,y,f}$
2. Substitute values into the x component equation.	$(2.00\,\text{kg})(8.00\,\text{m/s}) + 0 = 0 + (8.00\,\text{kg})v_8\cos\theta$ $v_8\cos\theta = 2.00\,\text{m/s}$
3. Substitute values into the y component equation.	$0 + 0 = (2.00\,\text{kg})(-6.00\,\text{m/s}) + (8.00\,\text{kg})v_8\sin\theta$ $v_8\sin\theta = 1.50\,\text{m/s}$
4. Simultaneously solve the two resulting equations from steps 2 and 3.	$v = \sqrt{(2.00\,\text{m/s})^2 + (1.50\,\text{m/s})^2} = 2.50\,\text{m/s}$ $\tan\theta = 1.50/2.00 \Rightarrow \theta = \tan^{-1}\left(\dfrac{3}{4}\right) = 37°$

| 5. Determine the initial and final kinetic energies of the system. If they are equal, the collision is elastic. If they are not equal, the collision is inelastic. | $K_i = \dfrac{1}{2}(2\,\text{kg})(8\,\text{m/s})^2 + 0 = 64\,\text{J}$

 $K_f = \dfrac{1}{2}(2\,\text{kg})(-6\,\text{m/s})^2 + \dfrac{1}{2}(8\,\text{kg})(2.5\,\text{m/s})^2 = 61\,\text{J}$ |

It seems the collision is not quite elastic. However, as you will learn in Chapter 9, this is not necessarily so. We have only calculated kinetic energies associated with the translational motion of the spheres. If the spheres have increased their rate of spinning, the missing kinetic energy may be accounted for in the kinetic energy associated with this increased spinning motion.

Example #14—Interactive. A 2.00-kg sphere with a velocity of $8.00\,\text{m/s}\,\hat{i}$ runs into a stationary 8.00-kg sphere. Following the collision the 2.00-kg sphere moves with a velocity of $3.00\,\text{m/s}\,\hat{i} - 4.00\,\text{m/s}\,\hat{j}$. Determine the velocity of the 8.00-kg sphere following the collision.

Picture the Problem. Sketch the motion of the spheres both before and after the collision. The momentum of the system is conserved in both the x and y directions. **Try it yourself.** Work the problem on your own, in the spaces provided, to get the final answer.

1. Sketch the motion of the spheres both before and after the collision	
2. Write out the law of conservation of momentum in both its vector and component forms.	
3. Substitute values into the x component equation.	
4. Substitute values into the y component equation.	
5. Simultaneously solve the two resulting equations from steps 2 and 3.	$\vec{v} = 1.25\,\text{m/s}\,\hat{i} + 1.00\,\text{m/s}\,\hat{j}$

Example #15. Rainwater is falling straight down with a velocity of 15.0 m/s at the rate of 10.0 cm of rain per hour. An electronic scale is placed in the rain. The scale's pan consists of a 25.0-cm-diameter circular disk. If the weight of the water on the platform is negligible, determine the reading of the scale.

Picture the Problem. This is an impulse-momentum problem. The impulse exerted on the rainwater by the pan equals the change in momentum of the rainwater. The average force exerted by the pan to stop the water will be the reading of the scale. Remember the density of water is 1000 kg/m^3.

1. Choosing the positive y direction as upward, find the change in momentum of one kilogram of water moving downward at a speed of 15 m/s that is brought to rest.	$\Delta\vec{p} = \vec{p}_f - \vec{p}_i = 0 - (1.00\,\text{kg})(15.0\,\text{m/s}\,\hat{j})$ $= 15.0\,\text{kg·m/s}\,\hat{j}$
2. Determine the length of time required for one kilogram of rain to fall on the balance. First find the time it takes for a certain height of water to fall. The height and volume of our cylinder of water falling on the pan are related by $V = \pi r^2 h$. Solve for h and substitute into the time expression.	$h = \left(\dfrac{dh}{dt}\right)\Delta t$ or $\Delta t = \dfrac{h}{dh/dt}$ $\Delta t = \dfrac{V}{\pi r^2 (dh/dt)}$
3. Knowing both the change in momentum and change in time, the average force can be calculated.	$\vec{F}_{av}\,\Delta t = \Delta\vec{p}$ or $\vec{F}_{av} = \dfrac{\Delta\vec{p}}{\Delta t} = \pi r^2 \dfrac{\Delta\vec{p}}{V}\dfrac{dh}{dt}$
4. Knowing that the volume of 1 kg of water is 10^{-3} m^3, we can solve for the average force required to stop the rain.	$\vec{F}_{av} = \pi(0.125\,\text{m})^2 \dfrac{15\,\text{kg·m/s}\,\hat{j}}{10^{-3}\,\text{m}^3}\left(\dfrac{0.1\,\text{m}}{\text{hr}}\right)\left(\dfrac{1\,\text{hr}}{3600\,\text{s}}\right)$ $= 2.05\times10^{-2}\,\text{N}\,\hat{j}$
5. According to Newton's 3rd Law, the rain pushes back down on the scale with the same average force, so the scale reading will be 2.05×10^{-2} N.	

Example #16—Interactive. An empty, open railroad car of mass M with frictionless wheels is rolling along a horizontal track with speed v_0 when it begins to rain. There is no wind so the rain falls straight down. The mass of the rainwater that accumulates in the car is m. (*a*) Determine the speed of the car after the rain has accumulated. (*b*) After the rain stops, but while the car is still moving, a worker opens a drain hole in the bottom of the car and lets the water out. Determine the speed of the car when it is once again empty.

Picture the Problem. This is a conservation of momentum problem in the horizontal direction, because the only external forces: weight and a normal force, are directed in the vertical direction.

The collision between the car and the rainwater is perfectly inelastic. The rain initially has no horizontal momentum, but it eventually gets some. For the second "collision", when the rainwater is let out, think carefully about the horizontal velocity of the water the instant it leaves the rail car. **Try it yourself.** Work the problem on your own, in the spaces provided, to get the final answer.

1. Draw a sketch illustrating the motion of the car and rainwater before and after each collision.	
2. Conserve momentum in the horizontal direction for the collision of the water landing in the car.	$$v = \frac{M}{M+m}v_0$$
3. Conserve momentum again while the water is being let out.	$$v = \frac{M}{M+m}v_0$$

Chapter **9**

Rotation

I.   Key Ideas

In this chapter we will continue to study the motion of many-particle systems emphasizing the rotational aspects of such motion. For rigid-body motion, we will look at the kinematic concepts of angular displacement, velocity, and acceleration and the dynamic concepts of torque and moment of inertia. This chapter treats the rotation of a rigid object either about a fixed rotation axis or, in the case of a rolling object, about a moving rotation axis—one that keeps its direction fixed while moving.

Section 9-1.   Rotational Kinematics: Angular Velocity and Angular Acceleration. Suppose a disk is constrained to rotate about a fixed axis through its center and perpendicular to the disk as shown in Figure 9-1. A radial line drawn on the disk will sweep out angle θ as the disk rotates. This angle between the drawn line and a reference direction is called the **angular position** of the disk. The change in the angular position $\Delta\theta$ is called the **angular displacement**, and the rate of change of the angular position, $d\theta/dt$, is the angular velocity.

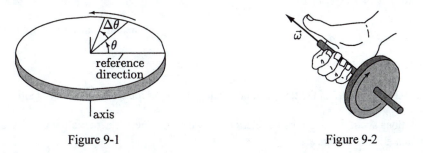

Figure 9-1   Figure 9-2

Angular velocity $\vec{\omega}$ is a vector parallel to the rotation axis. The magnitude of $\vec{\omega}$ is the rotation rate, and its direction is along the axis specified via a right-hand rule as shown in Figure 9-2. If you curl your fingers in the direction of rotation, $\vec{\omega}$ is in the same direction as your thumb—using your right hand, of course! **Angular speed** is the magnitude of the angular velocity vector. The **angular acceleration** vector is $\vec{\alpha} = d\vec{\omega}/dt$.

When a vector's direction remains parallel to an axis (not necessarily a rotation axis), the vector can be specified by a signed scalar. For example, if you make the z axis parallel to the given axis, the vector can be described by only its z component. One example of such a vector is the velocity vector of a particle undergoing one-dimensional linear motion (motion along a straight line). Another is the angular velocity vector of an object constrained to rotate such that the direction of its rotation axis remains fixed. This is called *one-dimensional rotational motion*.

For one-dimensional rotational motion, θ, ω, and α denote the angular position, velocity, and acceleration, respectively. Table 9-1 shows them with their one-dimensional linear-motion counterparts.

One-Dimensional (Fixed-Axis Direction) Linear and Rotational Motion		
Linear Motion	Rotational (Angular) Motion	
x	θ	Position
Δx	$\Delta \theta$	Displacement
$v_{av} = \dfrac{\Delta x}{\Delta t}$	$\omega_{av} = \dfrac{\Delta \theta}{\Delta t}$	Average velocity
$v = \dfrac{dx}{dt}$	$\omega = \dfrac{d\theta}{dt}$	Velocity
$a_{av} = \dfrac{\Delta v}{\Delta t}$	$\alpha_{av} = \dfrac{\Delta \omega}{\Delta t}$	Average acceleration
$a = \dfrac{dv}{dt} = \dfrac{d^2 x}{dt^2}$	$\alpha = \dfrac{d\omega}{dt} = \dfrac{d^2 \theta}{dt^2}$	Acceleration

Table 9-1

For constant angular acceleration, the relations between the rotational kinematic parameters are

$$\Delta \theta = \omega_0 \, \Delta t + \tfrac{1}{2} \alpha \left(\Delta t \right)^2$$
$$\Delta \omega = \alpha \, \Delta t$$
$$\omega^2 = \omega_0^2 + 2 \alpha \, \Delta \theta \qquad \text{Kinematic formulas for constant } \alpha$$
$$\omega_{av} = \tfrac{1}{2} \left(\omega + \omega_0 \right)$$

See Table 9-2 on page 287 of the text for more analogs between rotational and linear motion.

For rotations about a fixed axis the relations between the linear kinematic parameters and the angular kinematic parameters are

$$\Delta s = r \, \Delta \theta$$
$$v_t = r\omega$$
$$a_t = r\alpha \qquad \text{Relations between linear and angular kinematic parameters}$$
$$a_c = \frac{v^2}{r} = r\omega^2$$

where v_t and a_t are the tangential components of the linear velocity and acceleration, respectively, and a_c is the centripetal (radial) component of the acceleration.

The common units of angle are the degree, the radian, and the revolution. The radian is the SI unit of angle. By definition, one **radian (rad)** is the measure of the central angle of a circle whose intercepted arc length equals the radius. Some useful angular conversion factors are

$$\frac{2\pi \text{ rad}}{1 \text{ rev}}, \quad \frac{1 \text{ rev}}{360°}, \quad \text{and} \quad \frac{\pi \text{ rad}}{180°}$$ Angular conversion factors

Section 9-2. Rotational Kinetic Energy. The kinetic energy K of any object is merely the sum of the kinetic energies of the individual particles of the object. For an object rotating about a fixed axis, the velocity of the ith particle is $v_i = r_i \omega$, where r_i is the distance from the rotation axis to the ith particle. Summing over the elements, the total kinetic energy of a rotating object is

$$K = \sum \tfrac{1}{2} m_i v_i^2 = \sum \tfrac{1}{2} m_i \left(r_i \omega \right)^2 = \frac{1}{2} \left(\sum m_i r_i^2 \right) \omega^2 = \frac{1}{2} I \omega^2$$ Kinetic energy for rotations

where the **moment of inertia**, I, is defined to be

$$I = \sum m_i r_i^2$$ Moment of inertia

Section 9-3. Calculating the Moment of Inertia. The moment of inertia I is a measure of an object's inertial resistance to changes in angular velocity. It depends on the mass and the distribution of the mass about the axis. For a system of discrete particles, the moment of inertia can be calculated as above. For a continuous object, we imagine the object to consist of a continuum of very small mass elements, and as such the summation above becomes the integral

$$I = \int r^2 \, dm$$ Moment of inertia for continuous objects

where r is the radial distance from the axis of rotation to mass element dm. Table 9-1 on page 274 of the text lists moments of inertia for a number of objects about axes through their centers of mass. It can be shown that the moment of inertia I of an object about a given axis is equal to

$$I = Mh^2 + I_{cm}$$ Parallel-axis theorem

where M is the mass of the object, I_{cm} is its moment of inertia about a parallel axis that passes through the center of mass, and h is the distance between the two axes. This theorem greatly expands the usefulness of Table 9-1 on page 274 of the textbook. The moments of inertia about innumerable axes can be easily obtained using the parallel-axis theorem. Several examples at the end of this chapter provide practice calculating moments of inertia for a variety of objects.

Section 9-4. Newton's 2nd Law for Rotation. In motion, the quantity of fundamental dynamic importance is force. In rotational motion, the dynamic quantity associated with the force is the torque. *Torque τ* is the measure of a force's ability to produce a change in the rotational motion of an object. The torque τ produced by a force about an axis equals rF_t, where r, shown in Figure 9-3, is the distance from the rotation axis to the point of application of the force and F_t is the tangential component of the force. A little trigonometry will show that $\tau = rF_t = r(F\sin\theta) = (r\sin\theta)F = lF$, where the *lever arm l* is the perpendicular distance from the axis to the *line-of-action* of the force.

$$\tau = rF_t = rF \sin \theta = lF \qquad \text{Torque about an axis}$$

To calculate the torque due to gravity, assume that all the mass is located at the center of gravity of an object, and the gravitational force acts only at that point.

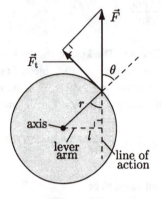

Figure 9-3

By mentally dividing up an object constrained to rotate about a fixed axis—a Ferris wheel, for example—into infinitely many infinitesimal pieces, and then applying Newton's laws for particle motion to each piece, it can be shown that

$$\tau_{\text{net, ext}} = \sum \tau_{\text{ext}} = I\alpha \qquad \text{Newton's 2nd Law for rotations}$$

where $\Sigma \tau_{\text{ext}}$ is the sum of the torques due to external forces, α is the angular acceleration, and I is the moment of inertia.

Section 9-5. Applications of Newton's 2nd Law for Rotation. When applying Newton's law for rotation, you should not draw the free-body diagram as a dot. Rather, you should show each object as a complete picture, and draw each force along the line of action of that force, with the tail of the vector at the point of application of the force. In addition to indicating the positive direction for the x and y axes, you should also indicate the positive direction of rotation.

Because of friction and inertia, if a string goes around a pulley wheel the tension in it is greater on one side of the wheel than the other. This is necessary for the string to exert a torque on the wheel. Use two different labels, e.g. T_1 and T_2, for these two tensions.

For any situation in which rotation occurs without slipping, for example if a string does not slip around the rim of the pulley, then the linear and angular velocities and accelerations can be related to each other:

$$v_t = R\omega$$
$$a_t = R\alpha \qquad \text{Nonslip conditions}$$

Power. Work is defined as $dW = F_t ds$, where dW is the work increment, ds is the magnitude of the displacement increment, and F_t is the component of the force in the direction of the displacement. The subscript t is used because the displacement points about the tangential direction. For an object rotating about a fixed axis $ds = r\,d\theta$, so

$$dW = F_t\,ds = F_t r\,d\theta = \tau\,d\theta \qquad\qquad \text{Work}$$

where the torque τ equals $F_t r$. The rate at which the torque does work is the power input of the torque:

$$P = \frac{dW}{dt} = \tau\frac{d\theta}{dt} = \tau\omega \qquad\qquad \text{Power}$$

Section 9-6. Rolling Objects. When a bicycle moves along a surface it can either follow a straight-line path or a curved path. In this section, discussion of rotation is restricted to the kind executed by the wheel of a bicycle traveling along a straight-line path. For this kind of rolling, the motion of each individual particle making up the wheel remains confined to a plane. It follows that the direction of the wheel's instantaneous rotation axis, which is perpendicular to the plane, remains fixed.

Rolling Without Slipping. At each instant, each point on the wheel, shown in Figure 9-4, is moving with speed $v = r\omega$, where r is its distance from the rotation axis and ω is the wheel's angular speed. The rotation axis perpendicular to the plane of the page passes through the point(s) of contact between the wheel and the road surface.

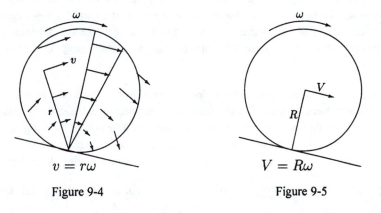

Figure 9-4   Figure 9-5

A special case of this relation, shown in Figure 9-5, is $V = R\omega$, where R is the wheel's radius and V is the speed of its center. *Note*: $V = R\omega$ only if the wheel rolls *without* slipping.

$$V = R\omega \qquad\qquad \text{Rolling-without-slipping condition}$$

By differentiating both sides of $v = r\omega$ with respect to time we get $a_t = r\alpha$, where a_t is the tangential acceleration and α is the angular acceleration, shown in Figure 9-6. A special case of this relation, shown in Figure 9-7, is $A_t = R\alpha$, where A_t is the tangential acceleration of the wheel's center.

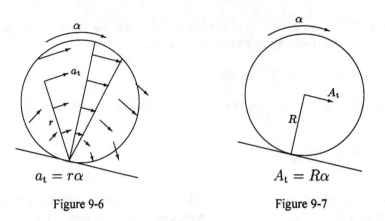

Figure 9-6   Figure 9-7

The most general method of solving for the motion of an object is to simultaneously apply both the linear and the rotational forms of Newton's 2nd Law. These are:

$$\sum_i F_{i,ext} = MA_{cm}$$

$$\sum_i \tau_{i,cm} = I_{cm}\alpha$$

Newton's 2nd Laws for translation and rotation

where each torque is calculated about an axis through the center of mass. When the center of mass is accelerating (a ball rolling down an incline, for example), its reference frame is not an inertial one. In a noninertial reference frame the equation $\Sigma\tau = I\alpha$ does not hold—except for certain specific rotation axes. The most general of these specific axes is the one through the center of mass. For this reason it is best to use the center of mass axis when applying $\Sigma\tau = I\alpha$ to rolling objects.

If an object rolls without slipping, the point of contact with the surface is instantaneously at rest. Any frictional force is that of static friction so no mechanical energy is dissipated. This suggests that many rolling-without-slipping problems can best be solved by using conservation of mechanical energy. The kinetic energy K of an object can be expressed as the sum of two terms, one for its translational motion K_{trans} and the other for its rotational motion K_{rot}. That is,

$$K = K_{trans} + K_{rot} = \tfrac{1}{2}MV_{cm}^2 + \tfrac{1}{2}I_{cm}\omega^2$$

Translational and rotational kinetic energy

where I_{cm} is the moment of inertia about an axis through the center of mass. By substituting the rolling-without-slipping condition $V = R\omega$ into this equation, ω can be eliminated and the kinetic energy can be expressed in terms of V_{cm} alone.

II. Physical Quantities and Key Equations

Physical Quantities

There are no new physical quantities in this chapter.

Key Equations

Kinematic formulas for constant α

$$\Delta\theta = \omega_0\,\Delta t + \tfrac{1}{2}\alpha(\Delta t)^2$$

$$\Delta\omega = \alpha\,\Delta t$$

$$\omega^2 = \omega_0^2 + 2\alpha\,\Delta\theta$$

$$\omega_{av} = \tfrac{1}{2}(\omega + \omega_0)$$

Relations between linear and angular kinematic parameters

$$\Delta s = r\,\Delta\theta$$

$$v_t = r\omega$$

$$a_t = r\alpha$$

$$a_c = \frac{v^2}{r} = r\omega^2$$

One-Dimensional (Fixed-Axis Direction) Linear and Rotational Motion

Linear Motion	Rotational (Angular) Motion	
x	θ	Position
Δx	$\Delta\theta$	Displacement
$v_{av} = \dfrac{\Delta x}{\Delta t}$	$\omega_{av} = \dfrac{\Delta\theta}{\Delta t}$	Average velocity
$v = \dfrac{dx}{dt}$	$\omega = \dfrac{d\theta}{dt}$	Velocity
$a_{av} = \dfrac{\Delta v}{\Delta t}$	$\alpha_{av} = \dfrac{\Delta\omega}{\Delta t}$	Average acceleration
$a = \dfrac{dv}{dt} = \dfrac{d^2x}{dt^2}$	$\alpha = \dfrac{d\omega}{dt} = \dfrac{d^2\theta}{dt^2}$	Acceleration

Angular conversion factors

$$\frac{2\pi\,\text{rad}}{1\,\text{rev}}, \quad \frac{1\,\text{rev}}{360°}, \quad \text{and} \quad \frac{\pi\,\text{rad}}{180°}$$

Torque about an axis

$$\tau = rF_t = rF\sin\theta = lF$$

Newton's 2nd Law for rotations

$$\sum\tau = I\alpha$$

Moment of inertia

$$I = \sum m_i r_i^2 \rightarrow \int r^2\,dm$$

Parallel-axis theorem

$$I = Mh^2 + I_{cm}$$

Kinetic energy of rotation

$$K = \sum \tfrac{1}{2}m_i v_i^2 = \frac{1}{2}I\omega^2$$

Work	$dW = F_t\, ds = F_t r\, d\theta = \tau\, d\theta$
Power	$P = \dfrac{dW}{dt} = \tau\,\dfrac{d\theta}{dt} = \tau\omega$
Newton's 2nd Laws for translation and rotation	$\sum_i F_{i,\text{ext}} = MA_{cm}$ $\sum_i \tau_{i,cm} = I_{cm}\alpha$
Rolling without slipping	$V = R\omega$
Translational and rotational kinetic energy	$K = K_{\text{trans}} + K_{\text{rot}} = \tfrac{1}{2}MV_{cm}^2 + \tfrac{1}{2}I_{cm}\omega^2$

III. Potential Pitfalls

What is a radian? If you cut a wedge of pumpkin pie and the length of one side of the wedge *equals* the length of the crust, the wedge angle equals one radian.

Don't think angular measure must be in radians in all the kinematic equations. Equations with *only* rotational kinematic parameters (e.g., θ, ω, and α) and time, such as $\Delta\theta = \omega\,\Delta t$, are valid with *any* consistent unit of angular measure. Thus, in the kinematic equations for rotational motion with constant α, any consistent unit for angular measure may be used. Equations with *both* linear and angular parameters such as $\Delta s = r\,\Delta\theta$, are valid *only* when the unit of angular measure is the radian.

Don't be mislead into thinking that linear acceleration means the acceleration component in the direction of the velocity vector. In general, linear acceleration $\vec{a}$ has both a centripetal, a_c, and a tangential, a_t, component. Linear position, velocity, and acceleration are the same old friends that we called position, velocity, and acceleration prior to introducing the corresponding angular terms. When referring to angular position, angular velocity, or angular acceleration, it is best to explicitly use the "angular" descriptor.

Don't think that angular velocity is a scalar. Angular velocity is a vector quantity. The direction of the angular velocity vector is found by the right-hand rule. Your right thumb points in the direction of the angular velocity vector if you curl the fingers of your right hand in the direction of the rotational motion. Alternatively, the direction of the angular velocity vector is the direction a common (or right-handed) screw would advance if rotated in the same direction as the rotating body. The scalar quantity ω represents either the angular velocity vector's magnitude (the angular speed) or its component in the direction of some axis.

For one-dimensional rotational motion, the magnitude of the angular velocity is the instantaneous rate at which an object rotates. The direction of rotation, clockwise or counterclockwise, is indicated by a plus or minus sign. Counterclockwise is usually the positive direction. When a particle is moving along the arc of a circle, the angular velocity of the particle is the rate at which the line from the particle to the center of the circle sweeps out the angle.

The rolling-without-slipping relation $V = R\omega$ is valid only for rolling on both flat and curved surfaces, like when cresting a hill.

For a rotating object, $r\omega$ is the linear velocity of a point at distance r from the rotation axis. Don't fall into the trap of thinking that for a rotating object $r\alpha$ is the linear acceleration of the point. This is only the tangential component of the linear acceleration vector. There is also a centripetal component given by $r\omega^2$.

Don't think the torque depends on the distance between the rotation axis and the point of application of the force. Rather, it depends on the length of the moment arm (the perpendicular distance between the rotation axis and the line of action of the force). If the line of action intersects the rotation axis then the torque must be zero.

IV. True or False Questions and Responses

True or False

_____ 1.   The acceleration of the center of mass of an object equals the net external force acting on it divided by its total mass unless the object rotates about the center of mass.

_____ 2.   Angular velocity and angular acceleration can be defined only in terms of angles measured in radians.

_____ 3.   A circular disk of radius r rotates about a fixed axis perpendicular to the disk and through its center. If the disk's angular acceleration is α, the linear acceleration of a point on its rim is equal to $r\alpha$.

_____ 4.   The second hand of a watch rotates with an angular velocity of roughly 0.1 rad/s.

_____ 5.   All parts of a rotating rigid object have the same angular velocity.

_____ 6.   All parts of a rotating rigid object have the same centripetal acceleration.

_____ 7.   The moment of inertia of a rotating rigid object about the rotational axis is independent of the rate of rotation.

_____ 8.   If the net external force acting on an object is zero, the net torque on it must also be zero.

_____ 9.   The net force acting on a rigid object rotating about a fixed axis through its center of mass is always zero.

_____ 10. The lever arm of a force is the perpendicular distance from the axis of rotation to the point at which the force acts.

_____ 11. The kinetic energy of a rigid object rotating about a fixed axis depends only on its angular speed and the total mass.

_____ 12. You cannot specify a moment of inertia unless you also specify an axis.

_____ 13. The moment of inertia of a rigid object about an axis through its center of mass is smaller than that about any other parallel axis.

_____ 14. For the purpose of calculating its moment of inertia, all the mass of a rigid object can be considered to be concentrated at its center of mass.

_____ 15. If a rigid object is rotating, it must be rotating about a fixed axis.

Responses to True or False

1. False. The center-of-mass acceleration equals the net external force divided by the total mass regardless of whether the object is rotating or not.

2. False. Angular velocity is often expressed in revolutions per minute, for example. However, many of the equations that include angular velocity or angular acceleration are valid only for angles measured in radians.

3. False. $r\alpha$ is only the tangential component of the linear acceleration of a point on the rim. The radial (or centripetal) component, which equals $r\omega^2$, also needs to be considered.

4. True. It moves through $2\pi = 6.28$ radians every 60 seconds, or through 0.105 radians each second.

5. True.

6. False. The centripetal acceleration of any part of the object is proportional to its distance from the axis of rotation.

7. True. The moment of inertia depends only upon the mass of the object and how it is distributed relative to the axis.

8. False. Consider two forces of equal magnitude and opposite direction. The forces add to zero, but the net torque is zero only if they act along the same line.

9. True. If its center of mass isn't accelerating, the net force must be zero.

10. False. The lever arm is the perpendicular distance from the axis of rotation to the line along which the force acts.

11. False. It also depends on the distribution of the mass relative to the rotation axis. The kinetic energy depends on the angular speed and the moment of inertia.

12. True.

13. True. This follows from the parallel-axis theorem.

14. False. The distribution of the object's mass in space plays an important role in determining the moment of inertia. If all the mass were at its center of mass, its moment of inertia about an axis through the center of mass would be zero.

15. False. In addition, the axis of rotation can change direction and undergo translations. Consider the motion of a bicycle wheel as the bike turns a corner.

V. Questions and Answers

Questions

1. Consider two points on a disk rotating with increasing speed about its axis, one at the rim and the other halfway from the rim to the center. Which point has the greater (*a*) angular acceleration, (*b*) tangential acceleration, (*c*) radial acceleration, and (*d*) centripetal accleration?

2. Can there be more than one value for the moment of inertia for a given rigid object?

3. In Figure 9-8, a man is hanging onto one side of a Ferris wheel, which swings him down toward the ground. (The Ferris wheel's motor is disengaged so it is free to rotate about its axis.) Can you use constant-angular-acceleration formulas to calculate the time it takes him to reach the bottom?

Figure 9-8

4. Does applying a positive (counterclockwise) net torque to an object necessarily increase its kinetic energy?

5. If the main rotor blade of the helicopter depicted in Figure 9-9 is rotating counterclockwise as seen from above, which way should the smaller tail rotor be pushing the air? Why?

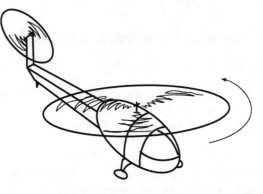

Figure 9-9

6. A solid ball, a solid disk, and a hoop, all with the same mass and the same radius, are set rolling without slipping up an incline, all with the same initial linear speed. Which goes farthest? If they are all set rolling with the same initial kinetic energy instead, which goes farthest?

Answers

1. (*a*) Both points have the same angular acceleration. This, of course, is the point of describing rotational motion in terms of angular quantities. (*b*) The tangential acceleration $a_t = r\alpha$ of the point on the rim is larger because it is farther from the axis. (*c*) The radial acceleration $a_{rad} = v^2/r = r\omega^2$ of the point at the rim is larger for the same reason. (*d*) Since the radial acceleration and centripetal acceleration are the same, the answer is the same as for (*c*).

Note: The centripetal direction is always defined as radially inward (toward the rotation axis). The radial direction is less definite. Sometimes radially inward is taken as the positive radial direction, but more often, radially outward is considered positive. Thus, $a_{rad} = \pm a_c$ depending on whether radially inward or outward is taken as the positive direction.

2. The moment of inertia may be different about different axes.

3. No, the gravitational torque on the system decreases. As shown in Figure 9-10, the weight force remains constant but its lever arm *l* decreases, as he swings down toward the bottom. This decrease in torque results in a decrease in the angular acceleration. The constant-angular-acceleration formulas are valid only if the angular acceleration remains constant.

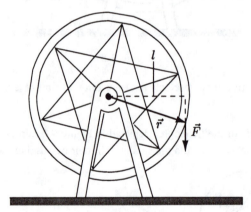

Figure 9-10

4. Not necessarily. It will increase the kinetic energy and therefore the angular speed of the object only if it does positive work, that is, if the motion is also counterclockwise. If the motion is clockwise then a counterclockwise torque will do negative work and reduce the object's kinetic energy.

5. The net torque about the helicopter's center of mass must equal zero, otherwise it will start to rotate. The rotational motion of the main rotor is counterclockwise. Thus the drag force of air, which opposes this motion, produces a clockwise torque on the helicopter. To counter this

clockwise torque, there must be a counterclockwise torque of equal magnitude acting on the helicopter. This is accomplished if the tail rotor pushes air to the helicopter's left (port). According to Newton's 3rd Law, if the tail rotor pushes air toward the left, then the air pushes the tail rotor to the helicopter's right (starboard), thus producing a counterclockwise torque on the helicopter. (The tail rotor's axis is oriented both horizontally and at right angles to the helicopter's forward direction.)

6. It is helpful to separately consider the translational $\left(\frac{1}{2}MV_{cm}^2\right)$ and rotational $\left(\frac{1}{2}I\omega^2\right)$ kinetic energies. All the objects have the same linear speed so they all have the same translational kinetic energy. Also, they all have the same angular speed so the larger the moment of inertia, the larger the rotational kinetic energy. The hoop $\left(I = MR^2\right)$ has the largest moment of inertia, the disk $\left(I = \frac{1}{2}MR^2\right)$ the next largest, and the sphere $\left(I = \frac{2}{5}MR^2\right)$ the smallest, so the hoop has the largest rotational kinetic energy and the sphere has the least. Each will roll up the hill until all its kinetic energy is transformed into gravitational potential energy, so the greater the kinetic energy, the higher up the hill it will roll. This means the hoop rolls the highest, with the disk second, and the sphere last. Of course, if all three start with the same kinetic energy instead of the same speed, they will all roll up to the same height.

VI. Problems, Solutions, and Answers

Example #1. Point P is located on a record turntable a distance of 10.0 cm from the axis. A dime is placed on the turntable directly over point P. The coefficient of static friction between the dime and the turntable is 0.210. If the turntable starts from rest with a constant angular acceleration of $1.20 \, \text{rad/s}^2$, how much time passes before the dime begins to slip?

Picture the Problem. The linear acceleration of P has both a tangential and a centripetal component. The tangential component is constant, but the centripetal component increases as the speed increases. The dime will slip when the frictional force can no longer provide the same acceleration to the dime that point P experiences.

1. Draw a figure to help visualize the forces and accelerations. Figure (a) shows the circular path of the turntable and dime, as well as the tangential and radial directions.	(a) 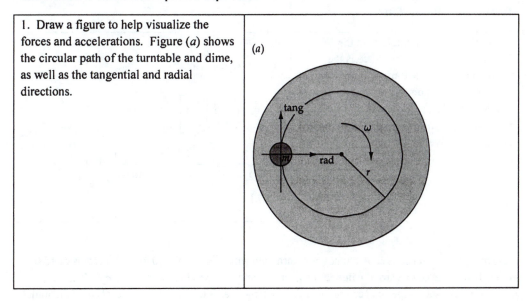

Figures (*b*) and (*c*) show the direction of the frictional force and the acceleration of the dime. Figure (*d*) shows an edge-on view and free-body diagram of the dime. The frictional force provides all the acceleration of the dime.	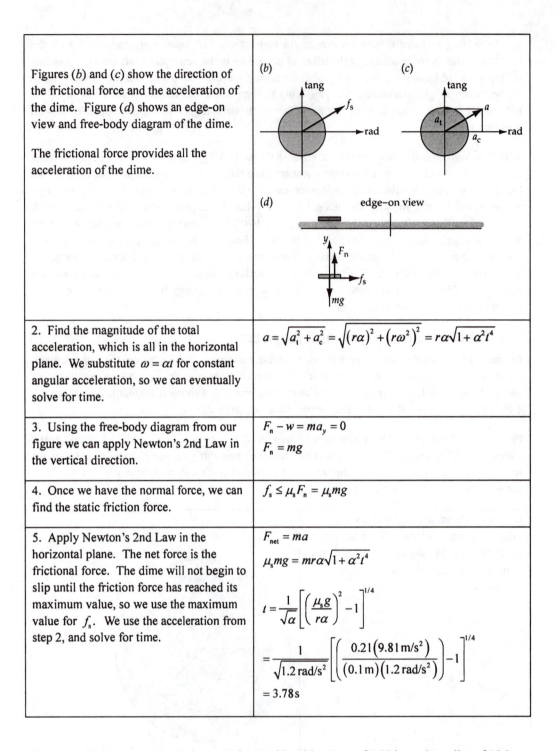
2. Find the magnitude of the total acceleration, which is all in the horizontal plane. We substitute $\omega = \alpha t$ for constant angular acceleration, so we can eventually solve for time.	$a = \sqrt{a_t^2 + a_c^2} = \sqrt{(r\alpha)^2 + (r\omega^2)^2} = r\alpha\sqrt{1 + \alpha^2 t^4}$
3. Using the free-body diagram from our figure we can apply Newton's 2nd Law in the vertical direction.	$F_n - w = ma_y = 0$ $F_n = mg$
4. Once we have the normal force, we can find the static friction force.	$f_s \le \mu_s F_n = \mu_s mg$
5. Apply Newton's 2nd Law in the horizontal plane. The net force is the frictional force. The dime will not begin to slip until the friction force has reached its maximum value, so we use the maximum value for f_s. We use the acceleration from step 2, and solve for time.	$F_{net} = ma$ $\mu_s mg = mr\alpha\sqrt{1 + \alpha^2 t^4}$ $t = \frac{1}{\sqrt{\alpha}}\left[\left(\frac{\mu_s g}{r\alpha}\right)^2 - 1\right]^{1/4}$ $= \frac{1}{\sqrt{1.2\,\text{rad/s}^2}}\left[\left(\frac{0.21(9.81\,\text{m/s}^2)}{(0.1\,\text{m})(1.2\,\text{rad/s}^2)}\right)^2 - 1\right]^{1/4}$ $= 3.78\,\text{s}$

Example #2—Interactive. A phonograph turntable with a mass of 1.80 kg and a radius of 15.0 cm is being braked to a stop. After 5.00 s its initial angular speed has decreased by 15%. Assuming constant angular acceleration, (*a*) how long does it take to come to rest? (*b*) If the initial angular speed is 33-1/3 rpm, through how many revolutions does it turn while coming to rest? (*c*) How much work is done on the turntable in order to bring it to rest?

Picture the Problem. This is a kinematics problem with constant angular acceleration. We are not given an initial angular speed, so we must assume the problem can be solved without it. Assign it the symbol ω_0. **Try it yourself.** Work the problem on your own, in the spaces provided, to get the final answer.

1. Draw a sketch to help visualize the situation.	
2. (*a*) Write out two constant angular acceleration equations. One will describe the fact that 85% of the angular speed remains after 5 s. The other will describe the fact that after some time *t*, the angular speed is zero. By dividing these equations, you can solve for *t*.	$t = 33\,\text{s}$
3. (*b*) Write out a constant angular acceleration equation for the position. Assume the initial position is zero. Knowing both ω_0 and *t*, you can find the angular acceleration, and then the final angular position.	9.3 rev
4. (*c*) The work required to bring the turntable to rest is equal to the kinetic energy loss of the turntable.	0.035 J

Example #3. A system consists of four point masses connected by rigid rods of negligible mass as shown in Figure 9-11. (*a*) Calculate the moment of inertia about the *A* and *y* axes by direct application of the formula $I = \Sigma m_i r_i^2$. (*b*) Use the parallel-axis theorem to relate the moments of inertia about the *A* and *y* axes and solve for the moment of inertia about the *y* axis. Compare this result with your result in part (*a*). (*c*) Calculate the moments of inertia about axes *x* and *z*.

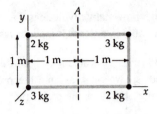

Figure 9-11

1. (a) Calculate the moment of inertia about the A axis. Each mass is a distance of 1 m from the axis.	$I_A = \sum m_i r_i^2$ $= (2\,kg)(1\,m)^2 + (3\,kg)(1\,m)^2 + (3\,kg)(1\,m)^2 + (2\,kg)(1\,m)^2$ $= 10\,kg \cdot m^2$
2. (a) Calculate the moment of inertia about the y axis. Two masses lie on the axis, so their distance from this axis is zero, and two masses are 2 m from this axis.	$I_y = \sum m_i r_i^2$ $= (2\,kg)(0\,m)^2 + (3\,kg)(0\,m)^2$ $+ (3\,kg)(2\,m)^2 + (2\,kg)(2\,m)^2$ $= 20\,kg \cdot m^2$
3. (b) Because the A axis goes through the center of mass of the system, we can use the parallel axis theorem to calculate I_y from I_A.	$I_y = I_A + Mh^2$ $= (10\,kg \cdot m^2) + (10\,kg)(1\,m)^2 = 20\,kg \cdot m^2$
4. (c) Calculate the moment of inertia about the x axis. Once again, two of the masses lie on the axis, so their distance from this axis are zero. The remaining two masses are each a distance of 1 m from the x axis.	$I_y = \sum m_i r_i^2$ $= (2\,kg)(0\,m)^2 + (3\,kg)(0\,m)^2$ $+ (3\,kg)(1\,m)^2 + (2\,kg)(1\,m)^2$ $= 5\,kg \cdot m^2$
5. (c) Calculate the moment of inertia about the z axis. Only the 3-kg mass at the origin lies on this axis. The 2-kg mass on the y axis is 1 m away. The other 2-kg mass is 2 m away, and the other 3-kg mass is a distance of $\sqrt{5}$ m away.	$I_z = \sum m_i r_i^2$ $= (3\,kg)(0\,m)^2 + (2\,kg)(1\,m)^2$ $+ (2\,kg)(2\,m)^2 + (3\,kg)(\sqrt{5}\,m)^2$ $= 25\,kg \cdot m^2$

Example #4—Interactive. A system consists of three point masses connected by rigid rods of negligible mass as shown in Figure 9-12. (a) Calculate the moment of inertia about the A and y

axes by direct application of the formula $I = \Sigma m_i r_i^2$. (b) Use the parallel axis theorem to relate the moments of inertia about the A and y axes and solve for the moment of inertia about the y axis. Compare this result with your result in part (a). (c) Calculate the moments of inertia about axes x and z. **Try it yourself.** Work the problem on your own, in the spaces provided, to get the final answer.

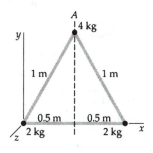

Figure 9-12

1. Calculate the moment of inertia about the A axis. Determine the distance of each mass from the axis, and use the moment of inertia equation.	$I_A = 1\,\text{kg·m}^2$
2. Calculate the moment of inertia about the y axis. Determine the distance of each mass from the axis, and use the moment of inertia equation.	$I_y = 3\,\text{kg·m}^2$
3. Because the A axis goes through the center of mass of the system, we can use the parallel axis theorem to calculate I_y from I_A.	
4. Calculate the moment of inertia about the x axis. Determine the distance of each mass from the axis, and use the moment of inertia equation.	$I_x = 3\,\text{kg·m}^2$
5. Calculate the moment of inertia about the z axis. Determine the distance of each mass from the axis, and use the moment of inertia equation.	$I_x = 6\,\text{kg·m}^2$

Example #5. A thin uniform rod of mass M and length L is free to pivot about an axis perpendicular to the rod a distance x from its center. (a) Find the moment of inertia about this axis as a function of x using the parallel-axis theorem and a formula, obtained from Table 9-1 on page

274 of the text, for the moment of inertia of a thin uniform rod about an axis perpendicular to it and passing through its center. (b) Check your result by setting x equal to $L/2$, the value for an axis perpendicular to the rod and passing through one end. Then compare your result with the expression obtained from Table 9-1 for the same axis.

Picture the Problem. This problem requires the use of the parallel axis theorem.

1. Draw a sketch to help visualize the problem.	
2. Use Table 9-1 to find the moment of inertia of a rod rotated about an axis perpendicular to the rod through its center.	$I = \dfrac{1}{12} ML^2$
3. Use the parallel axis theorem to find the moment of inertia about some axis a distance x from the center of the rod.	$I = Mh^2 + I_{cm}$ $= M\left(x^2 + \dfrac{1}{12} L^2\right)$
4. Setting $x = L/2$, solve for the moment of inertia for the rotation of this rod about its end. We get the same value as that in Table 9-1.	$I = M\left(\left(\dfrac{L}{2}\right)^2 + \dfrac{1}{12} L^2\right) = \dfrac{L^2}{4} + \dfrac{L^2}{12} = \dfrac{1}{3} L^2$

Example #6—Interactive. A uniform circular disk of mass M_d and radius R is free to pivot about an axis perpendicular to the plane of the disk a distance r from its center. (a) Find the moment of inertia about this axis as a function of r using the parallel-axis theorem and a formula from Table 9-1 in the text, for the moment of inertia of a solid cylinder about its symmetry axis. (b) Find an expression for the moment of inertia if the axis is halfway between the disk's center and its perimeter. **Try it yourself.** Work the problem on your own, in the spaces provided, to get the final answer.

1. Draw a sketch to help visualize the problem.	
2. Use Table 9-1 to find the moment of inertia of a disk rotated about an axis through its center perpendicular to the plane of the disk.	
3. Use the parallel axis theorem to find the moment of inertia about some axis a distance r from the center of the disk.	$I = M_d\left(r^2 + \tfrac{1}{2} R^2\right)$

4. Setting $r = R/2$, solve for the moment of inertia for the rotation of this disk about the new axis.	$I = \frac{3}{4} M_d R^2$

Example #7. As shown in Figure 9-13, a circular disk of mass M_1 and radius R is made from a uniform piece of thin sheet material. This disk is free to rotate about an axis perpendicular to it that passes through its center. A second disk with a radius of $R/2$ is cut from the same sheet and the two disks are glued together with the small disk's perimeter touching the perimeter of the large disk at a single point. Find an expression for the moment of inertia of the composite two-disk object about the given axis.

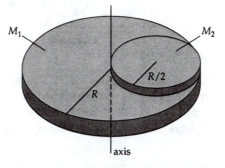

M_1 M_2

$R/2$

R

axis

Figure 9-13

Picture the Problem. The moment of inertia of the system will be the sum of the moments of inertia of each individual disk. Find the moment of inertia of the large disk about its center. Use the parallel axis theorem to find the moment of inertia of the small disk about the center of the large disk. Add them together.

1. Calculate the moment of inertia of the large disk about its central axis.	$I_1 = \frac{1}{2} M_1 R^2$
2. Calculate the moment of inertia of the small disk about the center of the large disk using the parallel axis theorem. The smaller disk will have ¼ the volume of the large disk, and will also have ¼ the mass.	$I = Mh^2 + I_{cm}$ $I_2 = M_2 \left(\frac{R}{2}\right)^2 + \frac{1}{2} M_2 \left(\frac{R}{2}\right)^2 = \frac{3}{8} M_2 R^2$ $ = \frac{3}{8}\left(\frac{M_1}{4}\right) R^2 = \frac{3}{32} M_1 R^2$
3. The total moment of inertia is the sum of the moments of inertia of the two disks.	$I_{tot} = I_1 + I_2$ $\phantom{I_{tot}} = \left(\frac{1}{2} + \frac{3}{32}\right) M_1 R^2 = \frac{19}{32} M_1 R^2$

Example #8—Interactive. A 0.500-kg, 0.250-m-radius disk is made from a uniform piece of thin sheet material. As shown in Figure 9-14, the disk is glued to a 0.200-kg meter stick, with the disk's center at the 75.0-cm mark. Find the moment of inertia of this composite object about an axis perpendicular to both the meter stick and the plane of the disk and passing through the stick's zero-cm mark.

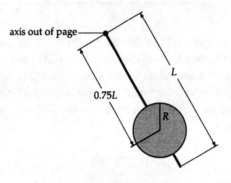

Figure 9-14

Picture the Problem. The moment of inertia of the system will be the sum of the moments of inertia of the disk and the meter stick. Find the moment of inertia of the meter stick about its end. Use the parallel axis theorem to find the moment of inertia of the disk about the rotation axis. Add them together. **Try it yourself.** Work the problem on your own, in the spaces provided, to get the final answer.

1. Calculate the moment of inertia of the stick rotated about its end.	$I_{stick} = 0.0667\,\text{kg} \cdot \text{m}^2$
2. Calculate the moment of inertia of the disk rotated about the end of the stick.	$I_{disk} = 0.300\,\text{kg} \cdot \text{m}^2$
3. The total moment of inertia is the sum of the two.	$I_{tot} = I_{stick} + I_{disk} = 0.367\,\text{kg} \cdot \text{m}^2$

Example #9. Calculate the moment of inertia about each of the axes A, B, and C for the thin solid half-disk of mass M and radius R, shown in Figure 9-15. The C axis is perpendicular to the plane of the half-disk.

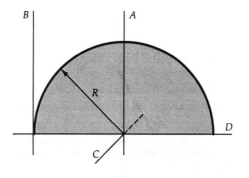

Figure 9-15

Picture the Problem. To find the moment of inertia about axis A, divide the half-disk into strips with a differential width in the direction of the D axis, and length in the direction of the A axis. Each differential strip will have a differential moment of inertia about the A axis that will vary with x, the distance from the A axis. The height of the strip will depend on x. The moment of inertia about axis B can be found with the parallel axis theorem.

To find the moment of inertia about axis C, divide the disk into semicircular strips of radius r, and thickness dr. The length of the strip depends on r, as will the differential moment of inertia of each semicircular strip.

1. Draw a sketch of the half-disk, showing a differential vertical strip of width dx.	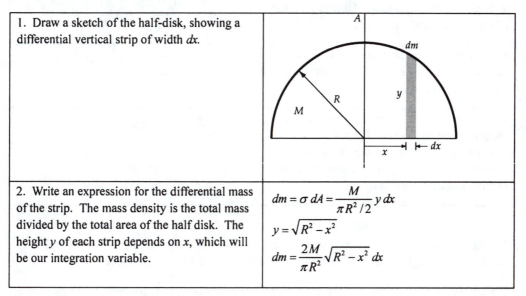
2. Write an expression for the differential mass of the strip. The mass density is the total mass divided by the total area of the half disk. The height y of each strip depends on x, which will be our integration variable.	$dm = \sigma\,dA = \dfrac{M}{\pi R^2/2}\, y\,dx$ $y = \sqrt{R^2 - x^2}$ $dm = \dfrac{2M}{\pi R^2}\sqrt{R^2 - x^2}\,dx$

3. Find the differential moment of inertia about the A axis due to a differential mass element calculated above. Each strip is a distance x from the axis.	$dI_A = x^2\, dm = \dfrac{2M}{\pi R^2} x^2 \sqrt{R^2 - x^2}\, dx$
4. Calculate the moment of inertia about the A axis by integrating. The variable x varies from $-R$ to R. The result is half of the moment of inertia of a solid cylinder about its diameter through its center shown in Table 9-1 in the text if we let $L \to 0$.	$I_A = \displaystyle\int_{-R}^{R} \dfrac{2M}{\pi R^2} x^2 \sqrt{R^2 - x^2}\, dx = \dfrac{1}{4} MR^2$
5. To find the moment of inertia about axis B, apply the parallel axis theorem. Since axis A goes through the center of mass, $I_A = I_{cm}$.	$I_B = Mh^2 + I_{cm} = MR^2 + I_A$ $= MR^2 + \dfrac{1}{4} MR^2 = \dfrac{5}{4} MR^2$
6. Draw a picture illustrating the semicircular disks to be used for calculating the moment of inertia about axis C.	
7. Find the differential mass of this strip. It has a length of πr, and a thickness dr.	$dm = \sigma\, dA = \dfrac{M}{\pi R^2 / 2} \pi r\, dr = \dfrac{2M}{R^2} r\, dr$
8. Find the differential moment of inertia of each strip above about axis C. Each strip is a distance r from axis C.	$dI_C = r^2\, dm = r^2 \dfrac{2M}{R^2} r\, dr = \dfrac{2M}{R^2} r^3\, dr$
9. The total moment of inertia about axis C is the sum of the moments of inertia from each differential strip. The sum of differential quantities is the integral of each strip, with the radius changing from 0 to R.	$I_C = \displaystyle\int_0^R \dfrac{2M}{R^2} r^3\, dr = \dfrac{1}{2} MR^2$

Example #10—Interactive. Calculate the moment of inertia about each of the axes A and B for the thin solid triangle of mass M, height a, and width $b=2a$, shown in Figure 9-16.

Picture the Problem. To find the moment of inertia about axis A, use vertical strips with a differential thickness dx, where x is the distance from axis A. To find the moment of inertia about axis B, use horizontal strips. **Try it yourself.** Work the problem on your own, in the spaces provided, to get the final answer.

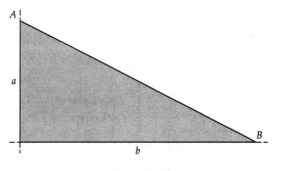

Figure 9-16

1. Find the differential mass of a vertical strip of thickness dx. The height of this strip should depend on x.	
2. Find the differential moment of inertia of this strip about axis A.	
3. Integrate all the differential moments of inertia to find the total moment of inertia about axis A.	$I_A = \frac{1}{6} Mb^2$
4. Find the differential mass of a horizontal strip of thickness dy. The length of this strip should depend on y.	
5. Find the differential moment of inertia of this strip about axis B.	
6. Integrate all the differential moments of inertia to find the total moment of inertia about axis B.	$I_B = \frac{1}{6} Ma^2 = \frac{1}{24} Mb^2$

Example #11. In Figure 9-17 a 0.400-kg pulley has a 6.00-cm radius and blocks 1 and 2 have masses of 0.800 kg and 1.60 kg, respectively. The pulley wheel is a uniform disk, friction in the pulley axle is negligible, and the string neither slips nor stretches. (a) Find the acceleration of block 2 and the angular acceleration of the pulley. (b) Find the tensions in the string segments on either side of the pulley. Check the validity of your results by verifying that Newton's 2nd Law for rotations holds for the pulley wheel.

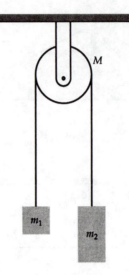

Figure 9-17

Picture the Problem. Draw separate free-body diagrams for each block and the pulley wheel. The tension of the string on either side of the pulley must be different if there is to be a net torque on the pulley. Apply Newton's 2nd Law to each block, and Newton's 2nd Law for rotations to the pulley. The magnitude of the acceleration of the blocks and the angular acceleration of the pulley are related.

1. Draw free-body diagrams of all three objects. Because rotations are involved, it is important to draw the forces exactly where they are applied. It is desirable for each object to accelerate in the "positive" direction at the same time. You can choose the coordinate system for each object so this will be the case.	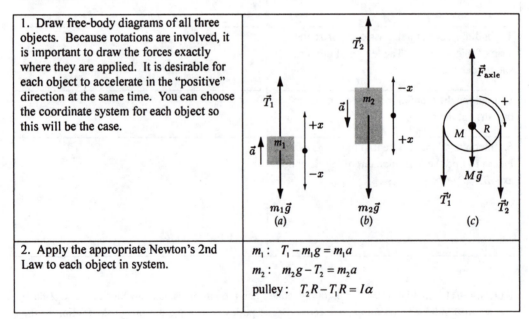
2. Apply the appropriate Newton's 2nd Law to each object in system.	$m_1:\quad T_1 - m_1 g = m_1 a$ $m_2:\quad m_2 g - T_2 = m_2 a$ $\text{pulley}:\quad T_2 R - T_1 R = I\alpha$

3. The three equations above have five unknowns, so more equations are needed. The linear and angular acceleration are geometrically related because the string does not slip on the pulley.	$a = R\alpha$
4. The pulley is a solid, uniform disk, so we know the moment of inertia.	$I = \dfrac{1}{2}MR^2$
5. Algebra: substitute the equation from step 3 into the pulley equation. Also, add the m_1 and m_2 equations. Add these two resulting equations, eliminating both tensions.	$T_2R - T_1R = I\dfrac{a}{R}$ or $T_2 - T_1 = \dfrac{I}{R^2}a$ $T_1 - T_2 = (m_1 - m_2)g + (m_1 + m_2)a$ $0 = (m_1 - m_2)g + (m_1 + m_2)a + \dfrac{I}{R^2}a$
6. Solve the resulting equation for the acceleration, substituting the moment of inertia from step 4.	$(m_2 - m_1)g = \left(m_1 + m_2 + \dfrac{I}{R^2}\right)a$ $a = \dfrac{(m_2 - m_1)g}{m_1 + m_2 + \dfrac{\frac{1}{2}MR^2}{R^2}}$ $a = \dfrac{((1.6\,\text{kg}) - (0.8\,\text{kg}))(9.81\,\text{m/s}^2)}{(0.8\,\text{kg}) + (1.6\,\text{kg}) + \frac{1}{2}(0.4\,\text{kg})} = 3\,\text{m/s}^2$
7. Now solve for the angular acceleration using the equation for step 3.	$\alpha = \dfrac{a}{R} = \dfrac{3.0\,\text{m/s}^2}{0.06\,\text{m}} = 50\,\text{rad/s}^2$
8. Finally we can solve for the tensions using our original Newton's 2nd Law equations for m_1 and m_2.	$T_1 = m_1(g - a)$ $= (0.8\,\text{kg})((9.81\,\text{m/s}^2) + (3\,\text{m/s}^2)) = 10\,\text{N}$ $T_2 = m_2(g + a)$ $= (1.6\,\text{kg})((9.81\,\text{m/s}^2) - (3\,\text{m/s}^2)) = 11\,\text{N}$

Example #12—Interactive. In Figure 9-18, the pulley wheel has a moment of inertia of 1.00×10^{-3} kg·m^2 and radii of 4.00 and 8.00 cm. Blocks 1 and 2 have masses of 1.60 and 1.00 kg, respectively. Assume that friction in the pulley axle is negligible and the strings neither slip nor stretch. (a) Find the acceleration of each block and the angular acceleration of the pulley. (b) Find the tension in each string.

Picture the problem. Draw a separate free-body diagram for each block and the pulley. Apply Newton's 2nd Law to each object. Relate the linear and angular accelerations. The accelerations of the blocks are not equal, so there are two relationships to write down. Do the algebra. **Try it yourself.** Work the problem on your own, in the spaces provided, to get the final answer.

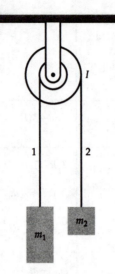

Figure 9-18

1. Draw free-body diagrams of all three objects. Because rotations are involved, it is important to draw the forces exactly where they are applied. It is desirable for each object to accelerate in the "positive" direction at the same time. You can choose the coordinate system for each object so this will be the case.	
2. Apply the appropriate Newton's 2nd Law to each object in system.	
3. Relate the linear accelerations to the angular acceleration.	
5. Algebra. You may need to use another sheet for this step.	$\alpha = 15.7 \, \text{rad/s}^2$ clockwise $a_1 = 0.63 \, \text{m/s}^2$ down $a_2 = 1.26 \, \text{m/s}^2$ up $T_1 = 16.7 \, \text{N}; T_2 = 8.55 \, \text{N}$

Example #13. A 2.00-kg 6.00-cm-radius wheel, shown in Figure 9-19, has a square hole through its center. The wheel's center of mass is at its geometric center. A string of negligible mass with one end attached to the ceiling is wrapped around its perimeter. As it falls, the string does not slip so the wheel rotates faster and faster as it gains speed. It is observed that the wheel is moving at 4.74 m/s after falling a distance of 2.00 m. Find its moment of inertia about an axis perpendicular to the page and passing through its center.

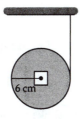

Figure 9-19

Picture the Problem. The string does not slip so no energy is dissipated via friction and the rolling-without-slipping condition holds. Choose a suitable system, apply the conservation of mechanical energy, and solve for the moment of inertia of the wheel.

1. Draw a simple sketch showing the initial and final conditions of the wheel.	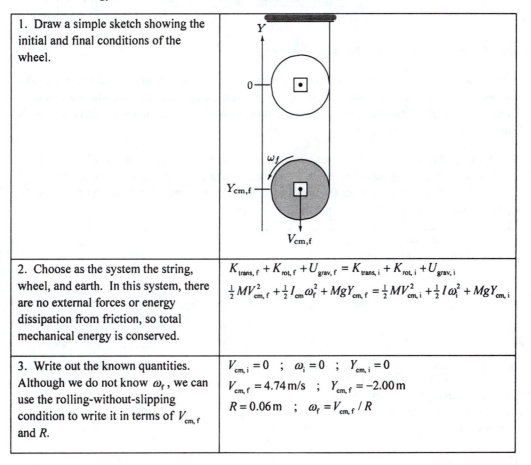
2. Choose as the system the string, wheel, and earth. In this system, there are no external forces or energy dissipation from friction, so total mechanical energy is conserved.	$K_{\text{trans, f}} + K_{\text{rot, f}} + U_{\text{grav, f}} = K_{\text{trans, i}} + K_{\text{rot, i}} + U_{\text{grav, i}}$ $\frac{1}{2}MV^2_{\text{cm, f}} + \frac{1}{2}I_{\text{cm}}\omega^2_f + MgY_{\text{cm, f}} = \frac{1}{2}MV^2_{\text{cm, i}} + \frac{1}{2}I\omega^2_i + MgY_{\text{cm, i}}$
3. Write out the known quantities. Although we do not know ω_f, we can use the rolling-without-slipping condition to write it in terms of $V_{\text{cm, f}}$ and R.	$V_{\text{cm, i}} = 0 \;\;;\;\; \omega_i = 0 \;\;;\;\; Y_{\text{cm, i}} = 0$ $V_{\text{cm, f}} = 4.74 \,\text{m/s} \;\;;\;\; Y_{\text{cm, f}} = -2.00\,\text{m}$ $R = 0.06\,\text{m} \;\;;\;\; \omega_f = V_{\text{cm, f}} / R$

4. Substitute the values in step 3 into the conservation of energy equation and solve for the moment of inertia.	$\frac{1}{2}MV_{cm,f}^2 + \frac{1}{2}I_{cm}\dfrac{V_{cm,f}^2}{R^2} + MgY_{cm,f} = 0+0+0$ $I_{cm} = MR^2\left(1+\dfrac{2gY_{cm,f}}{V_{cm,f}^2}\right)$ $= (2.00\,\text{kg})(0.06\,\text{m})^2\left(1+\dfrac{2(9.81\,\text{m/s}^2)(-2.00\,\text{m})}{4.74\,\text{m/s}^2}\right)$ $= 5.37\times10^{-3}\,\text{kg·m}^2$

Follow-up. The moment of inertia of a wheel like the one in this problem should be greater than that for a uniform solid disk of the same mass and radius, but smaller than that for a thin cylindrical shell. That is, its moment of inertia should be given by $I_{cm} = \beta MR^2$, where $\frac{1}{2} < \beta < 1$. For our result,

$$\beta = \frac{I_{cm}}{MR^2} = \frac{5.37\times10^{-3}\,\text{kg·m}^2}{(2.00\,\text{kg})(0.06\,\text{m})^2} = 0.747$$

which is close to the center of the acceptable range.

Example #14—Interactive. A uniform spherical ball is rolling along a horizontal road with speed v_0 when it comes to a hill. It rolls up the hill, just making it to the top, where it comes to rest. How high is the hill? Assume that it rolls without slipping and that air resistance is negligible.

Picture the Problem. The ball does not slip, so no energy is dissipated via friction and the rolling-without-slipping condition holds. Choose a suitable system, apply the conservation of mechanical energy, and solve for the increase in height of the ball. **Try it yourself.** Work the problem on your own, in the spaces provided, to get the final answer.

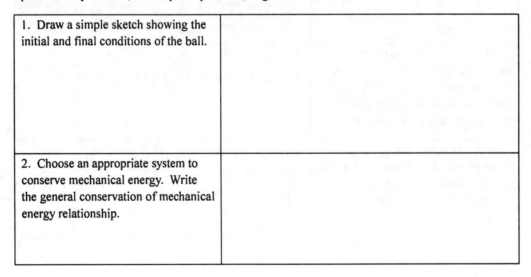

1. Draw a simple sketch showing the initial and final conditions of the ball.	
2. Choose an appropriate system to conserve mechanical energy. Write the general conservation of mechanical energy relationship.	

3. Write out the known quantities. Although we do not know ω_f, we can use the rolling-without-slipping condition to write it in terms of $V_{cm,f}$ and R.	
4. Substitute the values in step 3 into the conservation of energy equation and solve for the final height of the ball.	$\dfrac{7v_0^2}{10g}$

Example #15.   A bowling ball (mass 7.00 kg, diameter 20.0 cm) is set spinning about a horizontal axis at 120 rev/min. As the ball is set down on the floor, it is still spinning at 120 rev/min but its center of mass is at rest. If the coefficient of kinetic friction between ball and floor is 0.800, describe quantitatively the subsequent motion of the ball.

Picture the Problem. This is an example of rigid-body motion about a moving axis with a fixed direction. Because a finite force cannot produce an infinite linear acceleration, the initial motion is that of rolling *with* slipping. During the initial rolling-*with*-slipping stage calculate both the linear acceleration of the center of mass and the angular acceleration about the rotation axis through the center of mass. Using the kinematic formulas, calculate the time when slipping ceases.

1. Draw a free-body diagram of the bowling ball while it is rolling *with* slipping.	
2. Analyze the linear and rotational motion of the bowling ball. Let rotation in the clockwise direction be positive.	**Linear Motion** $\qquad$ **Rotational Motion** $$\sum F_x = ma_{cm,x} \qquad\qquad \sum \tau_{cm} = I_{cm}\alpha$$ $$f = \mu_k mg = ma_{cm} \qquad -\mu_k mgR = \tfrac{2}{5}mR^2\alpha$$ $$a_{cm} = \mu_k g \qquad\qquad\quad \alpha = -\dfrac{5\mu_k g}{2R}$$ $$v_{cm} = v_{0,cm} + a_{cm}t \qquad \omega = \omega_0 + \alpha t$$ $$v_{cm} = 0 + \mu_k gt \qquad\quad \omega = \omega_0 - \dfrac{5\mu_k g}{2R}t$$

3. As time increases, the velocity of the center of mass increases and the angular velocity decreases. The slipping will continue until $v_{cm} = R\omega$. To determine when this condition is reached we set up this equality and solve for the time.	$v_{cm} = R\omega$ $$\mu_k g t = R\left(\omega_0 - \frac{5\mu_k g}{2R}t\right)$$ $$t = \frac{2R\omega_0}{7\mu_k g}$$ $$t = \frac{2(0.10\,\text{m})(120\,\frac{\text{rev}}{\text{min}})\left(\frac{2\pi\,\text{rad}}{\text{rev}}\right)\left(\frac{1\,\text{min}}{60\,\text{s}}\right)}{7(0.80)(9.81\,\text{m/s}^2)} = 46\,\text{ms}$$
4. Substituting the time back into the velocity and angular velocity equations, we can determine the linear and angular speeds of the ball at the time it stops skidding.	$v_{cm} = (0.80)(9.81\,\text{m/s}^2)(0.046\,\text{s}) = 0.36\,\text{m/s}$ $\omega = v_{cm}/R = (0.36\,\text{m/s})/(0.10\,\text{m}) = 3.6\,\text{rad/s}$
5. Finally, we can also determine the distance the ball traveled and the angle through which it rotated before the skidding stopped.	$$\Delta x_{cm} = v_{avg}\,\Delta t = \frac{v_0 + v}{2}\Delta t$$ $$= \frac{0.36\,\text{m/s}}{2}(0.046\,\text{s}) = 0.0083\,\text{m or 8.3 mm}$$ $$\Delta\theta = \omega_{avg}\,\Delta t = \frac{\omega_0 + \omega}{2}\Delta t$$ $$= \frac{(12.6\,\text{rad/s}) + (3.6\,\text{rad/s})}{2}(0.046\,\text{s}) = 0.37\,\text{rad or 21.3°}$$

Example #16—Interactive. A solid spherical ball rolls down an incline. If the coefficient of static friction between the ball and the incline is 0.33, what is the steepest angle of the incline for which rolling without slipping will occur?

Picture the Problem. The ball will roll without slipping as long as the net torque about an axis through the center of mass is sufficient to produce the needed angular acceleration to prevent skidding. The steeper the incline, the larger the linear acceleration and the smaller the normal force. The smaller the normal force, the smaller the maximum static frictional force and, consequently, the smaller the torque exerted by the maximum static frictional force. **Try it yourself.** Work the problem on your own, in the spaces provided, to get the final answer.

1. Draw a free-body diagram of the situation to help with applying Newton's 2nd Laws.	

2. Apply Newton's 2nd Laws for both translation and rotation. Make sure to write the maximum static friction force in terms of the incline angle.	
3. Apply the condition of rolling-without-slipping to the accelerations in step 2, and solve for the angle above which this condition will no longer hold true.	$\theta = 49°$

Chapter **10**

Conservation of Angular Momentum

I.   Key Ideas

In this chapter we will continue to study the rotational motion of many-particle systems, emphasizing the vector aspects of such motion. The vector nature of angular velocity is presented in Chapter 9 of this Study Guide. Here we will introduce the vector nature of torque and of angular momentum. Conservation of angular momentum is a fundamental law of nature.

Section 10-1. The Vector Nature of Rotation. For a rotating object the angular velocity vector $\vec{\omega}$ is directed parallel to the rotation axis in the direction specified via the right-hand rule shown in Figure 9-2. (If you curl your fingers in the direction of rotation, $\vec{\omega}$ is in the same direction as your thumb, using your right hand of course!)

The Cross Product. Torque $\vec{\tau}$ is mathematically best expressed as the **cross product** (or **vector product**) of $\vec{r}$ and $\vec{F}$. That is, $\vec{\tau} = \vec{r} \times \vec{F}$ where $\vec{r}$ is the vector from the origin to the point of application of force $\vec{F}$.

The cross product of any two vectors $\vec{A}$ and $\vec{B}$ is perpendicular to both vectors. The mathematical convention for determining the direction of a cross product, illustrated in Figure 10-1, is called the *right-hand rule,* which is required to completely specify the direction of the cross product. To apply this rule, place the fingers of your right hand in the direction of the first vector ($\vec{A}$) such that when you make a fist they curl toward the direction of the second vector ($\vec{B}$). Your thumb then shows the direction of $\vec{A} \times \vec{B}$.

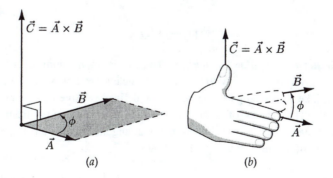

Figure 10-1

If ϕ is the angle between the directions of the two vectors, and $\hat{n}$ is the unit vector in the direction given by the right-hand rule, then

$$\vec{C} = \vec{A} \times \vec{B} = (AB\sin\varphi)\hat{n}$$ Cross (or vector) product

It follows that the cross product of any vector with itself is zero and that $\vec{A} \times \vec{B} = -\vec{B} \times \vec{A}$. Thus, for unit vectors $\hat{i}$, $\hat{j}$, and $\hat{k}$, we have $\hat{i} \times \hat{i} = \hat{j} \times \hat{j} = \hat{k} \times \hat{k} = 0$ and

$$\hat{i} \times \hat{j} = \hat{k} \qquad \hat{j} \times \hat{i} = -\hat{k}$$
$$\hat{j} \times \hat{k} = \hat{i} \qquad \hat{k} \times \hat{j} = -\hat{i}$$
$$\hat{k} \times \hat{i} = \hat{j} \qquad \hat{i} \times \hat{k} = -\hat{j}$$

The signs of cross products such as $\hat{k} \times \hat{j}$ can be found by using the graphic shown in Figure 10-2. If the order of the factors is such that you travel in the direction of the arrows (counterclockwise) around the circle then the sign of the product is positive. If you travel in the clockwise direction, then the sign of the product is negative. Using these rules for multiplying the unit vectors $\hat{i}$, $\hat{j}$, and $\hat{k}$, we can show that

$$\vec{C} = \vec{A} \times \vec{B} = (A_y B_z - A_z B_y)\hat{i}$$
$$+ (A_z B_x - A_x B_z)\hat{j} + (A_x B_y - A_y B_x)\hat{k}$$ Cross product using $\hat{i}$, $\hat{j}$, and $\hat{k}$

That is, $C_x = A_y B_z - A_z B_y$, $C_y = A_z B_x - A_x B_z$, and $C_z = A_x B_y - A_y B_x$. *Note:* In each of these three equations the subscript on the left side does not appear on the right side. Also, in each equation the unit vector corresponding to the first subscript equals the cross product of the unit vectors corresponding to the next two. For example, the first three subscripts in the equation $C_x = A_y B_z - A_z B_y$ are x, y, and z; the corresponding unit vectors are $\hat{i}$, $\hat{j}$,, and $\hat{k}$, and $\hat{i} = \hat{j} \times \hat{k}$. By remembering this you can quickly calculate the cross product of two vectors.

Figure 10-2

Section 10-2. Torque and Angular Momentum. The angular momentum $\vec{L}$ of a moving particle about the origin is given by $\vec{r} \times \vec{p}$, where $\vec{r}$ is the particle's position vector and $\vec{p}$ is its linear momentum $m\vec{v}$. Like torque, angular momentum is defined relative to a point in space, rather than an axis (line). The concept of angular momentum can be extended from that of a single particle to that of a system of particles, where the total angular momentum is simply the sum of the angular momenta of the individual particles. That is,

$$\vec{L} = \vec{r} \times \vec{p}$$ Angular momentum of a particle

$$\vec{\tau} = \vec{r} \times \vec{F}$$

Torque due to a force

$$\vec{L}_{\text{sys}} = \sum_i \vec{L}_i$$

Angular momentum of a system

$$\vec{\tau}_{\text{net,ext}} = \sum_i \vec{\tau}_i$$

Net torque on a system

The impulse-momentum theorem relates the change in a particle's linear momentum to the net force acting on it. The rotational analog of this theorem is the angular version

$$\Delta \vec{L}_{\text{sys}} = \int_{t_1}^{t_2} \vec{\tau}_{\text{net, ext}} \, dt$$

Angular impulse-angular momentum equation

Here torque and angular momentum are defined as vector quantities relative to a *point* rather than an axis. In the equations above, the $\Sigma\vec{\tau}$ terms refer to the sum of all torques, both internal and external. However, since the internal torques sum to zero, this sum can be replaced by a sum over only the external torques. It follows from Newton's laws that

$$\vec{\tau}_{\text{net, ext}} = \frac{d\vec{L}_{\text{sys}}}{dt}$$

Newton's 2nd Law for angular motion

It is often helpful to express the total angular momentum of a system as the sum of two terms: the spin angular momentum $\vec{L}_{\text{spin}}$, associated with any motion relative to the center of mass, and $\vec{L}_{\text{orbit}}$, which is the angular momentum that a point particle of mass M located at the center of mass and moving at the velocity of the center of mass would have. That is,

$$\vec{L} = \vec{L}_{\text{orbit}} + \vec{L}_{\text{spin}}$$
$$\vec{L}_{\text{orbit}} = \vec{r}_{\text{cm}} \times M\vec{v}_{\text{cm}}$$

Orbit and spin angular momenta

As an example, the total angular momentum of the earth can be express as the sum of its orbital angular momentum, due to is orbital motion around the sun, plus its spin angular momentum, due to its spinning motion about an axis through its center of mass.

For a symmetric rigid object that is rotating with angular velocity $\vec{\omega}$ about a fixed axis the angular momentum $\vec{L}$ is given by

$$\vec{L} = I\vec{\omega}$$

Angular momentum for a symmetric rigid object about a fixed axis

where I is the moment of inertia. Substituting this expression for angular momentum into the angular version of Newton's 2nd Law gives

$$\vec{\tau} = I\vec{\alpha}$$

Newton's 2nd Law for a rigid object about a fixed axis

For an object rotating about a fixed axis, the kinetic energy K is given by $K = \frac{1}{2}I\omega^2$. Substituting for ω in this equation gives

$$K = \frac{L^2}{2I}$$ Kinetic energy of a rigid object rotating about a fixed axis

Gyroscopic motion illustrates some of the nonintuitive vector properties of rotational motion. A spinning gyroscope moves so the change in its angular momentum vector is in the same direction as the net external torque vector.

Section 10-3. Conservation of Angular Momentum. If the net external torque $\vec{\tau}_{net, ext}$ remains zero, then the system's angular momentum $\vec{L}$ is constant. This follows from the angular version of Newton's 2nd Law and is called the *law of conservation of angular momentum*. Like the conservation of linear momentum this law is universally valid.

Section 10-4. Quantization of Angular Momentum. One of the strange discoveries in physics that took place during the early twentieth century is that, like energy, angular momentum is quantized with the fundamental unit of angular momentum $\hbar$ (read *h*-bar) being Planck's constant *h* divided by 2π.

II. Physical Quantities and Key Equations

Physical Quantities

Quantum of angular momentum $\hbar = \dfrac{h}{2\pi} = 1.05 \times 10^{-34}$ J$\cdot$s

Key Equations

Cross (or vector) product $\vec{C} = \vec{A} \times \vec{B} = (AB \sin \varphi)\hat{n}$

Cross product using $\hat{i}$, $\hat{j}$, and $\hat{k}$ $\vec{C} = \vec{A} \times \vec{B} = (A_y B_z - A_z B_y)\hat{i}$
$$+ (A_z B_x - A_x B_z)\hat{j} + (A_x B_y - A_y B_x)\hat{k}$$

Angular momentum of a particle $\vec{L} = \vec{r} \times \vec{p}$

Angular momentum of a system $\vec{L}_{sys} = \sum_i \vec{L}_i$

Torque due to a force $\vec{\tau} = \vec{r} \times \vec{F}$

Net torque on a system $\vec{\tau}_{net,ext} = \sum_i \vec{\tau}_i$

Orbital and spin angular momentum $\vec{L} = \vec{L}_{orbit} + \vec{L}_{spin}$
$$\vec{L}_{orbit} = \vec{r}_{cm} \times M\vec{v}_{cm}$$

Impulse-momentum theorem (angular version) $\Delta \vec{L}_{sys} = \int_{t_1}^{t_2} \vec{\tau}_{net, ext}\, dt$

Newton's 2nd Law (angular version) $\vec{\tau}_{net, ext} = \dfrac{d\vec{L}_{sys}}{dt}$

Angular momentum for symmetric rigid object about a fixed axis	$\vec{L} = I\vec{\omega}$
Newton's 2nd Law for a rigid body about a fixed axis	$\vec{\tau} = I\vec{\alpha}$
Kinetic energy of a rigid object rotating about a fixed axis	$K = \dfrac{L^2}{2I}$
Conservation of angular momentum	If $\vec{\tau}_{net,\ ext} = 0$, then $\vec{L}_i = \vec{L}_f$

III. Potential Pitfalls

Do not think the fact that angular velocity as a vector is too abstract to be useful. The direction of the vector specifies the direction of the rotation axis—often an important consideration.

Do not think that for an object rotating about a fixed axis, the angular velocity vector and angular momentum vector must be in the same direction. The angular momentum vector, unlike the angular velocity vector, depends on the mass distribution. Only for sufficiently symmetric mass distributions are the two in the same direction.

Do not think that conservation of angular momentum is useful only if the net torque on the system remains zero. For example, the z component of the angular momentum is conserved as long as the z component of the net torque remains zero, even if the other components of the net torque are nonzero.

IV. True or False Questions and Responses

True or False

_____ 1. The vector product of two vectors is always perpendicular to both vectors.

_____ 2. A force directed toward a point exerts zero torque about that point.

_____ 3. The vector product is not commutative.

_____ 4. Both angular momentum vectors and torque vectors are defined relative to some particular origin or reference point.

_____ 5. The angular momentum vector is always parallel to the net torque on the object.

_____ 6. The angular momentum of a system of particles depends on the motion of the center of mass, but not on the motion of the particles relative to the center of mass.

_____ 7. A particle of mass m moving in a circle of radius R at a speed v has an angular momentum with a magnitude of mvR about *any* point.

_____ 8. If the total linear momentum of an arbitrary system of particles is conserved, the total angular momentum must also be conserved.

_____ 9. A gyroscope precesses, rather than falls over, because its spin counteracts the force of gravity.

_____ 10. A gyroscope supported at its center of mass will not precess under the influence of gravity.

_____ 11. Both the angular momentum and the angular velocity of an object that is rotating about a symmetry axis are parallel.

_____ 12. A rotating wheel is balanced if the axis of rotation passes through its center of mass.

Responses to True or False

1. True.

2. True.

3. True. $\vec{A} \times \vec{B} \neq \vec{B} \times \vec{A}$

4. True.

5. False. The rate of change of the angular momentum vector is always parallel to the net torque.

6. False. This omits any angular momentum the system may have about the center of mass; that is, it omits the spin angular momentum.

7. False. Only about the center of a circle is the magnitude of the angular momentum equal to *mvR*.

8. False. A pair of non-colinear (not acting along the same line) forces that are equal in magnitude but opposite in direction produce a net torque but not a net force. These forces will change the system's angular momentum, but not its linear momentum.

9. False. Spin isn't a force. For a spinning gyroscope supported at a point, it is the force of the support exerted on the gyroscope that prevents it from falling. The force of the support is equal in magnitude and opposite in direction to the gravitational force. However, they do not necessarily act along the same line unless the gyroscope is supported at a point along the vertical line through its center of mass. The gyroscope precesses as a result of the torque due to the gravitational force.

10. True. In that case there is no gravitational torque.

11. True.

12. False. This does not guarantee dynamic balance—only static balance.

V. Questions and Answers

Questions

1. If the polar ice caps melted tomorrow, what would be the effect on the earth's rotation rate?

2. A basketball thrown against a closed door causes it to rotate open. The door is stationary before the collision and rotating after it, so during the collision it clearly gains angular momentum. Where did this angular momentum come from? The ball didn't lose any, or did it?

3. The propeller of a light airplane is rotating clockwise, as seen from someone observing it from behind the plane. If the pilot pulls back the stick, the flaps on the tail fins are raised. This results in the air pushing down on the plane's tail, thus causing the tail to lower and nose to raise. However, exerting a torque on an object with angular momentum causes it to precess. When the tail flaps are up, which way, port or starboard, will the plane precess? (Port is to the left and starboard is to the right of a person facing the same direction as the plane.)

Answers

1. If all the water now frozen at the earth's poles were distributed over the earth's oceans, the earth's moment of inertia would increase. Its rate of rotation would have to decrease for its angular momentum to remain constant.

2. The ball does lose angular momentum during the collision. Its angular momentum is given by $\vec{L} = \vec{r} \times m\vec{v}$, where $\vec{r}$ is the vector from the door's rotation axis to the ball, m is the ball's mass, and $\vec{v}$ is the velocity of its center of mass. During the collision $\vec{v}$ decreases so the angular momentum of the ball decreases. This decrease equals the increase in the door's angular momentum.

3. The tendency for the plane to turn sideways (yaw) is a gyroscopic effect. The propellers of single engine aircraft of U.S. manufacture rotate clockwise as seen from the rear. In accord with the right-hand rule, the angular momentum vector $\vec{L}$ of the propeller points in the forward direction, parallel to the propeller shaft. The downward force of the air on the tail of the plane results in a vector torque about the center of mass in the starboard direction. This causes the plane to precess, slowly turning to starboard so that $\Delta\vec{L}$ is in the same direction as the net torque. To prevent this rotation the pilot uses the rudder to deflect the air to the port. The air then exerts a starboard force on the rudder, which tends to turn the nose of the plane to port.

VI. Problems, Solutions, and Answers

Example #1.   Consider the two vector quantities $\vec{A} = 2.40\hat{i} + 3.20\hat{j} + 3.00\hat{k}$ and $\vec{B} = 4.50\hat{i} + 6.00\hat{j}$. (a) Find their vector product $\vec{C} = \vec{A} \times \vec{B}$ directly from the components of $\vec{A}$ and $\vec{B}$. (b) Use your part (a) result to verify that $|\vec{A} \times \vec{B}| = AB\sin\phi$.

Picture the Problem. For part (b), use the scalar (or dot) product to find the angle.

1. Use the definition of the cross product using $\hat{i}$, $\hat{j}$, and $\hat{k}$ for part (a).	$\vec{C} = \vec{A} \times \vec{B} = \left(A_y B_z - A_z B_y \right) \hat{i}$ $+ \left(A_z B_x - A_x B_z \right) \hat{j} + \left(A_x B_y - A_y B_x \right) \hat{k}$ $= \left(3.2 \times 0 - 3 \times 6 \right) \hat{i} + \left(3 \times 4.5 - 2.4 \times 0 \right) \hat{j}$ $+ \left(2.4 \times 6 - 3.2 \times 4.5 \right) \hat{k}$ $= -18.0 \hat{i} + 13.5 \hat{j}$
2. For part (b), first calculate $\left\| \vec{A} \times \vec{B} \right\|$.	$\left\| \vec{A} \times \vec{B} \right\| = \left\| -18 \hat{i} + 13.5 \hat{j} \right\|$ $= \sqrt{(-18)^2 + 13.5^2} = 22.5$
3. To show that the answer in step 2 is equal to $AB \sin \phi$, we need to find the magnitudes of $\vec{A}$ and $\vec{B}$, as well as the angle ϕ.	$A = \left\| \vec{A} \right\| = \sqrt{A_x^2 + A_y^2 + A_z^2} = \sqrt{2.4^2 + 3.2^2 + 3^2} = 5$ $B = \left\| \vec{B} \right\| = \sqrt{B_x^2 + B_y^2 + B_z^2} = \sqrt{4.5^2 + 6^2 + 0^2} = 7.5$
4. To find the angle, we need to use the two definitions of the dot product.	$\vec{A} \cdot \vec{B} = AB \cos \phi \Rightarrow \phi = \cos^{-1} \left(\dfrac{\vec{A} \cdot \vec{B}}{AB} \right)$ $\vec{A} \cdot \vec{B} = A_x B_x + A_y B_y + A_z B_z$ $= 2.4 \times 4.5 + 3.2 \times 6 + 3 \times 0 = 30$ $\phi = \cos^{-1} \left(\dfrac{30}{5 \times 7.5} \right) = 36.9°$
5. Finally, we can calculate $AB \sin \phi$ and compare it to the result in step 2.	$AB \sin \phi = (5)(7.5) \sin 36.9° = 22.5$

Example #2—Interactive. Consider the two vector quantities $\vec{A} = 2.4 \hat{i} + 3.2 \hat{j} - 3 \hat{k}$ and $\vec{B} = 3 \hat{i} + 4 \hat{j}$. Find the angle between $\vec{A} \times \vec{B}$ and $\vec{C} = 3 \hat{i} + 4 \hat{j}$.

Picture the Problem. Find the cross product, and then take the dot product of $\vec{A} \times \vec{B}$ and $\vec{C}$ to find the angle. **Try it yourself.** Work the problem on your own, in the spaces provided, to get the final answer.

1. Find the cross product $\vec{A} \times \vec{B}$.	
2. Use the two definitions of the dot product to find the angle between $\vec{A} \times \vec{B}$ and $\vec{C}$.	90°

Example #3. A particle released from rest falls freely under the influence of only gravity. Pick a coordinate system with an origin not on the line of fall and show explicitly that $\Sigma\vec{\tau} = d\vec{L}/dt$ holds for this motion.

Picture the Problem. Express the position, force, and linear momentum in terms of m, g, v, t, and the unit vectors $\hat{i}$, $\hat{j}$, and $\hat{k}$. Substitute into $\Sigma\vec{\tau} = d\vec{L}/dt$ and turn the math crank.

1. Draw a sketch to visualize the situation.	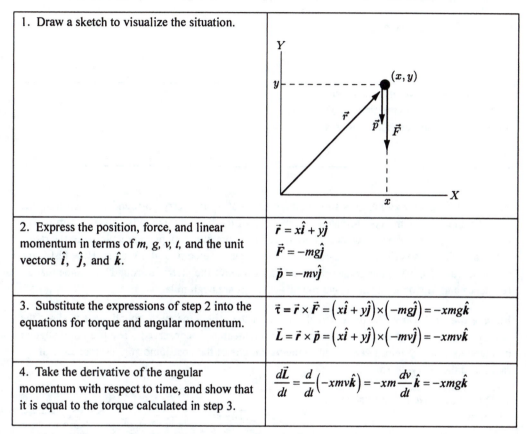
2. Express the position, force, and linear momentum in terms of m, g, v, t, and the unit vectors $\hat{i}$, $\hat{j}$, and $\hat{k}$.	$\vec{r} = x\hat{i} + y\hat{j}$ $\quad$ $\vec{F} = -mg\hat{j}$ $\quad$ $\vec{p} = -mv\hat{j}$
3. Substitute the expressions of step 2 into the equations for torque and angular momentum.	$\vec{\tau} = \vec{r} \times \vec{F} = \left(x\hat{i} + y\hat{j}\right) \times \left(-mg\hat{j}\right) = -xmg\hat{k}$ $\vec{L} = \vec{r} \times \vec{p} = \left(x\hat{i} + y\hat{j}\right) \times \left(-mv\hat{j}\right) = -xmv\hat{k}$
4. Take the derivative of the angular momentum with respect to time, and show that it is equal to the torque calculated in step 3.	$\dfrac{d\vec{L}}{dt} = \dfrac{d}{dt}\left(-xmv\hat{k}\right) = -xm\dfrac{dv}{dt}\hat{k} = -xmg\hat{k}$

Example #4—Interactive. A particle, projected with an initial horizontal velocity, falls under the influence of only gravity. Pick a coordinate system with an origin not on the line of fall and show explicitly that $\Sigma\vec{\tau} = d\vec{L}/dt$ holds for this motion.

Picture the Problem. The only force is the weight, which equals $m\vec{g}$. Draw the position vector $\vec{r}$ from the origin to the particle, write out expressions for the torque vector $\vec{\tau}$ and the angular momentum vector $\vec{L}$, and take the time derivative of the angular momentum vector. **Try it yourself.** Work the problem on your own, in the spaces provided, to get the final answer.

1. Draw a sketch to visualize the situation.	

2. Express the position, force, and linear momentum in terms of m, g, v, t, and the unit vectors $\hat{i}$, $\hat{j}$, and $\hat{k}$. Remember that your linear momentum should have both horizontal and vertical components.	
3. Substitute the expressions of step 2 into the equations for torque and angular momentum.	
4. Take the derivative of the angular momentum with respect to time, and show that it is equal to the torque calculated in step 3.	

Example #5. A 2.00-kg rabbit is sitting on a small, stationary merry-go-round 70.0 cm from the axis of rotation. The merry-go-round has a moment of inertia about the axis of 3.00 kg•m². The rabbit becomes startled and executes a horizontal jump in a direction perpendicular with the radial direction. This leaves the merry-go-round with an angular velocity of 4.00 rad/s. How fast is the rabbit moving relative to the ground just after her feet leave the merry-go-round? Assume that air resistance and friction in the merry-go-round's bearings are negligible.

Picture the Problem. The net external torque on the rabbit–merry-go-round system about the axis is zero. It follows that the system's angular momentum is conserved. The pre-jump angular momentum of the system is zero. Equate the magnitude of the post-jump angular momenta of the rabbit and the merry-go-round.

1. Draw a sketch to help visualize the situation.	
2. Determine the magnitude of the angular momentum of the rabbit just after she jumps.	$L_{\text{rabbit}} = \left\| \vec{r} \times \vec{p} \right\| = rmv \sin \phi$ $= mv(r \sin \phi) = mvr_{\perp}$
3. Determine the magnitude of the angular momentum of the merry-go-round just after the rabbit jumps.	$L = I\omega$

4. Equate the two angular momenta and solve for the speed of the rabbit.	$r_\perp mv = I\omega$ $v = \dfrac{I\omega}{r_\perp m} = \dfrac{\left(3.00\,\text{kg}\cdot\text{m}^2\right)\left(4.00\,\text{rad/s}\right)}{\left(0.70\,\text{m}\right)\left(2.00\,\text{kg}\right)} = 8.57\,\text{m/s}$

Example #6—Interactive. A merry-go-round consists of a circular platform mounted on a frictionless vertical axis. It has a 2.40-m radius and a 210-kg·m² moment of inertia and is rotating with an angular velocity of 2.40 rad/s with a 38.0-kg boy standing on it next to its edge. The boy then runs around its perimeter, running at a constant speed relative to the platform. As he comes up to speed the platform's angular speed decreases to 1.91 rad/s. How fast is the boy moving relative to the platform?

Picture the Problem. There are no external torques on the boy-platform system, so its angular momentum remains constant. Find the initial angular momentum of the system. Obtain an expression for the angular momentum of the system when the boy is running, and set it equal to the initial angular momentum. Solve for the speed of the boy relative to the ground. Calculate the speed of the edge of the platform. Knowing it and the boy's speed relative to the ground, find his speed relative to the platform. **Try it yourself.** Work the problem on your own, in the spaces provided, to get the final answer.

1. Draw a sketch to help visualize the problem.	
2. Find the initial angular momentum of the boy-platform system. Remember to include the angular momentum of both the boy and the platform.	
3. Write an expression for the angular momentum of the system while the boy is running. Equate this expression to the initial momentum to solve for the speed of the boy relative to the ground.	
4. Calculate the speed of the edge of the platform relative to the ground, and use this and your result from step 3 to calculate the speed of the boy relative to the platform.	2.31 m/s

Example #7.   Consider the sun as a uniform sphere of radius 1.39×10^9 m and mass 1.99×10^{30} kg, rotating on its axis with a period of 26.0 days. If, at the end of the sun's life, all the mass collapses inward to form a neutron star of radius 16.0 km, what will be its period of rotation?

Picture the Problem. The net external torque on the system is zero; thus angular momentum is conserved. Equate the values of the angular momentum of the system before and after the collapse. Recall the relationship between angular speed and period.

1. Write an expression for the initial and final angular momenta of the system. Recall the relationship between angular speed and period.	$L_i = I_i\omega_i = I_i\dfrac{2\pi}{T_i}$ $L_f = I_f\omega_f = I_f\dfrac{2\pi}{T_f}$
2. Equate the initial and final angular momenta and solve for the final period. The sun is approximately a sphere, both before and after collapse, so use the appropriate moment of inertia. Also, assume the mass of the sun remains constant during the collapse.	$I_i\dfrac{2\pi}{T_i} = I_f\dfrac{2\pi}{T_f}$ $T_f = \dfrac{I_f}{I_i}T_i = \dfrac{\frac{2}{5}MR_f^2}{\frac{2}{5}MR_i^2}T_i = \dfrac{R_f^2}{R_i^2}T_i$ $= \left(\dfrac{1.6\times10^4\text{ m}}{1.39\times10^9\text{ m}}\right)^2(26\text{ days})\left(\dfrac{24\text{ hr}}{\text{day}}\right)\left(\dfrac{3600\text{ s}}{\text{hr}}\right)$ $= 3.0\times10^{-4}$ s

Example #8—Interactive. A student standing on a frictionless, horizontal platform with her arms extended out from her sides is completing one rotation every 2.40 s. She raises her arms straight up over her head and her rate of rotation increases to once every 0.800 s.   Determine the percentage by which her moment of inertia decreased when she raised her arms.

Picture the Problem.   The net external torque about the vertical axis is zero; thus angular momentum about this axis is conserved. Equate the values of the angular momentum of the system before and after she raises her arms. **Try it yourself.** Work the problem on your own, in the spaces provided, to get the final answer.

1. Write an expression for the initial and final angular momenta of the system.	
2. Equate the initial and final angular momenta and solve for the ratio I_f/I_i.	
3. The percentage by which the moment of inertia has decreased is 1 minus the result from step 2.	67%

Example #9. In Figure 10-3, an 80.0-kg man runs up and jumps onto the bottom of a stationary Ferris wheel whose moment of inertia is $1300 \text{ kg} \cdot \text{m}^2$. The radius of the wheel is 2.40 m and friction on its axle is negligible. After the man jumps on, the wheel rotates freely, swinging him up above the ground. With what minimum initial speed v_0 must he run if it swings him over the top?

Figure 10-3

Picture the Problem. The man-Ferris wheel collision is inelastic, so the total mechanical energy of this system is not conserved. During the collision, the angular momentum of the man-Ferris wheel system is conserved. Equate the values of the angular momentum of the system immediately prior to and immediately following the collision. Following the collision, mechanical energy is conserved, so we can equate the final mechanical energy with the man at the top of the Ferris wheel with the mechanical energy immediately following the collision.

1. Equate the initial and final angular momenta of the collision. The instant before the collision is "instant 0" and the moment after the collision is "instant 1." Initially, only the man is moving, and he is a minimum distance R from the center of the Ferris wheel. After the collision, both the man and the wheel rotate. Solve for ω_1 to be used in the next step.	$L_0 = L_1$ $\left\| \vec{r}_{man} \times \vec{p}_{man} \right\| = I\omega$ $Rmv_0 = \left(I_{wheel} + I_{man} \right)\omega_1$ $\omega_1 = \dfrac{Rmv_0}{\left(I_{wheel} + I_{man} \right)}$
2. After the collision, while the wheel rotates, lifting the man, mechanical energy is conserved. In this step, consider the system to consist of the man, the Ferris wheel, and the earth. Let the bottom of the Ferris wheel be the zero point for the gravitational potential energy. Although the center-of-mass of	$K_1 + U_1 = K_2 + U_2$ $\frac{1}{2}\left(I_{wheel} + I_{man} \right)\omega_1^2 + mgh_1 = \frac{1}{2}\left(I_{wheel} + I_{man} \right)\omega_2^2 + mgh_2$ $\dfrac{1}{2}\left(I_{wheel} + I_{man} \right)\left(\dfrac{Rmv_0}{\left(I_{wheel} + I_{man} \right)} \right)^2 + 0 = 0 + mg2R$ $\dfrac{\left(Rmv_0 \right)^2}{2\left(I_{wheel} + I_{man} \right)} = 2mgR$

the Ferris wheel does not change, the man is lifted a height of $2R$ by the time he gets to the top. If the man just makes it around, the kinetic energy of the system is zero when the man is at the top of his ride. Let "instant 2" be the moment when the man reaches the top of his ride.	
3. Finally, we can use our result from step 2 to solve for the minimum initial velocity of the man. That would be one fast man!	$$v_0^2 = \frac{4g}{mR}\left(I_{wheel} + I_{man}\right)$$ $$v_0 = \sqrt{\frac{4g}{mR}\left(I_{wheel} + mR^2\right)}$$ $$= \sqrt{4gR\left(\frac{I_{wheel}}{mR^2} + 1\right)}$$ $$= \sqrt{4(9.81\,\text{m/s}^2)(2.4\,\text{m})\left(\frac{1300\,\text{kg}\cdot\text{m}^2}{(80\,\text{kg})(2.4\,\text{m}^2)} + 1\right)}$$ $$= 19.0\,\text{m/s}$$

Example #10—Interactive. Tarzan, the ape man, swings down from a tree branch to kick a bad guy in the head. In your calculations, model Tarzan as a uniform thin rod 1.90 m long with a mass of 91.0 kg. When he starts swinging, his body makes an angle of 40° with the vertical, as shown in Figure 10-4. At what speed are Tarzan's feet moving when they whack the back guy in the head, which is at the bottom of the swing?

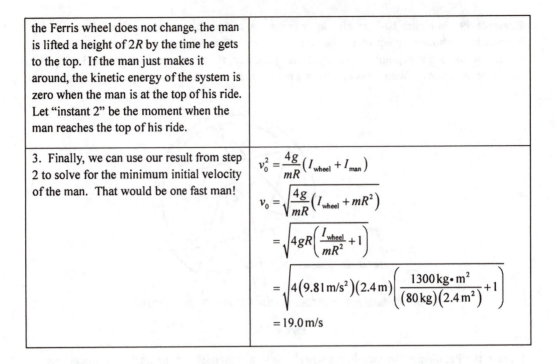

Figure 10-4

Picture the Problem. If you choose the system to be Tarzan, the rope, and the earth, there are no external forces, and the total mechanical energy of the system is conserved. Remember that the gravitational potential energy of the system depends on Tarzan's center of mass, and the kinetic energy depends on his moment of inertia and his angular speed. Once you have the angular speed of Tarzan's feet the instant before they hit the bad guy's head, you can determine their linear speed. **Try it yourself.** Work the problem on your own, in the spaces provided, to get the final answer.

1. Write out the basic conservation of energy equation as a guiding principle.	
2. Figure out Tarzan's moment of inertia about the proper axis.	
3. Determine the change in height of Tarzan's center of mass.	
4. Determine Tarzan's angular speed at the bottom of the swing from conservation of energy principles.	
5. Determine the linear speed of Tarzan's feet the instant before they hit the bad guy's head.	3.6 m/s

Chapter R

Special Relativity

I. Key Ideas

The concept of relative velocities was introduced in our discussion of one-dimensional mechanics in Chapter 2.   Inertial (nonaccelerated) and noninertial (accelerated) reference frames were discussed in Chapter 4.   Those discussions are limited to objects in reference frames that move very slowly compared to the speed of light, relative to each other.   When one reference frame moves at a speed near that of light compared to another reference frame, interesting things begin to happen. Clocks in the two reference frames no longer move at the same speed, the length of objects is not the same in the two frames, and simultaneous events in one reference frame do not appear as simultaneous events in the other reference frame.   These phenomena are described by the special theory of relativity (often referred to as *special relativity*), developed by Albert Einstein and others in 1905.   This chapter explores many of the consequences of the special theory of relativity.

Section R-1. The Principle of Relativity and the Constancy of the Speed of Light. The principle of relativity can be stated as follows:

> It is impossible to devise an experiment that determines whether you are at rest or moving uniformly.

Postulate I, The principle of relativity

Suppose you play a game of billiards in a rocket ship, and someone else in another rocket ship moving relative to you at a different very high constant speed is also playing billiards. When you and the other person each strike a billiard ball at the same angle and with the same speed, afterward each billiard ball will move in exactly the same manner.   That is, the laws of physics behave the same in each of these reference frames.

Newton believed that the principle of relativity applied only to mechanical motions such as colliding billiard balls and oscillating pendula. Einstein extended the principle of relativity to *all* areas of physics, including light and other electromagnetic phenomena. Indeed, the paper in which he started the whole notion of "relativity" was entitled "On the Electrodynamics of Moving Bodies."

Before Einstein put forth his special theory of relativity in 1905 it was generally believed that light and other electromagnetic phenomena did not obey a principle of relativity. This can be seen by looking at the speed of light, $c = 1/\sqrt{\varepsilon_0 \mu_0} = 3 \times 10^8$ m/s, which follows from Maxwell's equations that describe electromagnetic waves. According to Newtonian concepts this velocity can be measured only by a privileged observer who is at rest in "absolute space" in which there is a medium called the **ether** that supports the propagation of light and other electromagnetic waves. Other observers moving relative to this privileged observer will necessarily measure the speed of light to be different from the value c, and consequently will also find Maxwell's equations to have a form different from what the privileged observer obtains.

Michelson and Morley in 1887 developed an interferometer sensitive enough to measure the expected change of the speed of light as the earth moved in different directions through the hypothesized ether. The result of these and many other subsequent experiments was that the speed of light did not change, and no motion through the ether was ever detected. This is one of the most famous null experimental results ever obtained, proving that the ether did not exist. This led Albert Einstein to posulate

> The speed of light is independent of the speed of the light source.   Postulate II

The speed of light refers to the speed at which light travels through the vacuum of empty space, free of any ether.

A consequence of Postulate I and Postulate II is the constancy of the speed of light:

> The speed of light c is the same in any inertial reference frame.

> The constancy of the speed of light

This concept requires a bit of contemplation, as it runs counter to commonsense (Newtonian) concepts of how velocities add together. One thing that follows from the second postulate is an explanation of the null result of the Michelson–Morley experiment. If all observers measure the same value for the speed of light, it follows that no experiment will ever show two observers measuring different values for the speed of light.

Section R-2. Moving Sticks. Any stick moving perpendicular to its length has the same length as an identical stationary stick. This rule is independent of the material from which the two sticks are made. As a result, the rule is not a rule of sticks, specifically. Instead, it reflects a property of space. That is, not only could the "stick" be replaced by any object, it could also be replaced by a region of empty space with a specific length. The frame of reference in which a stick is at rest is called its **proper frame** or **rest frame**, and the length of a stick in its proper reference frame is called it **proper length** or **rest length**.

Section R-3. Moving Clocks. Clocks moving at high speeds relative to "stationary" clocks run slow. If a high-speed spaceship travels by us, we would observe that all the clocks on the ship run slower than our clocks. Since clocks measure time, this means that we would observe that time passes more slowly on the spaceship. However, thanks to the principle of relativity, the people on the ship can consider themselves at rest and us to be moving. From their perspective, our clocks appear to be running slow compared to the clocks on a spaceship. Both of these observations are consistent with the constancy of the speed of light and the principle of relativity.

If the time between ticks of a clock in a proper rest frame is T_0, then it can be shown that the time between ticks of an identical clock moving with speed v relative to the rest frame is given by

$$T = \frac{T_0}{\sqrt{1-\left(v^2/c^2\right)}}$$
Time dilation

In the moving frame of reference, more time passes between each tick of the clock, which means that time (as measured by the number of ticks of the identical clocks) passes more slowly in the moving frame of reference than in the stationary one.

A **spacetime event**, or more simply **event**, is something that occurs at a specific instant in time and at a specific location in space. Any two events that occur both at the same time and same place in one reference frame will also occur at the same time and same place in all reference frames. This is known as the **principle of invariance of coincidences**.

> If two events occur at the same time and at the same place in one reference frame, then they occur at the same time and at the same place in all reference frames.

Principle of invariance of coincidences

Why are coincidences invariant? Because coincidences can have lasting effects. Consider a person accidentally running into a lamppost. This collision requires two spacetime events to occur at the same time and place (a spacetime coincidence). The lamppost needs to be in a particular place at a particular time. The person also needs to be in that same place at the same time. Such a collision will result in the person being bruised after the collision. This bruise will exist regardless of the reference frame used to observe it. Therefore, the coincidence that caused the bruise (both the person and the lamppost being in the same place at the same time) must have occurred in every reference frame, as well.

Section R-4. Moving Sticks Again. Consider a stick at rest with a proper length L_0 and an identical stick moving with a speed v in a direction *parallel* to the length of the stick. In this case, an observer in the rest frame will measure the length of the moving stick to be shorter than the length of the stick at rest. The measured length of the moving stick will be

$$L = L_0\sqrt{1-\left(v^2/c^2\right)}$$
Length contraction

Length contraction only occurs along the direction parallel to the stick. This rule is independent of the material from which the two sticks are made. As a result, the rule is not a rule of sticks, specifically. Instead, it reflects a property of space and time. That is, not only could the "stick" be replaced by any object, it could also be replaced by a region of empty space with a specific length.

Section R-5. Distant Clocks and Simultaneity. Let us now consider two identical clocks in a single proper reference frame, at rest relative to each other, and separated by a distance of L_0. We can synchronize these clocks while they are at rest so that they display the same clock reading at the same instant of time in the proper frame. If these same two synchronized clocks now move at a speed v relative to the proper frame, in a direction parallel to L_0, the clock "in front" is actually behind the clock "in back." This loss of synchronicity is a direct result of length contraction and the constancy of the speed of light. It is referred to as the relativity of simultaneity.

> If two clocks are synchronized in their rest frame, then in a frame where they move with speed v parallel to the line joining them the clock in the rear is ahead of the clock in the front by vL_0 / c^2.

The relativity of simultaneity

Section R-6. Applying the Rules. See the example problems at the end of this chapter.

Section R-7. Relativistic Momentum, Mass, and Energy. In special relativity, the conservation laws for energy and momentum still apply, with some slight changes due to motion at high speeds. The most interesting of these is that the mass of a particle is no longer constant. If a particle has a rest mass m_0 in a proper frame, then when the particle is moving with speed v, its mass is larger than the rest mass and is given by

$$m = \frac{m_0}{\sqrt{1-\left(v^2/c^2\right)}}$$

Relativistic mass

Momentum is still defined such that $\vec{p} = m\vec{v}$, but the relativistic mass must now be used. As a result, the relativistic momentum is given by

$$\vec{p} = \frac{m_0\vec{v}}{\sqrt{1-\left(v^2/c^2\right)}}$$

Relativistic momentum

As in the mechanics you have already learned, the kinetic energy of a particle can be written as $K = \int v\,dp$. Using our new value for the relativistic momentum, this integral results in

$$K = \frac{m_0c^2}{\sqrt{1-\left(v^2/c^2\right)}} - m_0c^2$$

Relativistic kinetic energy

If we define the total relativistic energy as $m_0c^2 / \sqrt{1-\left(v^2/c^2\right)}$, we have

$$E = K + m_0c^2 = \frac{m_0c^2}{\sqrt{1-\left(v^2/c^2\right)}}$$

Total relativistic energy

Since the energy m_0c^2 is present even when the velocity of the particle is zero, it is referred to as the **rest energy** of the particle.

$$E_0 = m_0c^2$$

Rest energy

The total energy squared of a particle may also be written as

$$E^2 = p^2c^2 + m_0^2c^4 \qquad\qquad \text{Relativistic energy–momentum relationship}$$

That is, the total energy of a particle can be found from its rest mass and its momentum.

II. Physical Quantities and Key Equations

Physical Quantities

Speed of light $c = 1/\sqrt{\varepsilon_0\mu_0} = 3.00\times10^8$ m/s

Energy conversion $1\,\text{eV} = 1.6\times10^{-19}$ J

$1\,\text{MeV} = 10^6$ eV

Rest energy of an electron $m_0c^2 = 0.511\,\text{MeV}$

Rest energy of a proton $m_0c^2 = 938.3\,\text{MeV}$

Rest energy of a neutron $m_0c^2 = 939.6\,\text{MeV}$

Key Equations

Time dilation $T = \dfrac{T_0}{\sqrt{1-\left(v^2/c^2\right)}}$

Length contraction $L = L_0\sqrt{1-\left(v^2/c^2\right)}$

Relativistic mass $m = \dfrac{m_0}{\sqrt{1-\left(v^2/c^2\right)}}$

Relativistic momentum $\vec{p} = \dfrac{m_0\vec{v}}{\sqrt{1-\left(v^2/c^2\right)}}$

Rest energy $E_0 = m_0c^2$

Total relativistic energy $E = K + m_0c^2 = \dfrac{m_0c^2}{\sqrt{1-\left(v^2/c^2\right)}}$

Relativistic kinetic energy $K = \dfrac{m_0c^2}{\sqrt{1-\left(v^2/c^2\right)}} - m_0c^2$

Relativistic energy–momentum relationship $E^2 = p^2c^2 + m_0^2c^4$

III. Potential Pitfalls

When dealing with problems involving the relationship between space and time measurements made by various observers, it is important to keep in mind the concept of "event." An event could be the location of a lightning bolt striking a point at a given time. The location of each end of a moving rod at the same time constitutes two different events. Each event has four coordinates (x, y, z, t) assigned to it by an observer in reference frame S, and four other coordinates (x', y', z', t') assigned to it by an observer in reference frame S'.

Since two events that occur simultaneously for one observer will not in general be simultaneous for other observers, you must make sure you are clear about which observer determines that the two events are simultaneous.

Do not confuse the time at which you see an event take place with the time at which the event actually takes place. For example, suppose you see a distant supernova explode, and record the time on a clock at your location when you make your observation. To determine the actual time that the supernova explosion occurred, you must correct for the travel time that it took the light signal to reach your location.

Often, problems can be solved using the simple expression distance = velocity × time, provided that the distance, velocity, and time all refer to the same reference frame. For example, suppose you are a muon, moving at a speed $0.998c$ relative to the earth, that has to traverse a distance of 9000 m measured by an earth observer. In your frame of reference, the distance to be traversed is contracted and is only $(9000 \text{ m})\sqrt{1-0.998^2} = 600 \text{ m}$, so the time Δt you would need to traverse this distance is found from

$$\text{distance} = \text{velocity} \times \text{time}$$

$$600 \text{ m} = 0.998 (3 \times 10^8 \text{ m/s}) \Delta t$$

giving $\Delta t = 2 \times 10^{-6}$ s.

In relativistic energy problems do not mistakenly use the classical expression $K_{\text{classical}} = \frac{1}{2} m_0 v^2$. Whenever v is comparable to c, roughly $v \geq 0.1c$, you must use relativistic expressions.

Often, particles are described in terms of their kinetic energies rather than their velocities. For example, you may be told that a particle of charge e has been accelerated through a certain voltage, which is equal to the kinetic energy in electron volts that the particle acquires. Relativistic expressions must be used when the kinetic energy is comparable to the rest energy of an object. For example, you must use relativistic expressions for a 2-MeV electron because an electron's rest mass is about 0.5 MeV. For a 2-MeV proton, however, you can use nonrelativistic expressions to a good approximation because a proton's rest mass is about 938 MeV.

Do not mistakenly use the expression $\vec{p} = m_0 \vec{v}$ for an object's momentum in relativistic situations. The correct relativistic expression is

$$\vec{p} = \frac{m_0 \vec{v}}{\sqrt{1 - (v^2/c^2)}}$$

IV. True or False Questions and Responses

True or False

_____ 1. All observers in inertial reference frames will measure the speed of light in any direction to be 3×10^8 m/s.

_____ 2. The medium in which light was (mistakenly) assumed to propagate was called the ether.

_____ 3. The Michelson and Morley experiment was devised primarily to make very precise measurements of the speed of light.

_____ 4. An observer finds that the time interval between two events that take place at the same location is 2 s. Another observer moving relative to this observer will measure the time interval between the two events to be less than 2 s.

_____ 5. An observer in reference frame S finds that an event A takes place at her origin simultaneously with a second event B located at a distance along her $+x$ axis. Another observer in reference frame S' moving in the positive direction along the common x–x' axis finds that event A occurs after event B.

_____ 6. An observer in reference frame S finds that two lightning bolts strike the same place simultaneously. An observer in S' moving relative to S will determine that the two bolts strike at different times.

_____ 7. The reason length contraction occurs is that light from the front and light from the back of an object take different times to reach an observer.

_____ 8. You have a meter stick located at rest along your y axis. Another person moving along your x axis will measure the length of the meter stick to be less than 1 m.

_____ 9. A clock moving along the x–x' axis at constant speed is struck by a lightning bolt and is later struck by a second lightning bolt. The time interval recorded by this very sturdy clock between the two strikes is the proper time interval.

_____ 10. An object moving at a speed of $0.8c$ takes 5 yr to traverse a distance of 4c•yr.

_____ 11. You measure that a rocket ship moving at $0.8c$ takes 5 yr to travel from one galaxy to another. The time measured by the pilot of the rocket ship is 3 yr.

_____ 12. Two observers moving relative to each other measure the same value for their relative speed.

_____ 13. An electron and a proton are each accelerated through a potential difference of 50,000 V and are then injected into the magnetic field of a cyclotron. You must use relativistic expressions to analyze the motion of the electron, but the motion of the proton can be treated with classical expressions to a good approximation.

_____ 14. An object's kinetic energy is equal to the difference between its total energy and its rest energy.

_____ 15. The kinetic energy of a proton that has been accelerated through a potential difference of 300×10^6 V is 300 MeV.

_____ 16. To a good approximation, the momentum of a 500-MeV electron is 500 MeV/c.

_____ 17. To a good approximation, the momentum of a 500-MeV proton is 500 MeV/c.

Responses to True or False

1. True.

2. True.

3. False. The Michelson–Morley experiment was designed to measure the speed of the earth relative to the ether, which did involve high precision.

4. False. Since both events take place at the same location in the reference frame of the first observer, 2 s is the proper time interval between them. Any observer moving relative to the first observer will find a time interval larger than 2 s because of time dilation.

5. True.

6. False. Since both events occur at the same place, if one observer finds the two events to be simultaneous, then all other observers will also find them to be simultaneous. Disagreements about simultaneity arise only concerning events that are spatially separated in at least one of the reference frames.

7. False. The length of a moving object is determined by finding the spatial coordinates of its end points at the same time, and need not involve light coming from its end points.

8. False. All observers will agree about lengths along the y axis.

9. True.

10. True.

11. True.

12. True.

13. True.

14. True.

15. True.

16. True.

17. False. Since the proton's kinetic energy of 500 MeV is comparable to its rest energy of 938 MeV, you must use the relativistic relation between energy and momentum to calculate the proton's momentum. The result is $p = 1090$ MeV/c.

V. Questions and Answers

Questions

1. An observer stationary in reference frame S' measures a light signal emitted at $t' = 0$ to move from his origin directly along his $+y'$ axis at, of course, a speed c. Describe what an observer in S measures.

2. An observer in reference frame S' is moving relative to reference frame S at 0.8c. At a certain instant a red flash is emitted at the 10-m mark on the x' axis of S', and 5 s later, as determined by clocks in S', a blue flash is emitted at the same 10-m mark. What is the proper time interval between the red and blue flashes?

3. Is it possible for one observer to find that event A happens after event B and another observer to find that event A happens before event B?

4. Suppose that an observer finds that two events A and B occur simultaneously at the same place. Will the two events occur at the same place for all other observers?

5. Explain how the measurement of time enters into the determination of the length of an object.

6. You find that it takes a time interval Δt for a spaceship to move through a distance L. How does the time interval elapsed on the clock of the pilot of the spaceship compare to what you measure?

7. In terms of energies, when can you use $p = m_0 v$ to a good approximation?

8. In terms of energies, when can you use $p = E/c$ to a good approximation?

Answers

1. The observer in S also measures the light signal to move with speed c.

2. Since the red and blue flashes are both emitted at the same place as determined in reference frame S', the time interval of 5 s determined by a single clock in S' is the proper time interval between the two events.

3. Yes. The time ordering $t'_A - t'_B$ between the two events as determined in reference frame S' can be positive, zero, or negative depending on the spatial separation, $t_A - t_B$ in reference frame S, and the relative velocity v.

4. Yes.

5. If the object is moving relative to you, you measure its length by finding the difference between the coordinates of its end points at the same time.

6. The pilot measures the time interval with a single clock that records the proper time between the beginning and end points. This proper time interval will be smaller than your time interval Δt.

7. When the kinetic energy $K \ll m_0 c^2$ you can use the expression $p = m_0 v$ to a good approximation.

8. When the total energy $E \gg m_0 c^2$, or equivalently when the kinetic energy $K \gg m_0 c^2$, you can use $p = E/c$ to a good approximation.

VI. Problems, Solutions, and Answers

Example #1. A rocket car traverses a track 2.4×10^5 m long in 10^{-3} s as measured by an observer next to the track. How much time elapses on a clock in the rocket car when it traverses the track?

Picture the Problem. Keep clear in your mind who is measuring each distance and time interval. Because the two clocks are moving with respect to each other, we will need to use time dilation. However, in order to determine the amount of time dilation, we need to know the relative speed of the clocks.

1. Determine the speed of the car as measured by the observer at the side of the track.	$v = \dfrac{\text{distance}}{\text{time}} = \dfrac{2.4 \times 10^5 \text{ m}}{10^{-3} \text{ s}}$ $= 2.4 \times 10^8 \text{ m/s} = 0.8c$
2. The clock in the car measures the proper time required to travel that distance.	$T = \dfrac{T_0}{\sqrt{1 - \left(v^2 / c^2 \right)}}$ $T_0 = T\sqrt{1 - \left(v^2 / c^2 \right)}$ $= \left(10^{-3} \text{ s} \right) \sqrt{1 - 0.8^2} = 6 \times 10^{-4} \text{ s}$

Example #2—Interactive. How far does the driver of the rocket car in the above example determine she travels in traversing the track?

Picture the Problem. Keep clear in your mind who is measuring each distance and time interval. The rocket car will see a contracted length of the track. **Try it yourself.** Work the problem on your own, in the space provided, and check your answer.

1. Determine the contracted length of the track from the velocity found in the previous example.	 $L = 1.44 \times 10^5$ m

Example #3. What is the velocity of a 2.00-MeV electron?

Picture the Problem. If $K \ll m_0 c^2$, then you can use the nonrelativistic expression for kinetic energy. Otherwise, you must use the relativistic expression for kinetic energy to determine the electron's velocity.

1. Look up the rest mass of the electron.	$m_0 c^2 = 0.511\,\text{MeV}$
2. The given energy is much larger than the rest energy of the electron, so to determine its speed we must use the relativistic expression.	$$K = \frac{m_0 c^2}{\sqrt{1-\left(v^2/c^2\right)}} - m_0 c^2$$ $$2.00\,\text{MeV} = \frac{0.511\,\text{MeV}}{\sqrt{1-\left(v^2/c^2\right)}} - 0.511\,\text{MeV}$$ $$v = 0.979c$$

Example #4—Interactive. How fast is a 1,000,000-MeV proton moving?

Picture the Problem. If $K \ll m_0 c^2$, then you can use the nonrelativistic expression for kinetic energy. Otherwise, you must use the relativistic expression for kinetic energy to determine the proton's velocity. **Try it yourself.** Work the problem on your own, in the spaces provided, and check your answer.

1. Look up the rest mass of the proton.	
2. The given energy is much larger than the rest energy of the proton, so to determine its speed we must use the relativistic expression.	$v = 0.0462c$

Example #5. What is the kinetic energy of a proton whose momentum is 500 MeV/c ?

Picture the Problem. If $pc \gg m_0 c^2$, then you can use $E = pc$ or $K = pc$ to a good approximation. Otherwise, you will need to use the relativistic expression relating energy and momentum to find the proton's energy.

1. Look up the rest mass of the proton.	$m_0 c^2 = 938\,\text{MeV}$
2. The given energy is on the order of the rest energy, so the relativistic expression must be used.	$$E^2 = \left(K + m_0 c^2\right)^2 = p^2 c^2 + m_0^2 c^4$$ $$\left(K + 938\,\text{MeV}\right)^2 = \left(500\,\text{MeV}/c\right)^2 c^2 + \left(938\,\text{MeV}\right)^2$$ $$K = 125\,\text{MeV}$$

Example #6—Interactive. What is the kinetic energy of an electron whose momentum is 300 MeV/c ?

Picture the Problem. If $pc \gg m_0 c^2$, then you can use $E = pc$ or $K = pc$ to a good approximation. Otherwise, you will need to use the relativistic expression relating energy and momentum to find the electron's energy. **Try it yourself.** Work the problem on your own, in the spaces provided, and check your answer.

1. Look up the rest mass of the electron.	
2. The given energy is much larger than the rest energy, so an approximation can be used.	$K = 300\,\text{MeV}$

Chapter **11**

Gravity

I.   Key Ideas

Section 11-1. Kepler's Laws. Around 1600, Johannes Kepler inferred from many observations three empirical laws about planetary motion:

Law 1. All planets move in elliptical orbits with the sun at one focus.

Law 2. A line joining any planet to the sun sweeps out equal areas in equal times.

Law 3. The square of the period of any planet's orbit is proportional to the cube of its mean distance from the sun.

Section 11-2. Newton's Law of Gravity. Isaac Newton showed that a gravitational force law together with his own three laws of motion could be used to derive Kepler's empirically established laws. **Newton's law of gravity** states that every object in the universe attracts every other object with a force that is directly proportional to the product of the masses of the two objects and inversely proportional to the square of the distance between them:

$$\vec{F}_{1,2} = -\frac{Gm_1 m_2}{r_{1,2}^2}\hat{r}_{1,2} \hspace{3cm} \text{Newton's law of gravity}$$

Here $\vec{F}_{1,2}$ is the force exerted by object 1 on object 2, $r_{1,2}$ is the magnitude of the relative position vector $\vec{r}_{1,2}$ from object 1 to object 2, $\hat{r}_{1,2}$ is the unit vector in the direction of $\vec{r}_{1,2}$ (that is $\hat{r}_{1,2} = \vec{r}_{1,2} / r_{1,2}$), and G is the universal gravitational constant:

$$G = 6.67 \times 10^{-11}\,\text{N} \cdot \text{m}^2 / \text{kg}^2 \hspace{2cm} \text{Universal gravitational constant}$$

Newton showed that Kepler's 2nd law, the law of equal areas, is a direct consequence of the conservation of angular momentum and holds for any central force. He also showed that planets with circular orbits have a period given by

$$T = \sqrt{\frac{4\pi^2}{GM_S} r^3}$$ Kepler's 3rd law

where r is the radius of the planet's orbit and M_S is the mass of the sun.

Newton's law of gravity describes the gravitational force between particles. Although he had to invent calculus to do it, Newton was able to prove that the force of gravity exerted by any spherically symmetric object on a point mass on or outside its surface is the same as if all the mass of the object were concentrated at its center. Thus, the gravitational force exerted on a particle of mass m a distance r from the earth's center is

$$F = \frac{GM_E m}{r^2} \quad r \ge R_E$$ Gravitational force exerted by the earth

where M_E is the earth's mass and R_E is its radius. The acceleration of a particle due to the gravitational force of the earth is then

$$g(r) = \frac{F}{m} = \frac{GM_E}{r^2} \quad r \ge R_E$$ Acceleration due to the earth's gravity

To verify this equation, Newton compared the free-fall acceleration of objects falling here at the earth's surface, a distance of one earth radius from the earth's center, with the acceleration of the moon in its nearly circular orbit at a distance of approximately 60 earth radii from the earth's center.

Measurement of the Universal Gravitational Constant.   The universal gravitational constant G can be experimentally determined by measuring the force of gravitational attraction between two objects of known mass. This measurement is difficult because the force is extremely small for objects of laboratory (rather than planetary) size. Knowledge of G is of great scientific value because it enables the experimental determination of the mass of the earth and other objects in the solar system. The currently accepted value for the mass of the earth is approximately 5.98×10^{24} kg.

Gravitational and Inertial Mass. Newton's law of gravity makes it clear that the mass of an object is a measure of two distinct properties: (1) that property responsible for the gravitational force the object exerts on other objects, referred to as its **gravitational mass**, and (2) that property responsible for the object's resistance to acceleration, referred to as its **inertial mass**. The equivalence of these two properties is by no means obvious, but it has been confirmed experimentally to a very high precision. Therefore, we do not ordinarily distinguish between the two, and just use the term mass. This **principal of equivalence** is a very important concept in Einstein's theory of general relativity.

Section 11-3. Gravitational Potential Energy. Gravity is a conservative force, so it can be described in terms of potential energy. The **gravitational potential energy** of a particle at a location is defined as the negative of the work done by the force of gravity on the particle as it moves to that location from a reference location where its gravitational potential energy is zero.

The gravitational potential energy $U(r)$ of a particle in the earth's gravitational field at a distance r from the center of the earth is

$$U(r) = \frac{GM_Em}{R_E} - \frac{GM_Em}{r} \quad r \geq R_E \qquad \text{Gravitational potential energy } (U=0 \text{ at } r=R_E)$$

if the reference location, where its potential energy is chosen to be zero, is at the earth's surface. Remember that although we refer to this potential energy to be that of a particle near the earth, this potential energy is, in actuality, a property not of the particle but of the particle-earth system.

Consider an object of mass m leaving the earth's surface with some initial speed. At the earth's surface, where $r = R_E$, the potential energy of the object is zero. As the object rises, it gains potential energy and loses kinetic energy. As the object approaches an infinite distance from the earth, its potential energy approaches GM_Em/R_E, which we will call U_{max}. If the object leaves the earth's surface at sufficient speed so that its kinetic energy exceeds U_{max}, the earth's gravity will never be able to stop it and pull it back to earth. The object will still have some kinetic energy when it is an arbitrarily great distance away. When the object leaves the earth's surface with a kinetic energy equal to U_{max} its speed is called the **escape speed** of the earth, v_e. The escape speed, calculated by equating the initial kinetic energy and U_{max}, is

$$v_e = \sqrt{\frac{2GM_E}{R_E}} = \sqrt{2gR_E} \qquad \text{Escape speed}$$

which is about 11 km/s (or 7 mi/s).

Classification of Orbits by Energy. In the above equation for gravitational potential energy, the surface of the earth is the reference location where the potential energy was chosen to be zero. However, the reference location is arbitrary. We can choose it to be anywhere outside the earth. If we choose it to be an infinite distance from the earth, the equation for the gravitational potential energy becomes

$$U = -\frac{GM_Em}{r} \qquad r \geq R_E \qquad \text{Gravitational potential energy with } U=0 \text{ at } r=\infty$$

The reason for making this choice is, of course, that it results in a simpler expression for the potential energy. With this choice, the potential energy is negative for any finite separation and approaches zero as the separation approaches infinity. Any orbit in which the energy of the orbiting object is not sufficient for the object to escape is called a bound orbit. The total mechanical energy of an object is negative if the object is in a bound orbit (an ellipse) and zero or positive if the object is in an unbound orbit (a parabola or hyperbola).

Section 11-4. The Gravitational Field. The vector form of the gravitational field can be calculated at any point in space and is found by dividing the gravitational force experienced by an object of mass m by its mass. That is,

$$\vec{g} = \frac{\vec{F}}{m} \qquad \text{Gravitational field}$$

The magnitude of the gravitational field due to the earth is then $g(r) = GM_E / r^2$.

The gravitational field of a thin uniform spherical shell of mass M and radius R at a distance r from the center of the shell is

$$\vec{g}(r) = -\frac{GM}{r^2}\hat{r} \quad \text{for} \quad r > R$$

Gravitational field of a thin spherical shell

$$\vec{g}(r) = 0 \quad \text{for} \quad r < R$$

A uniform solid sphere can be thought of as a collection of concentric spherical shells. The gravitational field of a solid sphere of radius R and mass M a distance r from its center is

$$\vec{g}(r) = -\frac{GM}{r^2}\hat{r} \quad \text{for} \quad r \geq R$$

Gravitational field of a uniform solid sphere

$$\vec{g}(r) = -\frac{G}{r^2}\left(\frac{Mr^3}{R^3}\right)\hat{r} \quad \text{for} \quad r \leq R$$

where $\left(Mr^3 / R^3\right)$ is the mass inside a sphere of radius r.

Section 11-5. Finding the Gravitational Field of a Spherical Shell by Integration. By dividing a spherical shell into small rings, its gravitational field given above can be derived by integrating the gravitational field due to the sum of rings that constitute the shell.

II. Physical Quantities and Key Equations

Physical Quantities

Universal gravitational constant	$G = 6.67 \times 10^{-11}\ \text{N} \cdot \text{m}^2 / \text{kg}^2$
Radius of the earth	$R_E = 6.37 \times 10^6\ \text{m}$
Mean earth-sun distance	$r_E = 1.50 \times 10^{11}\ \text{m} = 1\,\text{AU}\ (\text{astronomical unit})$
Mass of the earth	$5.98 \times 10^{24}\ \text{kg}$
Mass of the sun	$M_S = 1.99 \times 10^{30}\ \text{kg}$
Length of a year	$1\,\text{yr} = 3.16 \times 10^7\ \text{s}$

Key Equations

Newton's law of gravity	$\vec{F}_{1,2} = -\dfrac{Gm_1 m_2}{r_{1,2}^2}\hat{r}_{1,2}$
Kepler's 3rd law	$T = \sqrt{\dfrac{4\pi^2}{GM_S}r^3}$

Gravitational field of the earth	$\vec{g} = \dfrac{\vec{F}}{m} = -\dfrac{GM_E}{r^2}\hat{r} \qquad r \geq R_E$
Escape speed	$v_e = \sqrt{\dfrac{2GM_E}{R_E}} = \sqrt{2gR_E}$
Gravitational potential energy $(U \to 0 \text{ as } r \to \infty)$	$U = -\dfrac{GM_E m}{r} \qquad r \geq R_E$
Gravitational field of a thin uniform spherical shell	$\vec{g}(r) = -\dfrac{GM}{r^2}\hat{r} \quad \text{for} \quad r > R$ $\vec{g}(r) = 0 \quad \text{for} \quad r < R$
Gravitational field of a uniform solid sphere	$\vec{g}(r) = -\dfrac{GM}{r^2}\hat{r} \quad \text{for} \quad r \geq R$ $\vec{g}(r) = -\dfrac{G}{r^2}\left(\dfrac{Mr^3}{R^3}\right)\hat{r} \quad \text{for} \quad r \leq R$

III. Potential Pitfalls

Every formula for gravitational potential energy (or for any potential energy, for that matter) *assumes* a particular zero location. Keep clearly in mind where this location is when working a particular problem. The expression *mgh* for gravitational potential energy means the potential energy is zero at $h = 0$.

The force of gravity exerted by a spherically symmetric object (a collection of uniform concentric spherical shells) on an object located outside the spherical object is the same as if the mass of the spherically symmetric object were concentrated at its geometric center. This is true *only* for objects where the mass is distributed with spherical symmetry.

In problems where the distances between planets and moons and such appear, the distances are measured between their *centers*.

An object moving freely in the earth's gravity is not necessarily moving toward the earth! However, such objects, even those in circular orbits, are always *accelerating* toward the earth. This is true no matter what direction they are moving.

The earth's escape speed is 11 km/s or 7 mi/s. Remember that this number refers only to escape from the earth's surface. An object starting from a higher altitude requires a lower initial velocity to escape.

Every planet, moon, or star has its own escape speed, determined by its mass and radius.

Be careful of signs when working problems that deal with the energy of orbital motion. Use the simplest form of the equation for gravitational potential energy, $-GM_E m/r$, for which the potential energy is always negative and approaches zero as the distance between the object and the earth approaches infinity. For objects in bound orbits, the total mechanical energy is negative, whereas for objects in unbound orbits, the total mechanical energy is never negative.

For the sake of illustration, we often do problems involving the earth's gravity in which we neglect air drag. Don't worry if the results differ widely from reality. Without air drag, an object falling from a very large distance would strike the earth's surface at a speed approaching the earth's escape speed. However, due to air drag, the actual impact speed is an order of magnitude or so smaller than the escape speed. (The object may also burn up in the atmosphere and never reach the earth's surface.)

IV. True or False Questions and Responses

True or False

_____ 1. Kepler's 2nd law, the law of equal areas, implies that the force of gravity varies inversely with the square of the distance.

_____ 2. Kepler's 2nd law, the law of equal areas, follows from the fact that an orbiting planet's angular momentum about the sun is constant.

_____ 3. According to Kepler's 3rd law, the period of a planet's orbital motion varies as the 3/2 power of its orbital radius.

_____ 4. For the special case of a circular orbit, the inverse-square force law can be derived from Kepler's laws using Newton's laws of motion.

_____ 5. A confirmation of Newton's law of gravity is the fact that the acceleration of the moon in its orbit is the same as that of objects falling freely near the surface of the earth.

_____ 6. Cavendish's experiment provides a direct measurement of the mass of the earth, from which the gravitational constant G can be inferred.

_____ 7. The gravitational field at a point is defined as the gravitational force on a unit mass at that point.

_____ 8. The gravitational field is a scalar quantity.

_____ 9. The gravitational field due to a uniform spherical shell of matter is zero throughout the region $r < R_{inner}$ where R_{inner} is the inner radius of the shell.

_____ 10. The gravitational field due to a spherically symmetric distribution of matter is identical to that of a point mass only at distances from its center that are *very large* compared to its radius.

Responses to True or False

1. False. The law of equal areas follows directly from the conservation of angular momentum. It holds for any and all central forces, not just those that vary inversely with the square of the distance.

2. True.

3. True; $(\text{period})^2 \propto (\text{radius})^3$.

4. True.

5. False. The ratio of the moon's acceleration to the acceleration of objects falling freely at the surface of the earth is R_E^2 / r^2, where R_E is the radius of the earth and r is the radius of the moon's orbit. (This is the confirmation of Newton's law of gravity.)

6. False. It's the other way around. Cavendish's experiment provides a direct measurement of the gravitational constant G from which the mass of the earth can be inferred.

7. True.

8. False. It is the force per unit mass, a vector.

9. True.

10. False. The gravitational field due to a spherical mass distribution is identical to that of a point mass at any distance from its center that is *larger than* its radius. For a nonspherical mass distribution, the distribution's gravitational field approaches that of a point mass as the distance from its center becomes large compared to its "radius." The larger this distance, the less impact the details of the actual mass distribution have on the gravitational field.

V. Questions and Answers

Questions

1. Some communications satellites remain stationary over one point on the earth. How is this accomplished?

2. Must the satellite orbit described in Question 1 be circular?

3. The two satellites of the planet "Krypton" have near-circular orbits with diameters in the ratio 1.7 to 1. What is the ratio of their periods?

4. Estimate the force of gravity exerted upon you by a mountain 2-km high. (You'll have to make some simplifying assumptions about the shape of the mountain, its density, and so forth. See Table 13-1 on page 396 of the text for the densities of various materials.)

5. The earth's orbit isn't a perfect circle; the earth is a little closer to the sun in January than it is in July. How can you tell this from the apparent motion of the objects in the heavens?

6. The drag force of the atmosphere on an orbiting satellite has a tendency to make the satellite's orbit more nearly circular. Why?

7. What does it mean to say that an astronaut in a satellite orbiting the earth is "weightless"?

Answers

1. If a satellite is in a circular equatorial orbit with a period equal to that of the earth's rotation, it will appear to be stationary relative to a single spot on the ground. Such an orbit is called a *geosynchronous orbit*. The altitude of the required orbit, which can be found using Kepler's 3rd law ($T^2 / r^3 = $ constant) using the satellite's orbital data and the radius of the earth, works out to be about 35,800 km, or 22,400 mi.

2. Yes, the orbit must be circular if the satellite is to appear truly stationary. If the satellite were in an elliptical orbit with a period equal to that of the earth's rotation, its angular speed would be faster than the earth's rotation rate part of the time and slower at other times—in accord with Kepler's law of equal areas.

3. By Kepler's 3rd law, the period T of an orbit is proportional to $R^{3/2}$, where R is the radius of the orbit. Thus the ratio of the periods is $(1.7)^{3/2}$ to 1, or 2.22 to 1.

4. The model I used for the mountain had the mass of a cone 2 km high and 10 km in diameter at the base. I assumed that I was standing at the base of the cone and that the gravity of the mountain acted as though from a point one-third of the way up its axis. These thoroughly arbitrary assumptions gave me a force of 2×10^{-4} N , which is about 2×10^{-7} times my weight. You may get something quite different, depending on the assumptions you make, but it will be small relative to your weight. Gravity is a weak force unless the mass of one of the objects is astronomically large!

5. From Kepler's 2nd law we know the line joining the earth and the sun sweeps out equal areas in equal times. In January, when this line is shortest, it sweeps out angles the fastest. It follows that viewed in January from the earth, the sun's apparent motion against the background of the "fixed" stars is greater than it is in July. This can be observed by comparing the sidereal day and the solar day. (The sidereal day is the time for each "fixed" star to reach its zenith on consecutive days, whereas the solar day is the time for the sun to reach its zenith on consecutive days. The zenith is the highest point reached.)

6. There is more residual atmosphere at lower altitudes. Thus, most of the effect of atmospheric drag on the satellite occurs when it is near perigee (its point of closest approach to the earth). Every time it comes around to perigee, it loses a little speed. As a result, it doesn't climb as far away from the Earth during the next orbit, so its path becomes a little more nearly circular each time.

7. Properly speaking, the astronaut isn't weightless since weight is the force that gravity exerts on him or her. The earth's gravity hasn't gone away. However, both the astronaut and scale are in free fall, so the scale exerts no net force against the astronaut and his or her apparent weight is zero.

VI. Problems, Solutions, and Answers

Example #1. An astronaut has landed on a 7940-km-diameter planet in another solar system. The astronaut drops a stone from the port of his landing craft and observes that it takes 1.78 s to fall the

4.50-m distance to the planet's surface. Using this data she calculates the mass of the planet. What value does she obtain?

Picture the Problem. From the distance the stone falls and the fall time, find the local free-fall acceleration. Use Newton's law of gravity to calculate its mass.

1. Calculate the local free-fall acceleration. We let the distance the stone falls be d.	$y - y_0 = v_{0y}t + \frac{1}{2}a_y t^2$ $d = \frac{1}{2}a_y t^2$
2. Apply Newton's 2nd Law to the stone, using Newton's law of gravity as the applied force. Let M be the mass of the planet, R its radius, and m the mass of the stone. The force and acceleration are in the same direction.	$\sum F = ma$ $\dfrac{GMm}{R^2} = ma$ $a = \dfrac{GM}{R^2}$
3. Substituting the acceleration from step 2 into the equation from step 1, the mass of the planet can be calculated.	$d = \dfrac{1}{2}\dfrac{GM}{R^2}t^2$ $M = \dfrac{2dR^2}{Gt^2} = \dfrac{2(4.5\,\text{m})(3.97\times10^6\,\text{m})^2}{(6.67\times10^{-11}\,\text{N}\cdot\text{m}^2/\text{kg}^2)(1.78\text{s})^2}$ $= 6.71\times10^{23}\,\text{kg}$

Example #2—Interactive. The moons of Mars are Deimos and Phobos ("Terror" and "Fear"). Deimos's orbital radius is 14,600 mi with a period of 30.0 hr, and Phobos's orbital period is 7.70 hr. (*a*) What is the radius of the orbit of Phobos? (*b*) What is the mass of Mars?

Picture the Problem. Since you know the radius of one orbit and both orbital periods, you can get the radius of Phobos's orbit using Kepler's 3rd law ($T^2/r^3 = \text{constant}$). To calculate the mass of Mars use Kepler's 3rd law again. Remember to convert units before calculating the mass of Mars. **Try it yourself.** Work the problem on your own, in the spaces provided, to get the final answer.

1. Use Kepler's 3rd law to find the radius of Phobos's orbit.	
	$r = 5900\,\text{mi}$
2. Knowing the mass of one planet and its orbital radius, calculate the mass of Mars.	
	$M = 6.59\times10^{23}\,\text{kg}$

Example #3. The average distance of Saturn from the sun is 1.40×10^9 km. Assuming a circular orbit, find Saturn's orbital speed in meters per second.

Picture the Problem. If Saturn's orbit is circular, then its orbital speed is constant. The orbital speed of Saturn will be its orbital circumference divided by its period. Determine the distance Saturn travels in one orbital period. Using Saturn's orbital radius and Earth's radius and period, calculate Saturn's orbital period using Kepler's 3rd law.

1. Write an equation for Saturn's speed as a guide for the rest of the problem.	$v = d/t$
2. Determine the distance Saturn travels in one period.	$d = 2\pi r = 2\pi \left(1.40 \times 10^9 \text{ km} \right) = 8.79 \times 10^9 \text{ km}$
3. Using Kepler's 3rd law, determine Saturn's orbital period.	$\dfrac{T_S^2}{R_S^3} = \dfrac{T_E^2}{R_E^3}$ $T_S = \left(\dfrac{R_S}{R_E} \right)^{3/2} T_E = \left(\dfrac{1.40 \times 10^9 \text{ km}}{1.5 \times 10^8 \text{ km}} \right) (1 \text{ yr})$ $T_S = 28.5 \text{ yr}$
4. Substitute the results from steps 2 and 3 into step 1 to find the orbital speed.	$v = \dfrac{8.79 \times 10^9 \text{ km}}{28.5 \text{ yr}} \dfrac{1 \text{ yr}}{3.16 \times 10 \text{ s/yr}} \dfrac{1000 \text{ m}}{\text{km}} = 9760 \text{ m/s}$

Example #4—Interactive. The planet Jupiter takes 11.9 years to orbit the sun. What is the radius of Jupiter's orbit? **Try it yourself.** Work the problem on your own, in the spaces provided, to get the final answer.

1. Knowing Earth's orbital period and radius, use Kepler's 3rd law to find the radius of Jupiter's orbit.	
	$r = 7.82 \times 10^{11} \text{ m} = 5.21 \text{ AU}$

Example #5. What is the escape speed from the surface of the moon? Take the mass of the moon to be 7.40×10^{22} kg and its radius to be 1.74×10^6 m.

1. Use conservation of energy. When the object is at the surface, the system has both kinetic and gravitational potential energy. If the object barely "escapes," the system has no kinetic or gravitational potential energy at the ending point.	$K_1 + U_1 = K_2 + U_2$ $\frac{1}{2}mv_e^2 - \frac{GM_M m}{R_M^2} = 0 + 0$ $v_e = \sqrt{\frac{GM_M}{R_M^2}} = \sqrt{\frac{\left(6.67\times10^{-11}\ \text{N·m}^2/\text{kg}^2\right)\left(7.40\times10^{22}\ \text{kg}\right)}{\left(1.74\times10^6\ \text{m}\right)^2}}$ $v_e = 2.38\times10^3\ \text{m/s}$

Example #6. If the resistance of the atmosphere is neglected, (*a*) at what speed would an object have to be projected directly upward from the earth's surface in order to reach an altitude of 200 km? (*b*) Once it reaches 200 km, how much more energy would an 800-kg object have to be given in order to be put into a circular orbit at this altitude?

Picture the Problem. Mechanical energy of the earth-object system is conserved because air resistance is negligible. For part (*a*), the object needs enough kinetic energy to reach the 200 km altitude. For part (*b*), the object needs to have enough kinetic energy added to maintain the appropriate orbital speed.

1. Use conservation of energy to determine the initial speed required for part (*a*).	$K_1 + U_1 = K_2 + U_2$ $\frac{1}{2}mv^2 - \frac{GM_E m}{R_E} = 0 - \frac{GM_E m}{R_E + \left(200\,\text{km}\right)}$ $v = \sqrt{\frac{2GM_E}{R_E} - \frac{2GM_E}{R_E + \left(200\,\text{km}\right)}} = \sqrt{\frac{2GM_E\left(200\,\text{km}\right)}{R_E\left(R_E + \left(200\,\text{km}\right)\right)}}$ $v = \sqrt{\frac{2\left(6.67\times10^{-11}\ \text{N·m}^2/\text{kg}^2\right)\left(5.98\times10^{24}\ \text{kg}\right)\left(2.00\times10^5\ \text{m}\right)}{\left(6.37\times10^6\ \text{m}\right)\left(\left(6.37\times10^6\ \text{m}\right)+\left(2.00\times10^5\ \text{m}\right)\right)}}$ $v = 1950\,\text{m/s}$
2. Find the speed required for a satellite in a circular orbit using Newton's 2nd law.	$\sum F = \frac{mv^2}{r}$ $\frac{GM_E m}{r^2} = \frac{mv^2}{r}$ $v^2 = \frac{GM_E}{r}$
3. The required additional energy is the kinetic energy needed to maintain the speed calculated in step 2.	$K = \frac{1}{2}mv^2 = \frac{GM_E m}{2r} = \frac{GM_E m}{2\left(R_E + 200\,\text{km}\right)}$ $= \left(\frac{\left(6.67\times10^{-11}\ \text{N·m}^2/\text{kg}^2\right)\left(5.98\times10^{24}\ \text{kg}\right)\left(800\,\text{kg}\right)}{2\left(\left(6.37\times10^6\ \text{m}\right)+\left(2.00\times10^5\ \text{m}\right)\right)}\right)$ $= 2.43\times10^{10}\ \text{J}$

Example #7—Interactive. An 1800-kg earth satellite is to be placed in a circular orbit at an altitude of 150 km above the earth's surface. (*a*) How much work must be done on the satellite to accomplish this? Neglect the resistance of the atmosphere. (*b*) How much of the work goes into increasing the potential energy of the satellite and how much into increasing its kinetic energy?

Picture the Problem. The work that must be done on the satellite is equal to the total change in the earth-satellite system's mechanical energy because air resistance is negligible. The change in the satellite's potential energy is due to its increased height, as in the previous problem. The change in its kinetic energy corresponds to its orbital speed. This problem is easiest if part (*b*) is done first. **Try it yourself.** Work the problem on your own, in the spaces provided, to get the final answer.

1. Knowing the change in the height of the satellite, find the increase in its potential energy.	$\Delta U = 2.6 \times 10^9 \text{ J}$
2. Find the increase in the satellite's kinetic energy. Assuming it starts at rest, its initial kinetic energy is zero. Its final kinetic energy is the kinetic energy required to maintain its orbital speed.	$\Delta K = 5.51 \times 10^{10} \text{ J}$
3. The total work done must be equal to the sum of the increases in kinetic and potential energy.	$W = 5.77 \times 10^{10} \text{ J}$

Example #8. A research satellite is in an elliptical orbit around the earth. Its closest approach to the earth's surface (perigee) is 100 km, and its greatest distance (apogee) is 1600 km. Find its speed at each of these two points.

Picture the Problem. The only force acting on the satellite is the gravitational force exerted by the earth. This force is directed toward the earth's center, so we know that the torque exerted on the satellite about the earth's center equals zero. If follows that the satellite's angular momentum about the earth's center is conserved. In addition, because the gravitational force is conservative, we know that the mechanical energy of the earth-satellite system is conserved. Equate expressions for the mechanical energy of the system at apogee and perigee to obtain a second equation. Turn the mathematical crank.

1. Draw a picture to help visualize the problem.	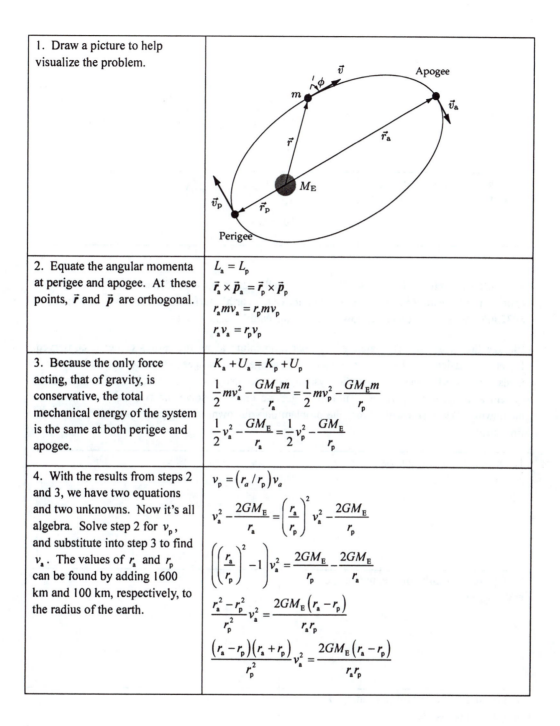
2. Equate the angular momenta at perigee and apogee. At these points, $\vec{r}$ and $\vec{p}$ are orthogonal.	$L_a = L_p$ $\vec{r}_a \times \vec{p}_a = \vec{r}_p \times \vec{p}_p$ $r_a m v_a = r_p m v_p$ $r_a v_a = r_p v_p$
3. Because the only force acting, that of gravity, is conservative, the total mechanical energy of the system is the same at both perigee and apogee.	$K_a + U_a = K_p + U_p$ $\dfrac{1}{2} m v_a^2 - \dfrac{GM_E m}{r_a} = \dfrac{1}{2} m v_p^2 - \dfrac{GM_E m}{r_p}$ $\dfrac{1}{2} v_a^2 - \dfrac{GM_E}{r_a} = \dfrac{1}{2} v_p^2 - \dfrac{GM_E}{r_p}$
4. With the results from steps 2 and 3, we have two equations and two unknowns. Now it's all algebra. Solve step 2 for v_p, and substitute into step 3 to find v_a. The values of r_a and r_p can be found by adding 1600 km and 100 km, respectively, to the radius of the earth.	$v_p = \left(r_a / r_p \right) v_a$ $v_a^2 - \dfrac{2GM_E}{r_a} = \left(\dfrac{r_a}{r_p} \right)^2 v_a^2 - \dfrac{2GM_E}{r_p}$ $\left(\left(\dfrac{r_a}{r_p} \right)^2 - 1 \right) v_a^2 = \dfrac{2GM_E}{r_p} - \dfrac{2GM_E}{r_a}$ $\dfrac{r_a^2 - r_p^2}{r_p^2} v_a^2 = \dfrac{2GM_E \left(r_a - r_p \right)}{r_a r_p}$ $\dfrac{\left(r_a - r_p \right)\left(r_a + r_p \right)}{r_p^2} v_a^2 = \dfrac{2GM_E \left(r_a - r_p \right)}{r_a r_p}$

	$v_a = \sqrt{\dfrac{2GM_E r_p}{r_a(r_a + r_p)}}$ $v_a = \sqrt{\dfrac{2(6.67 \times 10^{-11} \text{ N} \cdot \text{m}^2/\text{kg}^2)(5.98 \times 10^{24} \text{ kg})(6.47 \times 10^6 \text{ m})}{(7.97 \times 10^6 \text{ m})((6.47 \times 10^6 \text{ m}) + (7.97 \times 10^6 \text{ m}))}}$ $v_a = 6.70 \times 10^3 \text{ m/s}$
5. Substitute the result of step 4 back into step two to find v_p.	$v_p = \dfrac{7.97 \times 10^6 \text{ m}}{6.47 \times 10^6 \text{ m}}(6.70 \times 10^3 \text{ m/s})$ $= 8.25 \times 10^3 \text{ m/s}$

Example #9—Interactive. A satellite is in an elliptical orbit around the earth. At its closest approach to the earth (perigee) it is 112 km above the earth's surface and is moving at a speed of 8032 m/s. When it is at apogee, how far above the earth's surface is it?

Picture the Problem. The satellite's angular momentum about the earth's center is conserved. Equate expressions for the angular momentum at apogee and perigee. Because mechanical energy is also conserved in the earth-satellite system, you can also equate the mechanical energy of the system at apogee and perigee. Turn the algebraic crank to solve the two equations and two unknowns. **Try it yourself.** Work the problem on your own, in the spaces provided, to get the final answer.

1. Draw a picture to help visualize the problem.	
2. Equate the angular momentum at apogee and perigee.	
3. Equate the total mechanical energy at apogee and perigee.	

4. Turn the algebra crank.

$$r_a = 7.14 \times 10^6 \, \text{m} \, ; h = r_a - r_E = 7.71 \times 10^5 \, \text{m}$$

Chapter **12**

Static Equilibrium and Elasticity

I.   Key Ideas

If an object is stationary and remains stationary, it is said to be in **static equilibrium**. In this chapter we study the conditions necessary for static equilibrium to exist. The concept of center of gravity will be introduced, and the parameters affecting stability will be examined.

Section 12-1.   Conditions for Equilibrium. For an object to remain in static equilibrium, the acceleration of its center of mass must remain zero, which means that the net external force acting on the object must remain zero. However, the object can rotate even when the net external force remains zero. For an object to be in static equilibrium, the net external torque on the object about any point must also be zero. Thus the two necessary conditions for an object to be in static equilibrium are

$$\sum \vec{F}_{ext} = 0$$
$$\sum \vec{\tau}_{ext} = 0 \qquad \text{(about any point)}$$

Conditions for equilibrium

Section 12-2.   The Center of Gravity. In a gravitational field, a gravitational force is exerted on each constituent particle of an object. The vector sum of this collection of forces is considered a single force, the object's weight. In order to calculate the net torque exerted by the object's weight, the weight is best thought of as acting at a single point, the **center of gravity**. The center of gravity is defined so that the torque produced by the resultant weight force about any point equals the sum of the torques produced by the many individual gravitational forces about the same point. The x coordinate X_{cg} of the center of gravity of the object is related to the x coordinates of the particles that make up the object according to the equation

$$WX_{cg} = \sum_i w_i x_i = \sum_i m_i g x_i$$

Center of gravity

Where w_i is the weight of the ith particle, x_i is the x coordinate of the position of the ith particle, and $W = \Sigma w_i$. The center of mass is defined as $MX_{cm} = \Sigma m_i x_i$. If the gravitational field $\vec{g}$ is uniform then the above equation can be simplified to $WX_{cg} = g\Sigma m_i x_i = g\Sigma m_i x_i = gMX_{cm}$. Since $W = Mg$ in this case, we have

$$X_{cg} = X_{cm}$$ Centers of mass and gravity in uniform gravitational field

That is, in a uniform gravitational field the center of gravity coincides with the center of mass.

When calculating the torque about a point due to the force of gravity acting on an object we consider the weight as a single force that is applied at the object's center of gravity.

Section 12-3. Some Examples of Static Equilibrium. See Problems 1, 3, 5, and 7 and their solutions at the end of this chapter.

Section 12-4. Couples. A **couple** consists of a pair of oppositely directed forces of equal magnitude acting along different lines of action. Although the torque exerted by a single force depends on the axis about which the torque is calculated, the torque exerted by a couple is, perhaps surprisingly, independent of the point about which it is calculated. The magnitude of the torque exerted by a couple is

$$\tau = FD$$ Torque exerted by a couple

Where F is the magnitude of one of the forces and D is the perpendicular distance between the lines of action of the two forces.

Section 12-5. Static Equilibrium in an Accelerated Frame. If the forces on an object are analyzed in its rest frame (a frame of reference in which that object is at rest), and if the rest frame is an inertial reference frame, then the conditions for static equilibrium stated in Section 12-1 apply. Interestingly, if the rest frame is a noninertial reference frame, then the forces can be analyzed by a modified version of the conditions stated in Section 12-1. These modified conditions are:

$$\sum \vec{F} = m\vec{a}_{cm}$$
$$\sum \vec{\tau}_{cm} = 0$$ Conditions for static equilibrium in an accelerated frame

where $\vec{a}_{cm}$ is the acceleration of the center of mass in a noninertial reference frame. The sum of the torques *about the center of mass* must be zero. This second condition restricts us to computing torques only about axes passing through the center of mass.

Section 12-6. Stability of Rotational Equilibrium. An object is in **stable rotational equilibrium** if small angular displacements of the object away from equilibrium result in a torque (or torques) that tends to return the object to its equilibrium position. For example, the pendulum of a tall clock when it is at rest hanging vertically is in stable rotational equilibrium. If rotated slightly from the vertical and released, the torque due to the gravitational force will tend to rotate it back to vertical.

An object is in **unstable rotational equilibrium** if small angular displacements of the object away from equilibrium result in torques that tend to rotate the object further away from the equilibrium position. An example of an object in unstable rotational equilibrium is a pencil balanced on its point. If rotated slightly from the vertical and released, the torque due to the gravitational force will tend to rotate it further from the vertical.

An object is in **neutral rotational equilibrium** if both the net force and the net torque remain zero following small linear and/or angular displacements of the object from an equilibrium position. A meter stick suspended by a nail through a small hole drilled through its center of mass and driven into a wall is an example. The torque due to the gravitational force is zero, so if the stick is rotated slightly from an equilibrium position and released, there is no torque to either return it to or move it farther from its initial position.

When you rotate something slightly, if the potential energy increases, remains the same, or decreases, then the rotational equilibrium is stable, neutral, or unstable, respectively.

Section 12-7.   Indeterminate Problems.   The conditions for static equilibrium consist of two vector equations $\Sigma \vec{F} = 0$ and $\Sigma \vec{\tau} = 0$. By taking x, y, and z components of each of these we end up with six component equations: $\Sigma F_x = 0$, $\Sigma F_y = 0$, $\Sigma F_z = 0$ and $\Sigma \tau_x = 0$, $\Sigma \tau_y = 0$, $\Sigma \tau_z = 0$. However, with these six equations we can solve for, at most, six unknowns. Any problem that has more unknowns than available (independent) equations cannot be solved and is said to be indeterminate. However, for static equilibrium problems that concern a deformed object, additional equations can often be obtained by relating the deformations to the forces acting on it. Such equations are introduced in the following section.

Section 12-8.   Stress and Strain.   Most solids behave elastically. If stretched or deformed by external forces, they tend to return to their original size and shape when the deforming stress is removed. For a given material, this is true up to some **elastic limit;** exceeding this limit will cause permanent deformation.

If an applied force tends to stretch an object, the force per unit area perpendicular to the force is called the **tensile stress:**

$$\text{Stress} = F / A \hspace{4cm} \text{Tensile stress}$$

The resulting fractional change in the object's length is the **tensile strain:**

$$\text{Strain} = \Delta L / L \hspace{4cm} \text{Tensile strain}$$

Up to the proportional limit—a point somewhat lower than the elastic limit for a substance—stress is proportional to strain. The constant of proportionality, which is the ratio of stress to strain, is called **Young's modulus** Y:

$$Y = \frac{\text{stress}}{\text{strain}} = \frac{F / A}{\Delta L / L} \hspace{4cm} \text{Young's modulus}$$

The maximum tensile stress a material can withstand is called its **tensile strength.** Young's modulus for both the tensile stress and the tensile strength of a given material may or may not be the same as Young's modulus for the compressive stress and the compressive strength of that material.

A sideways deformation resulting from forces that act parallel to the surface of a material is called a **shear strain.** See Figure 12-1. Within the proportional limit of a material, **shear stress,** shear force per unit area, is proportional to **shear strain**. The constant of proportionality is the **shear modulus** M_s :

$$M_s = \frac{\text{shear stress}}{\text{shear strain}} = \frac{F_s / A}{\Delta X / L} = \frac{F_s / A}{\tan \theta} \qquad\qquad \text{Shear modulus}$$

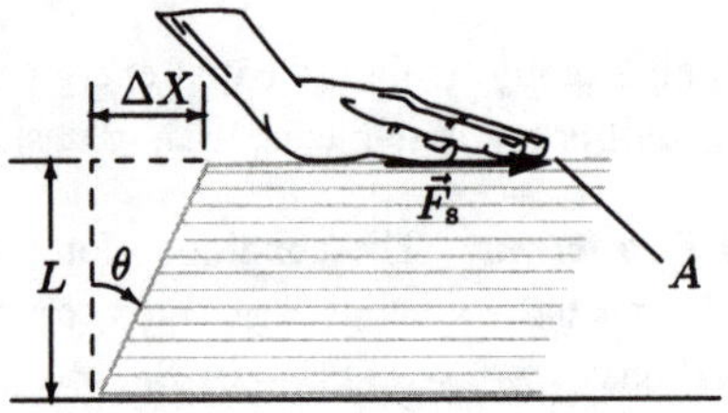

Figure 12-1

II.   Physical Quantities and Key Equations

Physical Quantities

Stress	N/m^2
Young's Modulus	N/m^2
Shear modulus	N/m^2

Key Equations

Conditions for equilibrium

$$\sum \vec{F}_{\text{ext}} = 0$$
$$\sum \vec{\tau}_{\text{ext}} = 0 \qquad \text{(about any point)}$$

Conditions for static equilibrium in an accelerated frame

$$\sum \vec{F} = m\vec{a}_{\text{cm}}$$
$$\sum \vec{\tau}_{\text{cm}} = 0$$

Center of gravity

$$WX_{\text{cg}} = \sum_i w_i x_i = \sum_i m_i g x_i$$

Torque exerted by a couple

$$\tau = FD$$

Tensile stress

$$\text{Stress} = F / A$$

Tensile strain

$$\text{Strain} = \Delta L / L$$

Young's Modulus

$$Y = \frac{\text{stress}}{\text{strain}} = \frac{F / A}{\Delta L / L}$$

Shear modulus

$$M_s = \frac{\text{shear stress}}{\text{shear strain}} = \frac{F_s / A}{\Delta x / L} = \frac{F_s / A}{\tan \theta}$$

III. Potential Pitfalls

Do not just draw force vectors anywhere. When drawing free-body diagrams, the line of action of a force must pass through the point of application of the force. The effective point of application of the gravitational force acting on an object is the object's center of gravity.

For an object to be in equilibrium, the net external torque must be zero about *any* axis.

For a particular problem, we are free to select the axis about which we calculate torques. Some axis choices are better than others because they result in simpler equations that are more easily solved. A useful rule of thumb is to choose an axis that passes through the point of application of the force that is the least specified. A force is completely specified when its point of application and both its magnitude and direction, or its components, are known.

Do not let signs confuse you. To determine whether the sign of a torque $\vec{\tau}$ associated with a force $\vec{F}$ is positive (counterclockwise) or negative (clockwise), imagine that the object is constrained to rotate about the selected axis and that $\vec{F}$ is the only force acting on it. The torque due to $\vec{F}$ is positive if the object would rotate in the positive direction (counterclockwise) and negative if it would rotate in the negative direction (clockwise). An alternative way to determine the sign of the torque $\vec{\tau}$ is to determine the direction of $\vec{r} \times \vec{F}$ where $\vec{r}$ is the vector from the axis to the point of application of the force. (The direction of a cross product is obtained by applying the right-hand rule.) If the direction of $\vec{r} \times \vec{F}$ is up out of the paper then $\vec{\tau}$ is positive (counterclockwise), and if the direction is into the paper, then it is negative.

Do not think that the elastic moduli are always constants. They are constants only for stresses that are less than the proportional limits of a material. They vary considerably for stresses approaching the elastic limit or the tensile or compressive strength of the material.

IV. True or False Questions and Responses

True or False

_____ 1. When an object is in static equilibrium, the net external force acting on it must equal zero.

_____ 2. When an object is in static equilibrium, the net external torque about *any* axis must equal zero.

_____ 3. The lever arm of a force about a point P always equals the distance from P to the point of application of the force.

_____ 4. The torque exerted by a couple about a point P is independent of the location of P, unless P is located between the lines of action of the two forces.

_____ 5. Strain has dimensions of length and the SI unit for it is the meter.

_____ 6. Young's modulus and the shear modulus have the same dimensions.

Responses to True or False

1. True.

2. True.

3. Not necessarily. The lever arm equals the perpendicular distance from P to the line of action of the force.

4. False. The torque exerted by a couple about a point P is always independent of the location of P.

5. False. Strain is dimensionless.

6. True. They have dimension of $[\text{M}]/([\text{L}][\text{T}]^2)$ and SI units of N/m^2.

V. Questions and Answers

Questions

1. Must the center of gravity of an object be located inside the material of the object?

2. If exactly three external forces act on an object in equilibrium and if two of their lines of action intersect, must the third line of action intersect at the same point?

Answers

1. No. For example, the center of gravity of a bowl is outside the material of the bowl.

2. Yes, if a point is on the line of action of a force, then the force's torque about that point must equal zero. Conversely, if the torque of a force about a point is zero, then the point must lie on the force's line of action. (Convince yourself that these assertions are true. Make some drawings, etc.) The intersection of point P of the lines of action of two of the forces lies on both lines, so the torques of both forces about P each equal zero. Since the sum of all three torques about P must equal zero, and the torques of two of the forces are both zero, the torque of the third force must also equal zero. It follows that the line of action of the third force must also pass through P.

VI. Problems, Solutions, and Answers

Example #1. A uniform, 0.100-kg meter stick is suspended by a pivot at the 50.0-cm mark. As shown in Figure 12-2, the system is balanced when a 0.300-kg mass is suspended from the 20.0-cm mark and an unknown mass *m* is suspended from the 70.0-cm mark. Determine the mass *m* and the force exerted on the stick by the pivot.

Picture the Problem. Sketch a free-body diagram of the stick. Because the suspended masses are each in equilibrium, each exerts a force equal to its weight on the stick. The weight of the stick is applied at the stick's center of gravity. Because the stick is uniform, the center of gravity of the stick is at the 50.0-cm mark. Because the stick is in equilibrium, both the net force and the net torque on the stick are equal to zero. To simplify the torque equations, choose the axis to be the point of application of the least-defined force, the force of the pivot on the meterstick. Use these two mathematical relationships to solve for the two unknown forces.

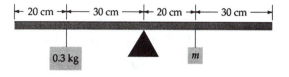

Figure 12-2

1. Draw a free-body diagram of the stick. There are four forces acting on it: the force of gravity, $\vec{w}_{ms}$, the contact force exerted by the pivot, $\vec{F}$, and the weight of the two suspended masses, $\vec{w}_1$ and $\vec{w}$. Actually, $\vec{w}_1$, the weight of the 3-kg mass, acts on the mass, not the meterstick. There are two forces acting on the 3-kg mass, the force exerted by the string and $\vec{w}_1$, the force exerted by gravity. Because the 3-kg mass is in equilibrium, we know these forces are equal and opposite. Because the force exerted by the meterstick on the string attached to the 3-kg mass and the force exerted by the string on the meterstick form an action-reaction pair, we know that these forces must be equal in magnitude and oppositely directed. Thus we conclude the force exerted by the string on the meter stick is equal to $\vec{w}_1$. Neither the magnitude nor the direction of $\vec{F}$ is known, but we do know its line of action passes through the axis.	
2. Sum the torques to zero, using the pivot point as the axis. Neither $\vec{F}$ nor $\vec{w}_{ms}$ will then have a lever arm. This equation can then be solved for the unknown mass m.	$\sum \tau = w_1 x_1 - w x_2 + w_{ms}(0) + F(0) = 0$ $m_1 g x_1 = m g x_2$ $m = \dfrac{m_1 x_1}{x_2} = \dfrac{(0.3\,\text{kg})(0.3\,\text{m})}{0.2\,\text{m}} = 0.45\,\text{kg}$

3. Sum the forces to zero, and solve for the force $\vec{F}$.	$\sum \vec{F} = \vec{F} + \vec{w}_1 + \vec{w} + \vec{w}_{ms} = 0$ $\vec{F} = -\left(\vec{w}_1 + \vec{w} + \vec{w}_{ms} \right)$ $\vec{F} = -g\left(m_1 + m + m_{ms} \right) \hat{j}$ $\vec{F} = \left(9.81 \,\text{m/s}^2 \right)\left(0.3\,\text{kg} + 0.45\,\text{kg} + 0.1\,\text{kg} \right) \hat{j}$ $\vec{F} = 8.34\,\text{N}\,\hat{j}$

Example #2—Interactive.   A uniform 0.300-kg meter stick is supported by a hinge and a vertical string. A 0.600-kg mass is supported by the stick, as shown in Figure 12-3. Determine the tension in the string and the force exerted by the hinge on the stick.

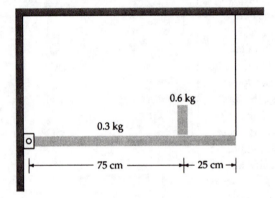

0.6 kg

0.3 kg

|←——————— 75 cm ———————|←— 25 cm —→|

Figure 12-3

Picture the Problem. Sketch a free-body diagram of the stick. There are four forces acting on it, the force of gravity and the contact forces exerted by the pivot, the string, and the mass. Because the 0.6-kg mass is in equilibrium, it exerts a force on the stick equal to its weight. The weight of the stick is applied at the stick's center of gravity. Because the stick is uniform, the center of gravity of the stick is 50 cm from either end. Because the stick is in equilibrium, both the net force and the net torque on the stick are equal to zero. To simplify the torque equations, choose the axis to be the point of application of the least-defined force, the force of the hinge on the meterstick. Sum the forces and torques to zero to solve for the two unknown forces. **Try it yourself.** Work the problem on your own, in the spaces provided, to get the final answer.

1. Draw a free-body diagram of the stick.	

2. Set the sum of the torques equal to zero and solve for the tension of the string.	$T = 5.89\,\text{N}$
3. Set the sum of the forces equal to zero to solve for the force of the hinge on the stick.	$\vec{F} = 2.94\,\text{N}$, upward

Example #3. A seesaw consists of a uniform plank, 4.00 m long, with a mass of 10.0 kg. As shown in Figure 12-4, a 40.0-kg boy sits 0.500 m from the short end of the seesaw and a girl of mass m_g sits 0.500 m from the other end. The plank is balanced when the pivot is located 25.0 cm from the center of the plank. Determine the mass of the girl.

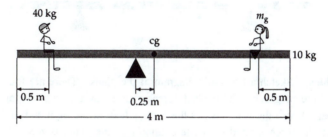

Figure 12-4

Picture the Problem. Draw a free-body diagram of the plank. Because the children are each in equilibrium, each child pushes on the plank with a force equal to the weight of that child. The weight of the plank is applied at the plank's center of gravity. Because the plank is uniform its center of gravity is at its geometric center. Select an axis that passes through the point of application of the force that is the least specified. Because we know neither its magnitude nor its direction, the least specified force is that exerted by the pivot. Because the stick is in equilibrium, both the sum of the forces and the sum of the torques are equal to zero.

1. Draw a free-body diagram of the plank.	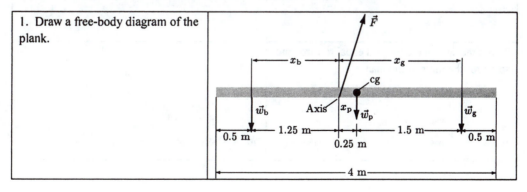

2. Set the torques equal to zero, and solve for the mass of the girl.	$\sum \tau = w_b x_b - w_p x_p - w_g x_g + F(0) = 0$ $m_b g x_b - m_p g x_p - m_g g x_g = 0$ $m_g = \dfrac{m_b x_b - m_p x_p}{x_g}$ $m_g = \dfrac{(40\,\text{kg})(1.25\,\text{m}) - (10\,\text{kg})(0.25\,\text{m})}{1.75\,\text{m}} = 27.1\,\text{kg}$

Example #4 –Interactive. A uniform 30.0-kg plank, 10.0 ft long, is supported at points 2.00 ft from each end as shown in Figure 12-5. A man walks along the plank toward the right. When he reaches a point 1.00 ft from the right end of the plank, the plank starts to tilt. Determine the mass of the man.

Figure 12-5

Picture the Problem. Sketch a free-body diagram of the plank. There are four forces acting on the plank: The force of gravity, the contact forces exerted by the two supports, and the force exerted by the man. When the man reaches the point 1 ft from the right end of the plank and the plank starts to tilt, the force exerted by the left support on the plank is zero. Select an axis that passes through the point of application of the force that is the least specified. Because we know neither its magnitude nor its direction, the least specified force is the force exerted by the right support. As the man approaches the 1-ft mark, the plank remains in equilibrium, which means the sum of the forces and the sum of the torques are both equal to zero. **Try it yourself.** Work the problem on your own, in the spaces provided, to get the final answer.

1. Sketch a free-body diagram of the plank.	
2. Set the sum of the torques equal to zero, and solve for the mass of the man.	$M = 90\,\text{kg}$

Example #5. A uniform 20.0-kg strut supporting a 30.0-kg object is suspended by a string and a hinge as shown in Figure 12-6. Determine the tension in the string and the force exerted on the strut by the hinge.

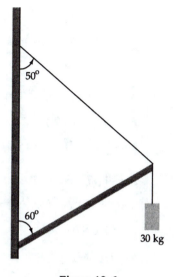

Figure 12-6

Picture the Problem. Draw a free-body diagram of the strut. Select an axis that passes through the point of application of the force that is the least specified. In this case, the least specified force is the force exerted by the hinge. We know neither its magnitude nor its direction. Indicate the lever arm for each force on your diagram. Remember that the lever arm is the perpendicular distance from the axis to the line of action of the force. Because the strut is in equilibrium, the sum of the forces and the sum of the torques are both equal to zero.

| 1. Draw a free-body diagram of the strut. There are four forces acting on the strut: the force of gravity, the force of the hinge, the force of the upper string, and the force of the lower string. Because the 30-kg mass is in equilibrium, the force of the lower string is equal to the weight of the 30-kg mass. The strut makes an angle of 60° with the y axis and thus makes an angle of 30° with the x axis. Because the sum of the angles in a triangle equals 180°, we know the angle between the strut and the string supporting the strut is 70°. | |

2. Set the sum of the torques equal to zero. Solve for the tension in the support string.	$\sum \tau = F(0) - w_s(L/2)\cos 30° - wL\cos 30° + TL\sin 70° = 0$ $m_s g(L/2)\cos 30° + mgL\cos 30° = TL\sin 70°$ $T = \dfrac{((m_s/2)+m)g\cos 30°}{\sin 70°}$ $T = \dfrac{(40\,\text{kg})(9.81\,\text{m/s}^2)\cos 30°}{\sin 70°} = 362\,\text{N}$
3. Set the sum of the forces equal to zero and solve for the force of the hinge on the strut. The tension has both x and y components that can be determined from the illustration to the right of the free-body diagram.	$\sum \vec{F}_{ext} = \vec{F} + \vec{w}_s + \vec{w} + \vec{T} = 0$ $\vec{F} = -(\vec{w}_s + \vec{w} + \vec{T})$ $\vec{T} = -T\sin 50°\hat{i} + T\cos 50°\hat{j}$ $\vec{F} = -\left[(m_s + m)\vec{g} + \vec{T}\right]$ $= T\sin 50°\hat{i} + \left[(m_s + m)g - T\cos 50°\right]\hat{j}$ $= (326\,\text{N})\sin 50°\hat{i}$ $\quad + \left[(50\,\text{kg})(9.81\,\text{m/s}^2) - (362\,\text{N})\cos 50°\right]\hat{j}$ $= (277\,\text{N})\hat{i} + (258\,\text{N})\hat{j}$

Example #6—Interactive. A sign is constructed from a uniform 15.0-kg sheet of plywood 80.0 cm high and 120 cm wide. As shown in Figure 12-7, the sign is suspended via a pin through its lower left corner and a horizontal cord attached to the upper left corner. Determine the tension in the cord and the force exerted by the pin on the sign.

Figure 12-7

Picture the Problem. Sketch a free-body diagram of the sign. Select an axis that passes through the point of application of the least specified force. This is the force exerted by the pin, since we know neither its magnitude nor its direction. Show the lever arm for each force on your diagram. Recall that the lever arm is the perpendicular distance from the axis to the line of action of the force. Because the sign is in equilibrium, the net force and net torque are equal to zero. **Try it yourself.** Work the problem on your own, in the spaces provided, to get the final answer.

1. Sketch a free-body diagram of the sign. There are three forces acting on the sign: the force of gravity, the force of the pin, and the force of the cord.	
2. Set the sum of the torques equal to zero, and solve for the tension in the cord.	$T = 110\,\text{N}$
3. Set the vector sum of the forces equal to zero and solve for the force exerted by the pin on the sign.	$\vec{F} = \left(110\,\text{N}\right)\hat{i} + \left(147\,\text{N}\right)\hat{j}$

Example #7. A thin 30.0-cm rod of nonuniform composition is supported in a horizontal position by a string that passes over two massless, frictionless pulleys as shown in Figure 12-8. The vertical segments of the string are attached to the rod at points 1.00 cm and 2.00 cm from the ends of the rod. Determine the location of the rod's center of gravity.

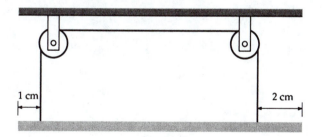

Figure 12-8

Picture the Problem. Sketch a free-body diagram of the rod. Select an axis that passes through the point of application of the least specified force. The weight is the least specified force, so choose the point of application of the weight force as the axis. Indicate the lever arm for each force on your diagram. The stick is in equilibrium, so the net force and the net torque acting on the stick are equal to zero. Because the pulleys are frictionless, we know that the tension T is the same in the two vertical string segments.

1. Sketch a free-body diagram of the rod. There are three forces acting on the rod: the force of gravity acting some distance x from its left end, and the tension forces exerted on the rod by the opposite ends of the string.	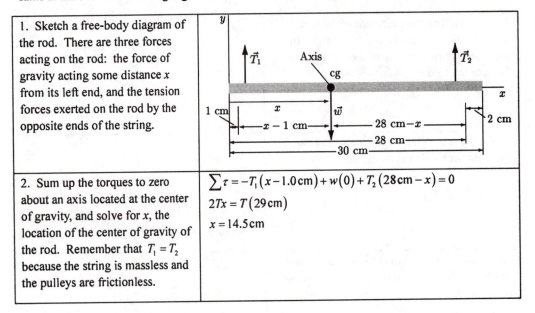
2. Sum up the torques to zero about an axis located at the center of gravity, and solve for x, the location of the center of gravity of the rod. Remember that $T_1 = T_2$ because the string is massless and the pulleys are frictionless.	$\sum \tau = -T_1(x-1.0\,\text{cm}) + w(0) + T_2(28\,\text{cm} - x) = 0$ $2Tx = T(29\,\text{cm})$ $x = 14.5\,\text{cm}$

Example #8—Interactive. A nonuniform meter stick is suspended from two light strings attached to the 10.0 and 90.0 cm marks. Masses of 68.4 g and 100 g are suspended from the strings as shown in Figure 12-9. The friction in the pulleys is negligible. Determine both the mass of the meter stick and the location of its center of gravity.

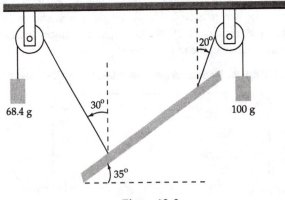

Figure 12-9

Picture the Problem. Draw a free-body diagram of the stick. Choose a point on the stick an arbitrary distance x from the left end of the stick as the center of gravity. Because the stick is in equilibrium, the net force and the net torque acting on the stick are equal to zero. Choose as a rotation axis the point of application of the gravitational force, since this force is the least specified. **Try it yourself.** Work the problem on your own, in the spaces provided, to get the final answer.

1. Draw a free-body diagram of the stick. There are three forces acting on the stick: the force of gravity and the forces exerted on the stick by the two strings.	
2. Set the vector sum of the forces equal to zero, and solve for the mass of the stick.	$m = 153.2\,\mathrm{g}$
3. Set the sum of the torques equal to zero and solve for x. Use the formula $|\vec{\tau}| = |\vec{r} \times \vec{F}| = rF\sin\phi$ where $\vec{r}$ is the vector from the axis to the point of application of the force and ϕ is the angle between the directions of $\vec{r}$ and $\vec{F}$. You may need to construct a diagram in which the angles and lengths are carefully drawn and labeled.	$x = 46.6\,\mathrm{cm}$

Example #9. A piece of tungsten wire 0.800 m long and a 0.500 m length of steel wire, each of diameter 1.00 mm, are joined end to end, making a total length of 1.30 m. If a mass of 22.0 kg is suspended from the ceiling by this vertical wire, how much does the wire stretch? Neglect the weight of the wire. (Young's modulus is 2.0×10^{11} N/m² for steel, 3.6×10^{11} N/m² for tungsten.)

Picture the Problem. Draw a sketch of the situation. The tensile stress is the same throughout both wires since they have the same diameter and are connected end to end. Obtain expressions for the strain in each wire using the appropriate Young's modulus, and add the extensions for each wire to obtain the net extension.

1. Draw a sketch of the situation.	
2. The stress and strain are related by Young's modulus. Use this relationship to find an expression for the extension of the two materials.	$Y_s = \dfrac{\text{stress}}{\text{strain}} = \dfrac{F/A}{\Delta L_s / L_s}$ $\Delta L_s = \dfrac{F}{A}\dfrac{L_s}{Y_s}$ $\Delta L_t = \dfrac{F}{A}\dfrac{L_t}{Y_t}$
3. The total extension of the wire will be equal to the sum of the extensions of the two materials. Both have the same force F exerted, as well as the same cross-sectional area A.	$\Delta L = \Delta L_s + \Delta L_t$ $= \dfrac{F}{A}\left(\dfrac{L_s}{Y_s} + \dfrac{L_t}{Y_t}\right) = \dfrac{4mg}{\pi d^2}\left(\dfrac{L_s}{Y_s} + \dfrac{L_t}{Y_t}\right)$ $= \dfrac{4(22\,\text{kg})(9.81\,\text{m/s}^2)}{\pi(1.0\times10^{-3}\,\text{m})^2}\left(\dfrac{0.5\,\text{m}}{2\times10^{11}\,\text{N/m}^2} + \dfrac{0.8\,\text{m}}{3.6\times10^{11}\,\text{N/m}^2}\right)$ $= 1.30\,\text{mm}$

Example #10—Interactive. If a mass of 40.0 kg is hung from a 2.00-m copper wire of diameter 1.50 mm, by how much does the wire stretch? Young's modulus for copper is $1.1\times10^{11}\,\text{N/m}^2$.

Picture the Problem. The change in length can be found by relating the stress, strain, and Young's modulus. **Try it yourself.** Work the problem on your own, in the spaces provided, to get the final answer.

1. Relate the stress and strain to Young's modulus.	
2. Using your equation from step 1, solve for the increased length.	$\Delta L = 4.03\,\text{mm}$

Chapter 13

Fluids

I. Key Ideas

Matter in bulk can be either solid or fluid. Solids are more or less rigid and tend to maintain a fixed shape. Fluids (liquids and gases) flow more or less freely and assume the shape of their container.

Section 13-1. Density. The **density** ρ of a substance is its mass per unit volume:

$$\rho = \frac{m}{V} \qquad \text{Density}$$

One cubic centimeter of water at 4°C has a mass of 1 gram (this was the original definition of the gram), so the density of water is 1 g/cm^3. An object with a density lower than that of water floats in water; an object of greater density sinks. The ratio of the density of a substance to that of water is the **specific gravity** of the substance. Solids and liquids have densities that are roughly independent of external conditions, but the density of a gas is strongly affected by its pressure and temperature. Thus, the temperature and pressure must be specified when stating the density of a gas. Standard conditions are atmospheric pressure at sea level and a temperature of 0°C.

Section 13-2. Pressure in a Fluid. The distinction between fluids and solids is that fluids are unable to support any shear stress and so conform to the shape of their container. The force per unit area exerted by a fluid is called the **pressure** P:

$$P = \frac{F}{A} \qquad \text{Pressure}$$

The unit of pressure in the SI system is the **pascal**. An increase in the pressure on a material tends to compress it in all directions at once. The ratio of the increase in pressure ΔP to the resulting fractional decrease in the volume $-\Delta V/V$ of the material is called the **bulk modulus** B:

$$B = -\frac{\Delta P}{\Delta V / V} \qquad \text{Bulk modulus}$$

The **compressibility** k of the material is the reciprocal of the bulk modulus:

$$k = \frac{1}{B}$$
Compressibility

The pressure at any point in a fluid is the same in all directions, so it is a scalar quantity. Also, it increases with increasing depth because of the weight of the fluid above. For a static fluid the pressure is the same at all points at the same depth (in any horizontal plane). Any change in pressure is transmitted undiminished throughout the fluid. This is a statement of **Pascal's principle.**

Liquids are highly incompressible. As a result, the pressure in a liquid increases linearly with depth:

$$P = P_0 + \rho g h$$
Pressure in a static liquid

where P_0 is the pressure at the top of the liquid and h is the depth below the top. Gases, on the other hand, are highly compressible, so they have densities that are approximately proportional to the pressure. Therefore, the pressure of the atmosphere decreases more or less exponentially with increasing altitude—that is, the pressure decreases by a constant fraction for a given increase in height.

Many pressure gauges measure the difference between the "absolute" pressure P and the atmospheric pressure P_{at}. This difference between absolute pressure and atmospheric pressure is called the gauge pressure P_{gauge}.

$$P_{gauge} = P - P_{at}$$
Gauge pressure

Section 13-3. Buoyancy and Archimedes' Principle. There is an upward, **buoyant force, $\vec{B}$,** on an object submerged in a fluid that is equal to the weight of the fluid displaced by the object. An object will float if it displaces a quantity of fluid equal to or greater in weight than the weight of the object. The specific gravity of an object can then be determined by the ratio of the weight of the object divided by the weight of an equal volume of water:

$$\text{Specific gravity} = \frac{\text{weight}}{\text{buoyant force when submerged in water}} = \frac{w}{B_{water}} \qquad \text{Specific Gravity}$$

If an object has a specific gravity equal to or greater than 1, then it will float in water.

Section 13-4. Fluids in Motion. The general flow of a fluid can be very complicated, so we restrict our study here to *steady, nonturbulent flow.* If the fluid is incompressible (a liquid), then the **volume flow rate** I_V (volume per unit time) in a tube or pipe is the same throughout its length. That is,

$$I_V = Av = \text{constant}$$
Continuity equation for incompressible flow

where A is the cross-sectional area of the pipe and v is the speed of the fluid. The mass flow rate equals ρI_V (mass per unit volume × volume per unit time). For the steady flow of both compressible and incompressible fluids, the mass flow rate is the same throughout the length of the tube. By applying the work–energy theorem and neglecting any dissipation of energy, we obtain **Bernoulli's equation,** the fundamental dynamic equation for the steady, nonviscous flow of an incompressible fluid:

$$P + \rho gh + \tfrac{1}{2}\rho v^2 = \text{constant} \qquad \text{Bernoulli's equation}$$

A pressure gradient is required to accelerate a nonviscous fluid; the acceleration is in the direction of decreasing pressure. Thus, in accordance with Newton's 2nd Law, a fluid speeds up when it flows into a region of lower pressure and slows down when it flows into a higher pressure region. These remarks apply to horizontal flow, for which the height h is a constant; we speak here of acceleration due to a pressure gradient, not that due to gravity. The pressure drop associated with the increase in the speed of a fluid is known as the **Venturi effect.**

Viscous Flow. Nonviscous flow is an idealization. When real fluids flow, an internal shear stress is produced that opposes the flow. This shear stress increases with the speed of the flow. This property of fluid flow is known as **viscosity.** Because of viscosity, a pressure gradient is required to cause a fluid to flow through a horizontal pipe at constant speed. The pressure drop, which occurs along the direction of flow, is proportional to the flow rate:

$$\Delta P = \frac{8\eta L}{\pi r^4} I_v \qquad \text{Poiseuille's law}$$

where L and r are the length and radius of the pipe and η, the **coefficient of viscosity,** is a measure of a fluid's resistance to flow.

Poiseuille's law and Bernoulli's equation apply only to steady, laminar (nonturbulent) flow. However, laminar flow ceases and turbulence sets in when the velocity of a fluid flowing through a tube becomes sufficiently great. The **Reynolds number** N_R is a dimensionless parameter that characterizes the degree of turbulence of the flow of a fluid. It is defined by

$$N_R = \frac{2r\rho v}{\eta} \qquad \textit{Reynolds number}$$

where r is the radius of the tube. The transition from laminar to turbulent flow occurs as the Reynolds number increases from 2000 to 3000.

II. Physical Quantities and Key Equations

Physical Quantities

Volume $\qquad\qquad\qquad\qquad 1\,\mathrm{L} = 10^3\,\mathrm{cm}^3 = 10^{-3}\,\mathrm{m}^3$

Pressure $\qquad\qquad\qquad\quad 1\ \text{pascal (Pa)} = 1\ \mathrm{N/m^2} = 1\ \mathrm{kg/(m\cdot s^2)}$

$$1\ \text{atmosphere (atm)} \begin{matrix} = 760\ \text{mmHg} \\ = 1.013\times10^5\ \text{Pa} \end{matrix}$$

$$1 \text{ bar} = 10^3 \text{ millibars (mbar)} = 10^5 \text{ Pa}$$

$$1 \text{ torr} = 1 \text{ mmHg} = 133.3 \text{ Pa}$$

Viscosity $1 \text{ poise} = 0.1 \text{ Pa} \cdot \text{s}$

Key Equations

Density $\rho = \dfrac{m}{V}$

Pressure $P = \dfrac{F}{A}$

Bulk modulus $B = -\dfrac{\Delta P}{\Delta V / V}$

Compressibility $k = \dfrac{1}{B}$

Gauge pressure $P_{\text{gauge}} = P - P_{\text{at}}$

Pressure in a static liquid $P = P_0 + \rho g h$

Continuity equation for incompressible flow $I_V = Av = \text{constant}$

Bernoulli's equation $P + \rho g h + \tfrac{1}{2} \rho v^2 = \text{constant}$

Poiseuille's law $\Delta P = \dfrac{8 \eta L}{\pi r^4} I_v$

Reynolds number $N_R = \dfrac{2 r \rho v}{\eta}$

III. Potential Pitfalls

It is easy to confuse density specific gravity. Density is the mass per unit volume of a substance. Contrary to the way it may sound, specific gravity is not defined in terms of weight. It is the ratio of the density of a substance to the density of water. It is thus a dimensionless number that is approximately numerically identical to the density expressed in grams per cubic centimeter.

Remember that the decrease of pressure with height (or the increase with depth) is linear only for a fluid that is incompressible and therefore has a constant density. The decrease of atmospheric pressure with height is not at all linear. This is because air is compressible so its density decreases with increasing height.

When solving problems that deal with the pressure at some depth in a fluid, remember that ρgh is the pressure difference over a vertical distance h. It is not necessarily the pressure at either the top or bottom.

The equilibrium condition for an object floating in a fluid is that the weight of the object is equal to the weight of the volume of fluid that the submerged portion of the object displaces. The volume of the fluid displaced is, of course, less than or equal to the volume of the object.

Just because an object sinks in a liquid, don't think that there is no buoyant force acting on it. The buoyant force is simply less than the object's weight.

Bernoulli's equation and the continuity equation (Av = constant) are applicable only for the flow of incompressible fluids, so they are of limited use for studying gas flow. The Bernoulli equation applies only for nonviscous laminar flow and Poiseuille's law applies only for laminar flow.

You can really mess up calculations with Bernoulli's equation if you are not careful to use consistent units for all the quantities. It is best to put everything into SI units.

When doing numerical problems, be sure to write out all units and determine the units of your answer. Use the conversions $1 \text{ Pa} = 1 \text{ N/m}^2$, $10 \text{ poise} = 1 \text{ Pa} \cdot \text{s} = 1 \text{ N} \cdot \text{s/m}^2$, and the like.

IV. True or False Questions and Responses

True or False

_____ 1. Fluids flow—changing their shape—freely, but they maintain a nearly constant volume.

_____ 2. Solids tend to be rigid (to maintain their shape and volume).

_____ 3. The bulk modulus and Young's modulus have the same dimensions.

_____ 4. The bulk modulus is the negative of the compressibility.

_____ 5. The density of a substance is the weight per unit volume of a substance.

_____ 6. The density of a substance can be expressed as the product of its specific gravity and the density of water.

_____ 7. An elastic material is one that is easily deformed.

_____ 8. The ratio of the increase in the pressure on a material to the fractional decrease in its volume is called its compressibility.

_____ 9. The pressure in a static fluid is the same at every point throughout the fluid.

_____ 10. The buoyant force on an object completely submerged in a fluid depends on the density of the fluid and the volume of the object but not on the shape or composition of the object.

_____ 11. A body that floats in a liquid does so at a depth at which it displaces an amount of fluid equal in weight to its own weight.

_____ 12. Among units of pressure, 1 bar is equal to 1 atmosphere.

_____ 13. Buoyancy arises from the increase in pressure with depth in a fluid.

_____ 14. Shearing forces cannot exist in a fluid.

_____ 15. Bernoulli's equation applies to any flow of an incompressible fluid.

_____ 16. Bernoulli's equation gives incorrect results when viscosity is significant.

_____ 17. Bernoulli's equation is derived by requiring that the total mechanical energy of an element of the fluid remains constant.

Responses to True or False

1. False. The statement is true for a liquid, but the term fluid also refers to gases.

2. True.

3. True.

4. False. The bulk modulus is the reciprocal of the compressibility.

5. False. Density is the mass per unit volume.

6. True. This is the definition of specific gravity.

7. False. Easily deformed materials are not necessarily elastic. What "elastic" means is that deformations of the material are reversible. Elastic materials can be stiff or spongy.

8. False. The statement describes the material's bulk modulus, which is the reciprocal of its compressibility.

9. False. The pressure varies with depth because of the weight of the fluid lying above. However, any change in pressure at one point in a fluid results in an equal change at all points in it.

10. True.

11. True.

12. Not exactly: 1 atm = 1.013 bar.

13. True.

14. False. A fluid cannot sustain a static shearing stress, but shearing stresses (called viscous forces) do exist in a real flowing fluid. They do not exist in an *ideal* fluid.

15. False. It applies only to the steady, nonturbulent flow of a nonviscous, incompressible fluid.

16. True.

17. False. It is derived by applying the work–energy theorem to account for the change in mechanical energy.

V. Questions and Answers

Questions

1. Assuming that you float in fresh water with about 5% of your body above water, estimate the volume of your body.

2. The "fog" produced by evaporating "dry ice" (frozen CO_2) stays at ground level and spreads along the ground. Why is this?

3. Figure 13-1 shows a small vessel upended in water, which is contained in a larger vessel, so as to trap some air inside. This gadget is known as a Cartesian diver. If the air pressure above the water is increased (for example, by pressing on a flexible lid on the larger container), the small vessel sinks to the bottom. Why does it sink? How can it be made to rise again?

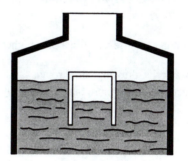

Figure 13-1

4. Once, in the course of an experiment, I needed to know the volume within a large, irregular, thin-walled aluminum container. The volume was of the order of 1 m³. Someone suggested the obvious, which was to fill the container with water from vessels of known volume. The only drawback was that a cubic meter of water weighs about a ton and would have deformed the thin aluminum walls of the container. Can you suggest a way to solve this problem?

5. A colleague of mine likes to do the following demonstration. He pours some red-dyed water into a glass U-tube that is open at both ends to show that the fluid "seeks its own level" (that is, it rises to the same height on both sides of the U-tube). He then pours in some more red liquid, and the fluid in the U-tube comes to rest as shown in Figure 13-2. What's the trick here?

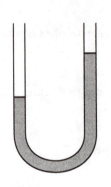

Figure 13-2

6. An ice cube floats in a glass of water. As the ice melts, does the water level rise or fall? Explain.

7. Aircraft carriers steam into the wind at an angle so that the planes take off headed directly into the wind. Why? In what direction should they head (into or with the wind) when the planes return? Assume that relative to the flight deck the planes travel in the same direction during recovery as during launch.

8. The fluid pressure in a pipe decreases as you go downstream, even when the pipe is level and the fluid is incompressible. Why?

9. No pump that works by suction can raise water in a pipe higher than about 34 feet. Why not?

10. You can't breathe underwater by drawing air through a tube stretching from the water's surface (a "snorkel") to a depth of much more than a foot. Why not?

Answers

1. You can answer this question only if you know your mass. Suppose your mass is 70 kg. For you to float, the buoyant force of the water must equal your weight. Thus by Archimedes' principle, you must displace 70 kg of water. A 70-kg mass of water has a volume of 70 L or 0.070 m^3. The submerged 95% of your body must therefore have a volume of 0.070 m^3, so the total volume of your body must be $0.070 \text{ m}^3 / 0.95 = 0.074 \text{ m}^3$.

2. The "fog" that you see is water vapor condensing out of the air locally due to the presence of the cold CO_2 vapor. Although the CO_2 vapor itself is transparent, you can see where it is by means of the condensed water vapor droplets. The CO_2 vapor stays at ground level because it is substantially denser than air.

3. The increased pressure on the water surface is transmitted to the trapped air inside the small vessel; this is in accordance with Pascal's principle. The trapped air is compressed and therefore displaces less water, so in accordance with Archimedes' principle the buoyant force holding the small vessel up is reduced. Increase the pressure enough, and the small vessel sinks. Decrease it enough, and the trapped air expands enough to raise the sunken vessel again.

4. We floated the container in a swimming pool and filled it with water from a calibrated vessel. The water pressure inside and outside the container was then about the same, so there was little or no stress on its fragile walls.

5. The fluid he added the second time wasn't water but some liquid that is less dense than water and doesn't mix readily with it. Thus, a taller column of the second fluid was required to produce the same pressure in the liquid at the bottom of the U-tube.

6. When the ice is floating, the part of it that is underwater is displacing a volume of water that is equal in weight to the whole weight of the ice. This is exactly the volume of the water produced by the melted ice, so the water level goes neither up nor down but stays the same. (Note that water shrinks on melting.)

7. To handle both launch and recovery most easily, you want a plane's speed, relative to the flight deck, to be as slow as safety permits. Adequate lift for takeoff requires a certain air speed. By taking off with the flight deck moving into the wind, the plane acquires the required air speed while minimizing the speed of the plane relative to the deck. During touchdown, the plane also requires a certain air speed (not necessarily the same as the air speed required for takeoff). To minimize the speed of the plane relative to the flight deck, the carrier heads into the wind during recovery as during launch.

8. The pressure decreases because the fluid has viscosity. If a steady flow is to be maintained, there must be a pressure difference to maintain the flow of the fluid against this friction.

9. A suction pump raises water by creating a pressure in the pipe that is lower than atmospheric pressure; it is the pressure of the atmosphere that raises the water. Since the absolute pressure at the pump cannot be less than zero, the largest pressure available to raise the water is equal to atmospheric pressure, which corresponds to a height of about 34 ft of water.

10. To draw air into your lungs, your diaphragm must lower the pressure in your lungs to less than atmospheric pressure. At a depth of a foot, the water outside your body is already at a pressure around 3 kPa higher than atmospheric. Because 4 to 7 kPa is the maximum external pressure difference your diaphragm can handle, you can't suck air through a tube at a depth of much more than a foot.

VI. Problems, Solutions, and Answers

Example #1.   The density of water is 1.000 g/cm³ at 4°C. Suppose that a 500-mL flask is filled exactly full of water at a temperature of 60°C, where the density of water is 0.980 g/cm³. When the flask of water is cooled to 4°C, how much more water must be added to fill the flask again? Neglect the thermal expansion or contraction of the flask itself.

Picture the Problem. Determine the mass of the water in the flask at 60°C, and determine the volume of this mass of water at 4°C.

1. Using the known density and volume of the water, calculate its mass at 60°C.	$m = \rho_{60} V_{60} = \left(0.980\,\text{g/cm}^3\right)\left(500\,\text{cm}^3\right) = 490\,\text{g}$

2. Determine the volume of this water at 4°C.	$V_4 = m / \rho_4 = (490\,\text{g}) / (1.000\,\text{g/cm}^3) = 490\,\text{cm}^3$
3. The amount of water to be added is the difference in the two volumes.	$V_{\text{added}} = V_{60} - V_4 = 10\,\text{cm}^3 = 10\,\text{mL}$

Example #2—Interactive. The mass of a small analytic flask is 15.2 g. When the flask is filled with water, the mass of the flask and water total 119.0 g; when it is filled with an unknown fluid, the total mass is 96.7 g. What is the specific gravity of the unknown fluid?

Picture the Problem. You can determine the mass of the unknown volume of water and the unknown fluid. This allows you to use the definition of specific gravity. **Try it yourself.** Work the problem on your own, in the spaces provided, to get the final answer.

1. Determine the mass of the water.	
2. Determine the mass of the unknown liquid.	
3. Since equal volumes of liquid are involved, the ratio of the mass of the unknown fluid to that of water will equal the specific gravity.	Specific gravity $= 0.785$

Example #3. A lead brick measuring 5.00 × 10.0 × 20.0 cm is dropped into a swimming pool 3.5 m deep. By how much does the volume of the brick change due to the pressure of the water?

Picture the Problem. The pressure at the bottom of the pool will cause the volume of the brick to shrink. Use Table 13-2 on page 398 of the text to find the bulk modulus.

1. Find the difference in pressure on the brick at the bottom of the pool compared to atmospheric pressure. The change in pressure is due entirely to the depth of water in the pool.	$\Delta P = \rho g h$
2. Using the definition of bulk modulus, determine the change in volume of the brick. The volume is calculated by multiplying the three dimensions of the brick.	$B = -\dfrac{\Delta P}{\Delta V / V}$ $\Delta V = -\dfrac{V\,\Delta P}{B} = -\dfrac{V \rho g h}{B}$ $= -\dfrac{(0.001\,\text{m}^3)(1000\,\text{kg/m}^3)(9.81\,\text{m/s}^2)(3.5\,\text{m})}{7.7 \times 10^9\ \text{N/m}^2}$ $= -4.46 \times 10^{-9}\ \text{m}^3 = -4.46\,\text{mm}^3$

Example #4—Interactive.   Blood pressure is normally measured on a patient's arm at approximately the level of the heart. If it were measured instead on the leg of a standing patient, how significantly would the reading be affected? Normal blood pressure is in the range of 70 to 140 torr, and the specific gravity of normal blood is 1.06.

Picture the Problem. The question is really asking simply: how much does pressure change with height in blood? You will need to estimate the height difference between arm and leg (1m). Because the blood is flowing, and is viscous, your answer won't be exact, but it will give you an idea of how significant the difference is. **Try it yourself.** Work the problem on your own, in the spaces provided, to get the final answer.

1. Write an equation for the pressure difference, as a guide.	$\Delta P = \rho g h$
2. The density of blood can be calculated using its specific gravity.	
3. Calculate the change in pressure, watching your units, and converting to torr. The resulting increase in pressure is significant. $\Delta P \approx 70\, torr$	

Example #5. Some colleagues and I once had to carry out an experiment on a lake. To do this we built a raft out of Styrofoam, 1 ft (0.305 m) thick and 8 ft (2.44 m) square. Take the specific gravity of Styrofoam to be 0.035. (a) How deep in the water did the unloaded raft float? (b) With three men (whose average mass was 88 kg each) and 120 kg of experimental equipment on the raft, how deep did it float?

Picture the Problem. When the raft is floating in equilibrium, the total weight of the raft and load will be equal to the weight of the water displaced.

1. Part (a). The weight of the raft must equal the weight of the displaced water. Let A be the area of the raft, t be its thickness, and h the depth to which the raft sank. The density of the Styrofoam is the density of water times the specific gravity of the Styrofoam.	$m_s g = m_w g$ $At\rho_w (\text{s.g.})_s\, g = Ah\rho_w g$ $h = t(\text{s.g.})$ $h = (0.305\,\text{m})(0.035) = 1.07\,\text{cm}$

2. For part (b), instead of just the mass of the Styrofoam, we need to include the mass of the load. The raft was still not in danger of sinking, since there was still 22.5 cm above the water.	$\left(m_s + 3m_m + m_e\right)g = m_w g$ $\left(At\rho_w\left(\text{s.g.}\right)_s + 3m_m + m_e\right)g = Ah\rho_w g$ $h = t\left(\text{s.g.}\right) + \dfrac{3m_m + m_e}{A\rho_w}$ $h = \left(0.305\,\text{m}\right)\left(0.035\right) + \dfrac{3\left(88\,\text{kg}\right) + \left(120\,\text{kg}\right)}{\left(2.44\,\text{m}\right)^2\left(1000\,\text{kg/m}^3\right)}$ $h = 7.52\,\text{cm}$

Example #6—Interactive. The density of air under "ordinary" (not standard) conditions is about 1.2 kg/m^3, whereas that of helium is 0.17 kg/m^3. What must the radius of a spherical helium-filled balloon be if it is to lift a total load of 350 kg, not including the helium itself? Assume the mass of the balloon to be negligible.

Picture the Problem. The balloon will rise if the mass of the helium and load is equal to the air it displaces. **Try it yourself.** Work the problem on your own, in the spaces provided, to get the final answer.

1. Express the weight of a sphere in terms of its density and the radius of the sphere.	
2. Equate the weight of a displaced sphere of air to the weight of the sphere of helium plus the weight of its load, and solve for the radius.	$r = 4.32\,\text{m}$

Example #7. In Figure 13-3, oil with a specific gravity of 0.68 is floating on top of some water in a container. A wooden object is floating at the fluid boundary such that one-third of its volume lies below the boundary. What is the density of the wood?

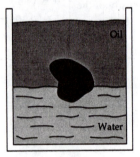

Figure 13-3

Picture the Problem. The wooden object will float such that its weight will equal the total weight of the fluid it displaces. One-third of the fluid by volume is water, and the remaining two-thirds is oil. Since the weights are proportional to g, the total mass of the displaced fluid must equal the mass of the of the wooden object.

1. Write an expression for the total mass of the displaced fluid, which will also be the mass of the wood. The total volume is V.	$m = \rho_{water} V_{water} + \rho_{oil} V_{oil}$ $= \rho_{water} \dfrac{V}{3} + (s.g.)_{oil}\, \rho_{water} \dfrac{2V}{3}$ $= \left[1 + 2(s.g.)_{oil}\right] \dfrac{\rho_{water} V}{3}$
2. The density of the wood can now be found.	$\rho_{wood} = \dfrac{m}{V} = \left[1 + 2(s.g.)\right] \dfrac{\rho_{water}}{3}$ $= \dfrac{1 + 2(0.68)}{3} (1000\,\text{kg/m}^3) = 787\,\text{kg/m}^3$

Example #8. Water is flowing smoothly at 15 ft/s in a horizontal pipe with a 2-in inside diameter at an absolute pressure of 40 lb/in^2. Neglect the viscosity of water. At a certain point the diameter of the pipe necks down to 1.05 in. (a) How fast does the water flow in the narrow section? (b) What is its pressure there (in lb/in^2)?

Picture the Problem. First, convert all quantities to SI units to avoid future complications. The continuity equation can be used to find the water's speed in the narrow section. Bernoulli's equation can be used to find the pressure difference to cause this change in speed.

1. Convert all quantities to SI units.	$15\,\text{ft/s} = 4.6\,\text{m/s}$ $2\,\text{in} = .05\,\text{m}$ $1.05\,\text{in} = 0.027\,\text{m}$ $40\,\text{lb/in}^2 = 276 \times 10^3\,\text{Pa}$
2. Use the continuity equation to find the water's speed at the narrow section.	$A_2 v_2 = A_{1.05} v_{1.05}$ $v_{1.05} = \dfrac{A_2 v_2}{A_{1.05}} = \dfrac{\pi r_2^2 v_2}{\pi r_{1.05}^2}$ $= \dfrac{(0.05\,\text{m})^2 (4.6\,\text{m/s})}{(0.027\,\text{m})^2} = 16.7\,\text{m/s}$
3. Use Bernoulli's equation to find the new pressure in the narrow section.	$P_2 + \rho g h + \tfrac{1}{2}\rho v_2^2 = P_{1.05} + \rho g h + \tfrac{1}{2}\rho v_{1.05}^2$ $P_{1.05} = P_2 + \tfrac{1}{2}\rho\left(v_2^2 - v_{1.05}^2\right)$ $P_{1.05} = 276 \times 10^3\,\text{Pa} + \tfrac{1}{2}(1000\,\text{kg/m}^3)\left((4.6\,\text{m/s})^2 - (16.7\,\text{m/s})^2\right)$ $P_{1.05} = 147 \times 10^3\,\text{Pa} = 21.3\,\text{lb/in}^2$

Example #9. A town's water tank is supported above ground on posts. Its diameter is 18 m, and the water in the tank is 7.5 m deep. The top is open to the atmosphere. If a hole is punched in the bottom of the tank and the water flows out in a stream 1 cm in diameter, how long does it take for the water level to drop by 10 cm? Neglect the viscosity of the water.

Picture the Problem. The flow at the top of the tank has to equal the flow at the bottom of the tank. The two equations that relate flow and pressure are Bernoulli's equation and the continuity equation.

1. Write out Bernoulli's equation using the top and the bottom of the tank as the places of interest. Let the bottom of the tank be at $h = 0$. Furthermore, because both the top of the tank and the hole are open to the air, $P_1 = P_2 = P_{at}$.	$P_t + \rho g h_t + \frac{1}{2}\rho v_t^2 = P_b + \rho g h_b + \frac{1}{2}\rho v_b^2$ $P_{at} + \rho g h + \frac{1}{2}\rho v_t^2 = P_{at} + \frac{1}{2}\rho v_b^2$ $v_b^2 = 2gh + v_t^2$
2. Write out the continuity equation for the top and bottom of the tank.	$A_t v_t = A_b v_b$ $v_t = \dfrac{A_b v_b}{A_t}$
3. Substitute into the relation in step 1 to solve for v_b. Since the hole at the bottom is so much smaller than the surface area of the top of the tank, we can neglect $\left(A_b / A_t\right)^2$ compared to 1.	$v_b^2 = 2gh + \left(\dfrac{A_b}{A_t}\right)^2 v_b^2$ $v_b^2 = 2gh\left(1 - \left(\dfrac{A_b}{A_t}\right)^2\right)^{-1} \approx 2gh$
4. Use the continuity equation to determine the volume flow rate. The volume of water that must flow is equal to the area of the tank times 10 cm, which is also equal to the flow rate multiplied by time. By equating the two equations for the flow rate, we can solve for time.	$I_V = A_b v_b = A_b\sqrt{2gh}$ $I_V t = A_t(0.1\,\text{m})$ $t = \dfrac{A_t(0.1\,\text{m})}{I_V} = \dfrac{A_t(0.1\,\text{m})}{A_b\sqrt{2gh}}$ $t = \left(\dfrac{18\,\text{m}}{0.01\,\text{m}}\right)^2 \dfrac{0.01\,\text{m}}{\sqrt{2(9.81\,\text{m/s}^2)(7.5\,\text{m})}}$ $= 2.67 \times 10^4\,\text{s} = 7.42\,\text{hr}$

Example #10—Interactive. At a certain point in a sloping pipe 2.5 cm in diameter, water is flowing at 10 m/s. From there the pipe gradually descends vertically by 7 m, and over the same distance its diameter increases to 3.5 cm. What is the pressure difference between the upper and lower points?

Picture the Problem. The two equations that relate flow and pressure are Bernoulli's equation and the continuity equation. **Try it yourself.** Work the problem on your own, in the spaces provided, to get the final answer.

1. Use the continuity equation to determine the speed of the water where the pipe diameter is 3.5 cm.	
2. Use Bernoulli's equation to calculate the pressure difference between the upper and lower points.	$\Delta P = 106\,\text{kPa}$

Example #11. Water is supplied to an outlet from a pumping station 5 km away. From the pumping station to the outlet there is a net vertical rise of 19 m. Take the coefficient of viscosity of water to be 0.01 poise. The pipe leading from the pumping station to the outlet is 1 cm in diameter, and the gauge pressure in the pipe at the point where it exits the pumping station is 520 kPa. At what volume flow rate does water flow from the outlet?

Picture the Problem. Bernoulli's equation will not work because the water is viscous. There will be a pressure drop due to viscous flow in the pipe and a drop due to the gain in height. You know the total pressure drop from the pumping station to the outlet and can calculate the drop due to the vertical rise. The difference will be due to the viscous drag in the pipe. Assume the flow is laminar and use Poiseuille's law to find the volume flow rate at the outlet.

1. Find the pressure drop due to the vertical rise.	$P = \rho g h = \left(1000\,\text{kg/m}^3\right)\left(9.81\,\text{m/s}^2\right)\left(19\,\text{m}\right)$ $= 186\,\text{kPa}$
2. Find the pressure drop due to the viscosity.	$\Delta P_v = \Delta P_{\text{total}} - \Delta P_h = 520\,\text{kPa} - 186\,\text{kPa} = 334\,\text{kPa}$
3. Use Poiseuille's law to determine the volume flow rate.	$\Delta P = \dfrac{8\eta L}{\pi r^4} I_v$ $I_v = \dfrac{\Delta P \pi r^4}{8\eta L}$ $= \dfrac{\left(334\times10^5\,\text{Pa}\right)\pi\left(0.005\,\text{m}\right)^4}{8\left(0.001\,\text{Pa·s}\right)\left(5\times10^3\,\text{m}\right)}$ $= 1.64\times10^{-5}\,\text{m}^3/\text{s} = 16\,\text{cm}^3/\text{s}$

Example #12—Interactive. Blood flows in the finer capillaries of the body at a rate of around 1 mm/s. If blood flows at this rate through a capillary $8\,\mu\text{m}$ in diameter, what is the pressure difference required to move blood at this rate through a capillary 1.8 mm long? Assume that the coefficient of viscosity of blood is $4.0\,\text{mPa·s}$.

Picture the Problem. If you get all the units correct, once you calculate the volume flow rate, just substitute into Poiseuille's law. Compare this pressure difference to normal blood pressure in the range of 70 to 140 torr. **Try it yourself.** Work the problem on your own, in the spaces provided, to get the final answer.

1. Calculate the volume flow rate.	
2. Use Poiseuille's law to find the pressure difference.	$\Delta P = 3.6\,\text{kPa} = 27\,\text{torr}$

Example #13. Water ($\eta = 0.01$ poise) is pumped into one end of a level 50-ft-long 1-in-diameter pipe at a gauge pressure of 40 lb/in². (a) If the flow is nonturbulent, how much water can be pumped through the pipe in 1 min if its other end is open to the atmosphere? (b) Calculate the Reynolds number for this flow to see whether or not your assumption was justified.

Picture the Problem. Use Poiseuille's law to get the volume flow rate in the pipe. From the volume flow rate, calculate the velocity at which the water is flowing. From the velocity and other given information, calculate the Reynolds number.

1. Calculate the volume flow rate using Poiseuille's law.	$I_V = \dfrac{\pi r^4}{8\eta L}\Delta P$ $= \dfrac{\pi\left(1.27\times10^{-2}\,\text{m}\right)^4}{8\left(0.001\,\text{Pa·s}\right)\left(15.2\,\text{m}\right)}\left(2.76\times10^5\,\text{Pa}\right)$ $= 0.185\,\text{m}^3/\text{s} = 11.1\,\text{m}^3/\text{min} = 2930\,\text{gal/min}!$
2. Find the velocity of the water leaving the pipe outlet.	$v = \dfrac{I_V}{A} = \dfrac{\left(0.185\,\text{m}^3/\text{s}\right)}{\pi\left(1.27\times10^{-2}\,\text{m}\right)^2} = 365\,\text{m/s}$
3. Calculate the Reynolds number. This is about three orders of magnitude larger than the transition into turbulent flow. Our assumption was not correct.	$N_R = \dfrac{2r\rho v}{\eta} = \dfrac{2\left(1.27\times10^{-2}\,\text{m}\right)\left(1000\,\text{kg/m}^3\right)\left(365\,\text{m/s}\right)}{0.001\,\text{Pa·s}}$ $= 9.3\times10^6$

Example #14—Interactive. In Example #11, Poiseuille's law, which holds only for nonturbulent flow, predicts that water ($\eta = 0.01$ poise) flows through the 1-cm-diameter pipe at a rate of 1.64×10^{-5} m³/s. Determine the Reynolds number for this flowing water, and state whether or not the assumption of nonturbulent flow is appropriate. **Try it yourself.** Work the problem on your own, in the spaces provided, to get the final answer.

1. Determine the velocity of the water through the pipe outlet.	
2. Calculate the Reynolds number. This value is near the low end of the laminar flow to turbulent flow transition range, so the use of Poiseuille's law is probably, but not necessarily, valid.	$N_R = 2090$

Chapter **14**

Oscillations

I.   Key Ideas

Oscillatory, periodic motions are very important in nature. Oscillation occurs when a system is disturbed from a stable equilibrium; a restoring force causes oscillation around the equilibrium position. Wave motion is a closely related phenomenon.

Section 14-1. Simple Harmonic Motion. Simple harmonic motion is the oscillatory motion that occurs under a restoring force that is proportional to the displacement of the system from equilibrium. That is,

$$F_x = -kx$$

Linear restoring force (Hooke's law)

The force of a spring is a classic example of a linear restoring force. The resulting motion has acceleration that is proportional to the position.

$$a_x = -\frac{k}{m}x = -\omega^2 x$$

Acceleration proportional to displacement

Whenever the acceleration of an object is proportional to its displacement and is oppositely directed, the object will move with simple harmonic motion. The motion is a sinusoidal oscillation; that is, a graph of the displacement versus time has the form of a sine function

Period and Frequency.   The frequency f of an oscillating system is the number of cycles completed per unit of time. The period T of the motion is the time required for one cycle of the motion. These are related by the equation

$$T = \frac{1}{f}$$

Period

The period is related to the force constant k of the spring and to the mass m of the oscillating object by the equation

$$T = 2\pi\sqrt{\frac{m}{k}}$$

Period for particle on a linear spring

For simple harmonic motion, the period is independent of amplitude.

Circular Motion and Simple Harmonic Motion.   There is a very close connection between simple harmonic motion and circular motion with constant speed. The projection on one coordinate axis of a point undergoing uniform circular motion is a simple harmonic motion. The frequency is related to the angular velocity ω of the circular motion by the equation

$$f = \frac{\omega}{2\pi}$$

Frequency–angular velocity

For simple harmonic motion the position and velocity of the oscillating object are expressed as functions of time by the relations

$$x = A\cos(\omega t + \delta)$$

Position in simple harmonic motion

$$v = \frac{dx}{dt} = -\omega A\sin(\omega t + \delta)$$

Velocity in simple harmonic motion

where A is the amplitude of the motion and δ is the phase constant. The argument of the trigonometric functions $(\omega t + \delta)$ is the phase of the motion.

Section 14-2. Energy in Simple Harmonic Motion.   An object undergoing simple harmonic motion has a constant total energy, but its potential energy and its kinetic energy vary with time. For a linear restoring force (Hooke's law) the potential energy is $(1/2)kx^2$. At the turning points, the kinetic energy is zero and the potential energy has its maximum value. As the system passes through its equilibrium point, the reverse is true. That is,

$$E_{\text{total}} = \tfrac{1}{2}kA^2 = \tfrac{1}{2}kx^2 + \tfrac{1}{2}mv^2 = \tfrac{1}{2}mv_{\text{max}}^2$$

Conservation of energy

Section 14-3. Some Oscillating Systems.   A familiar oscillating system is a *physical pendulum*— an object supported so it freely swings back and forth under the influence of gravity. The motion is not, in general, simple harmonic motion, but it approaches simple harmonic motion for small amplitudes. The period of the pendulum is independent of its mass, but depends on its moment of inertia I, the distance D between its rotation axis and the center of mass, and g, the gravitational field strength. The period T for small-amplitude oscillations is

$$T = 2\pi\sqrt{\frac{I}{MgD}}$$

Period of a physical pendulum

Upon first glance at this formula, it appears that the period depends upon the mass M. However, this is not the case. Since the moment of inertia is proportional to the mass, the ratio I/M does not depend on M, and neither does T.

For a *simple pendulum* (a point mass attached to the end of a massless string) $I = ML^2$ and $D = L$, where L is the strength of the string. Substituting these expressions for I and D into the formula for the period gives

$$T = 2\pi\sqrt{\frac{L}{g}}$$

Period of a simple pendulum

Section 14-4. Damped Oscillations. In real oscillations, the motion is not conservative; it is always damped by frictional forces. As a result, the energy and amplitude of the motion decrease with time. If damping forces are small, the energy decreases exponentially, that is, by a constant fraction in a given time interval. That is,

$$x = A_0 e^{-(b/2m)t}\cos(\omega't + \delta)$$

Position as a function of time for a damped oscillator

where

$$\omega' = \omega_0\sqrt{1 - \left(\frac{b}{2m\omega_0}\right)^2}$$

Frequency of a damped oscillator

ω_0 is the natural (undamped) frequency of the oscillator, and b is the damping constant. (The damping force is $-bv$, a linear drag force.)

The quality factor Q of a damped oscillation expresses the amount of damping; a high Q means the oscillation takes a long time to die out.

$$Q = \omega_0\tau = 2\pi\frac{E}{|\Delta E|}$$

Q factor for a damped oscillator

where ω_0 is the oscillator's natural frequency, $\tau = 2m/b$, and $E/|E|$ is the ratio of the energy at the beginning of a cycle to the energy dissipated during the cycle. Expressed in terms of the damping constant b, $Q = b/(8m\omega_0)$. If the damping is strong enough, there is no oscillation; the displacement just dies out without crossing the equilibrium position.

Section 14-5. Driven Oscillations and Resonance. Here we consider the motion of an oscillator that is driven by a repetitive (periodic) driving force. Consideration is restricted to the steady-state motion that results when the energy per cycle transferred by the driving force equals that dissipated by the frictional forces. This is simple harmonic motion at the frequency of the driving force. This is expressed as

$$x = A\cos(\omega t - \delta)$$

Position as a function of time for a driven damped oscillator

where

$$A = \frac{F_0}{\sqrt{m^2(\omega_0^2 - \omega^2)^2 + b^2\omega^2}}$$

Amplitude of a driven damped oscillator

and

$$\delta = \tan^{-1} \frac{b\omega}{m(\omega_0^2 - \omega^2)}$$

Phase constant of a driven damped oscillator

If the frequency of the driving force is equal or almost equal to the oscillator's natural frequency, the amplitude of the driven motion will be large. This enthusiastic response to the driving force is called resonance. If the oscillator has high Q, the resonance response is strong, but only over a narrow frequency range.

II. Physical Quantities and Key Equations

Physical Quantities

Frequency   1 hertz (Hz) = 1 cycle/s

Key Equations

Linear restoring force (Hooke's law) $F_x = -kx$

Acceleration proportional to displacement $a_x = -\omega^2 x$

Frequency $f = \dfrac{\omega}{2\pi}$

Period $T = \dfrac{1}{f}$

Period for particle on a linear spring $T = 2\pi\sqrt{\dfrac{m}{k}}$

Period of a physical pendulum $T = 2\pi\sqrt{\dfrac{I}{MgD}}$

Period of a simple pendulum $T = 2\pi\sqrt{\dfrac{L}{g}}$

Position in simple harmonic motion $x = A\cos(\omega t + \delta)$

Velocity in simple harmonic motion $v = -\omega A\sin(\omega t + \delta)$

Conservation of energy $E_{\text{total}} = \frac{1}{2}kA^2 = \frac{1}{2}kx^2 + \frac{1}{2}mv^2 = \frac{1}{2}mv_{\text{max}}^2$

Q factor for a damped oscillator $Q = \omega_0 \tau = 2\pi\dfrac{E}{|\Delta E|}$

Position as a function of time for a damped oscillator $x = A_0 e^{-(b/2m)t} \cos(\omega' t + \delta)$

Frequency of a damped oscillator $\omega' = \omega_0 \sqrt{1 - \left(\dfrac{b}{2m\omega_0}\right)^2} = \omega_0 \sqrt{1 - \left(\dfrac{1}{4Q^2}\right)}$

Q factor for a driven damped oscillator $Q = \dfrac{\omega_0}{\Delta\omega} = \dfrac{f_0}{\Delta f}$

Position as a function of time for a driven damped oscillator $x = A\cos(\omega t - \delta)$

Amplitude of a driven damped oscillator $A = \dfrac{F_0}{\sqrt{m^2(\omega_0^2 - \omega^2)^2 + b^2\omega^2}}$

Phase constant of a driven damped oscillator $\delta = \tan^{-1} \dfrac{b\omega}{m(\omega_0^2 - \omega2)}$

III. Potential Pitfalls

Do not be careless and use any of the constant-acceleration formulas for the motion of a harmonic oscillator! The force is proportional to the displacement from equilibrium, so the acceleration is definitely not constant.

In the formulas involving sine and cosine functions, the argument of the trigonometric function (that is, the quantity whose sine or cosine is taken) must be a dimensionless number; if thought of as an angle, it must be an angle in radians.

In almost all cases, it is best to put the origin of the x axis at the equilibrium position of the oscillating particle since that is the zero point of the displacement of, and thus of the force on, the oscillating particle.

Remember that the formula for the period of a simple pendulum applies only to the simple pendulum. If the actual system is something other than a particle on the end of a massless string, then the period will depend on how the mass is distributed in space.

The formulas we write for simple harmonic motion apply to any motion under a restoring force that is directly proportional to the displacement from an equilibrium position. The "spring constant k" used in the formulas is whatever quantity occupies that position in the force equation, $F = -kx$, whether or not there are springs involved.

The potential and kinetic energy of a harmonic oscillator both vary with time; it is only the total energy that is constant. The total energy may be all kinetic, all potential, or anything in between at different points in the cycle.

The motion of a mass on a spring hanging vertically is simple harmonic motion even though a second force—gravity—is involved. The weight simply changes the equilibrium point about which the mass oscillates.

IV. True or False Questions and Responses

True or False

_____ 1. Periodic motion is any motion that repeats itself cyclically.

_____ 2. Simple harmonic motion is periodic motion in which the position of a particle varies sinusoidally with the time.

_____ 3. The time it takes for one full cycle of simple harmonic motion to be completed is called the *phase constant* of the motion.

_____ 4. The frequency of simple harmonic motion is the number of completed cycles per unit time.

_____ 5. At any instant, the acceleration of a moving particle undergoing simple harmonic motion is directed opposite to its displacement from equilibrium.

_____ 6. When the speed of a particle undergoing simple harmonic motion is a maximum, the magnitude of its acceleration is a minimum.

_____ 7. If a particle is undergoing simple harmonic motion due to the action of a force $F_x = -kx$, its kinetic energy is $\frac{1}{2}kA^2$, where A is the amplitude of its oscillation.

_____ 8. The total mechanical energy of a particle undergoing simple harmonic motion is constant.

_____ 9. An object hanging on the end of a spring oscillates vertically with a period that is independent of its mass.

_____ 10. An object swinging on the end of a string as a simple pendulum oscillates with a period that is independent of its mass.

_____ 11. The motion of a simple pendulum is necessarily a simple harmonic motion.

_____ 12. The motion of real oscillatory systems can, at best, be only approximately described as simple harmonic motion.

_____ 13. The motion of a weakly damped harmonic oscillator is very nearly simple harmonic motion with a decreasing amplitude.

_____ 14. For a damping force that is proportional to the velocity, the energy of a damped oscillator diminishes linearly with time.

_____ 15. The period of a damped oscillator is always shorter than it would be if there were no damping.

_____ 16. Critical damping is the condition in which the displaced particle returns to equilibrium most rapidly, without oscillating.

____ 17. For $b = 0$, a driven harmonic oscillator has a maximum amplitude given by F_0/m when the frequency of the driving force is equal to the natural frequency of the oscillator.

____ 18. The fact that a driven harmonic oscillator absorbs maximum power from the driving force when the frequency of the driving force is nearly equal to the natural frequency of the oscillator is referred to as *resonance*.

____ 19. The width (in frequency) of the resonance curve increases as the damping constant b increases.

Responses to True or False

1. True.

2. True.

3. False. This time is the *period* of the motion.

4. True.

5. True.

6. True. In fact, if v is a maximum or minimum, $a = dv/dt$ is zero.

7. False. This is its total mechanical energy; its kinetic energy isn't constant.

8. True.

9. False; $T = 2\pi/\omega = 2\pi\sqrt{m/k}$.

10. True; $T = 2\pi\sqrt{L/g}$.

11. False. It is simple harmonic motion only if the amplitude is small.

12. True. There is some dissipation of energy, and therefore some damping of the motion, in any real oscillating system.

13. True.

14. False. It diminishes exponentially.

15. False. The frequency is always a little less, so the period is a little longer. The difference is pretty small unless the damping is near critical.

16. True.

17. False. With $b = 0$ the amplitude at resonance would go to infinity. Of course, the proportional limit of the spring will be exceeded, or something will break, before that happens.

18. True.

19. True.

V. Questions and Answers

Questions

1. A particle is undergoing simple harmonic motion. How far does it move in one full period?

2. A mass oscillates on the end of a certain spring at a frequency f. The spring is cut in half, and the same mass is set oscillating on one of the pieces. What is its new frequency of oscillation?

3. The mass of the string is usually neglected in treating the motion of a simple pendulum. If the mass of the string is not completely negligible, how is the motion of the pendulum affected?

4. A mass on the end of a string, which is hung over a nail, is set swinging as a simple pendulum of length L (see Figure 14-1). While it is swinging, the string is paid out until the free-swinging length is $2L$. What happens to the pendulum's frequency? Can we use the conservation of energy to find out what happens to its amplitude? Neglect friction at the nail.

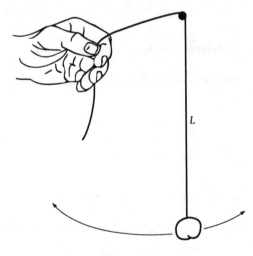

Figure 14-1

5. If a mass-and-spring system is set oscillating, why does its motion eventually stop?

6. A mass-and-spring system is undergoing simple harmonic motion with an amplitude A. If the mass is decreased by half but the amplitude is unchanged, how does the total energy of the oscillation change? How does the total energy change when the original mass is set oscillating with half the initial amplitude?

7. Would it have made any difference in our discussion in this chapter if we had defined *simple harmonic motion* as motion that obeys $x = A \sin(\omega t + \phi)$ rather than defining it in terms of the cosine?

Answers

1. The particle's net displacement in one period is, of course, zero. The total distance it covers, however, is from $z = 0$ (its initial position) to $x = A$, from there to $x = -A$, and from $x = -A$ back to its initial position. It therefore covers a total distance of $4A$.

2. What we have to find first is the force constant of half a spring. The same tension stretches one-half of the original spring half as much as it would the whole. Therefore, if the force constant of the whole spring is k, that of one of the halves is $2k$. Since the frequency is proportional to the square root of k, cutting the spring in half increases the frequency by $\sqrt{2}$.

3. If the string has mass, a small fraction of the total mass that is swinging is closer to the pivot than the pendulum length L, so the effective length is a little less than L. Consequently, the motion of the pendulum has a slightly shorter period than it would if the string were massless.

4. The period of a simple pendulum is proportional to $\sqrt{L}$, so the period increases by a factor of $\sqrt{2}$ and the frequency decreases to $1/\sqrt{2}$ its original value. The energy of the swinging pendulum is not a constant because (negative) work is done on it by the hand that pays out the string. Since we don't know how to evaluate this work, we can't apply the conservation of energy to this problem. (We know the displacement of the hand, but the tension in the string is variable.)

5. The motion dies out because real oscillations are damped. The drag of the air and some of the internal forces in the material of the spring dissipate the energy of the oscillating mass.

6. When the mass is halved, the frequency of the oscillation increases, but the total energy, $(1/2)kA^2$, is unaffected. When the amplitude is halved, the total energy decreases to $\frac{1}{4}$ of its original value.

7. It would have made no difference whatever. Defining *simple harmonic motion* in terms of the sine does change the phase constant by 90°:

$$x = A\sin(\omega t + \phi)$$
$$= A\cos(\omega t + \phi - \pi/2)$$
$$= A\cos(\omega t + \delta)$$

where $\delta = \phi - \pi/2$.

VI. Problems, Solutions, and Answers

Example #1. A 0.100-kg mass is suspended from a spring of negligible mass with a spring constant of 40.0 N/m. The mass oscillates vertically with an amplitude of 0.0600 m. (*a*) Determine the angular frequency of the motion. Express the height y of the mass above the equilibrium position as a function of time so that initially (at $t = 0$) the mass is (*b*) at its highest point; (*c*) 0.03 m above the equilibrium position and moving downward.

Picture the Problem. To find the position as a function of time requires that the amplitude A and phase constant δ be determined from the initial conditions for the position and velocity. In each part, simultaneously solve the position and velocity equations at time $t = 0$ for the amplitude and phase constant.

1. First find the angular frequency.	$\omega = \sqrt{k/m} = \sqrt{(40\,\text{N/m})/(0.1\,\text{kg})} = 20\,\text{rad/s}$
2. Solve the formulas for position and velocity simultaneously to get the amplitude and phase constant. For all parts, we are given the initial position and velocity, so we set $t = 0$. We also know both ω and A.	$y(t) = A\cos(\omega t + \delta)$ $v(t) = -\omega A\sin(\omega t + \delta)$ $0.06\,\text{m} = A\cos(\delta) = (0.06\,\text{m})\cos(\delta)$ $0 = -\omega A\sin(\delta) = (20\,\text{rad/s})(0.06\,\text{m})\sin(\delta)$
3. Solving for part (b), we get what's shown. For simplicity, we choose $\delta = 0$, and end up with the final expression for position as a function of time.	$\cos(\delta) = 1 \Rightarrow \delta = 0, \pm 2\pi, \pm 4\pi$ $\sin(\delta) = 0 \Rightarrow \delta = 0, \pm 2\pi, \pm 4\pi$ $y(t) = (0.06\,\text{m})\cos\left[(20\,\text{rad/s})t\right]$
4. We take the same steps to solve for part (c).	$0.03\,\text{m} = A\cos(\delta) = (0.06\,\text{m})\cos(\delta)$ $0 > -\omega A\sin(\delta) = (20\,\text{rad/s})(0.06\,\text{m})\sin(\delta)$
5. The equations in step 4 put different conditions on the phase constant. The only angle that satisfies both conditions is $\delta = \pi/3$, if we restrict ourselves to having $0 \le \delta \le 2\pi$.	$\cos(\delta) = 0.5 \Rightarrow \delta = \pi/3, 5\pi/3$ $\sin(\delta) > 0 \Rightarrow 0 < \delta < \pi$ $\therefore \ \delta = \pi/3$
6. The equation for the height y is then	$y(t) = (0.06\,\text{m})\cos\left[(20\,\text{rad/s})t + \pi/3\right]$

Example #2—Interactive. The motion of a mass oscillating on the end of a spring is graphed in Figure 14-2. What are (a) the amplitude, (b) the period, and (c) the frequency of its motion? (d) If the mass is 600 g, what is the force constant of the spring?

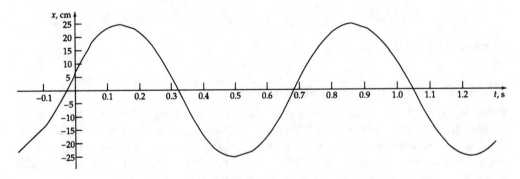

Figure 14-2

Picture the Problem. (*a*) and (*b*) can be read directly from the graph, (*c*) can be calculated from (*b*), and the force constant can be calculated from the mass and period. **Try it yourself.** Work the problem on your own, in the spaces provided, to get the final answer.

1. The amplitude is the maximum deviation from equilibrium.	$A = 25\,\text{cm}$
2. The period is the time required for the motion to repeat itself.	$T = 0.72\,\text{s}$
3. Calculate the frequency from the period.	$f = 1.4\,\text{Hz}$
4. Calculate the spring constant from the mass and period.	$k = 46\,\text{N/m}$

Example #3. A certain mass hanging in equilibrium on the end of a vertical spring stretches the spring by 2.00 cm. (*a*) If the mass is then pulled down 5.00 cm farther and released, at what frequency does it oscillate? (*b*) At what speed is it moving as it passes through its equilibrium position?

Picture the Problem. The hanging mass is in equilibrium when the spring is stretched 2.00 cm. Use this to calculate k/m, which can then be used to calculate the oscillation frequency. When passing through equilibrium, the mass has its maximum speed.

1. In equilibrium, the net force on the mass is zero. Here, y_0 is the extension of the spring when the mass is at its equilibrium position. See Section 14-3 on page 437 of the text for more details	$F_{net} = ky_0 - mg = 0$ $\dfrac{k}{m} = \dfrac{g}{y_0}$
2. Calculate the frequency of oscillation.	$f = \dfrac{\omega}{2\pi} = \dfrac{\sqrt{k/m}}{2\pi} = \dfrac{\sqrt{g/y_0}}{2\pi}$ $= \dfrac{\sqrt{(9.81\,\text{m/s}^2)/(0.02\,\text{m})}}{2\pi} = 3.52\,\text{Hz}$
3. The velocity of a particle experiencing simple harmonic motion is the time derivative of its position. In order for the speed to be a maximum, $\sin(\omega t) = \pm 1$.	$y = A\cos(\omega t)$ $v = \dfrac{dy}{dt} = -A\omega \sin(\omega t)$ $v_{max} = A\omega = A2\pi f = (0.05\,\text{m})\,2\pi(3.52\,\text{Hz})$ $= 1.11\,\text{m/s}$

Example #4—Interactive. A 5.00-kg block on the end of a spring oscillates at a frequency of 5.00 Hz with an amplitude of 5.00 cm. (*a*) What is the force constant of the spring? (*b*) What is the period of the block's motion? (*c*) What is the maximum speed at which the block moves?

Picture the Problem. The force constant can be calculated from the frequency and mass. The period can be found from the frequency, and the maximum speed is given by taking the derivative of the position function. **Try it yourself.** Work the problem on your own, in the spaces provided, to get the final answer.

1. Calculate the force constant from the frequency and mass.	$k = 4930\,\text{N/m}$
2. The period is the reciprocal of the frequency.	$T = 0.2\,\text{s}$
3. Find the maximum speed by taking the derivative of the position function.	$v_{max} = 1.57\,\text{m/s}$

Example #5. You are riding on a 10.0-m diameter Ferris wheel that is rotating at a constant rate of 1.00 rev every 10.0 s. (*a*) Determine your angular speed in rad/s. Express your height y above the rotation axis as a function of time t so that initially (at $t = 0$) you are (*b*) 3.00 m above the rotation axis and moving downward, and (*c*) 2.00 m below the rotation axis and moving upward.

Picture the Problem. Draw a sketch to help visualize the situation. Angular speed is total angle divided by total time. Your height y will have the functional form of cosine. The argument of the cosine function will be the angle $\theta = \omega t + \delta$ which is a function of time. Use the conditions given to solve for δ.

1. Draw a sketch to help visualize the situation.	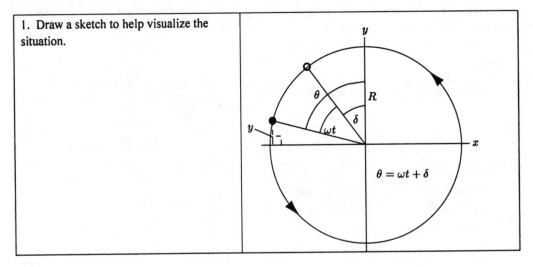

2. Find the angular frequency.	$\omega = \dfrac{\Delta\theta}{\Delta t} = \dfrac{1\,\text{rev}}{10.0\,\text{s}}\left(\dfrac{2\pi\,\text{rad}}{\text{rev}}\right) = \dfrac{\pi\,\text{rad}}{5.00\,\text{s}} = 0.628\,\text{rad/s}$
3. Write a generic expression for the height above the rotation axis as a function of time.	$y(t) = R\cos\theta = R\cos(\omega t + \delta)$
4. For part (b), you start 3 m above the axis at time $t = 0$.	$y_0 = R\cos(\delta)$ $\delta = \cos^{-1}\left(\dfrac{y_0}{R}\right) = \cos^{-1}\left(\dfrac{3.00\,\text{m}}{5.00\,\text{m}}\right) = 0.927\,\text{rad}$ $y(t) = (5.00\,\text{m})\cos\left[(0.628\,\text{rad/s})t + 0.927\,\text{rad}\right]$
5. For part (c), you start 2 m below the rotation axis, moving upward. In order to be moving upward, the angle must be greater than 180°.	$\delta = \cos^{-1}\left(\dfrac{y_0}{R}\right) = \cos^{-1}\left(\dfrac{-2.00\,\text{m}}{5.00\,\text{m}}\right) = 114°\ \text{or}\ 246°$ $246° = 4.30\,\text{rad}$ $y(t) = (5.00\,\text{m})\cos\left[(0.628\,\text{rad/s})t + 4.30\,\text{rad}\right]$

Example #6—Interactive. A phonograph record 30.0 cm in diameter revolves once every 1.80 s. (a) What is the linear speed of a point on its rim? (b) What is its angular velocity ω? (c) Write an equation for the y component of the point's position as a function of time, given that it is at its maximum value on the y axis when $t = 0$.

Picture the Problem. You know the diameter of the record. How far does a point on its rim move in 1.80 s? The angular velocity of rotation is the linear speed divided by the radius of the disk. The amplitude of the oscillation is the radius of the record. **Try it yourself.** Work the problem on your own, in the spaces provided, to get the final answer.

(a) $v = 0.524\,\text{m/s}$ (b) $\omega = 3.49\,\text{rad/s}$ (c) $y(t) = (0.15\,\text{m})\cos\left[(3.49\,\text{s}^{-1})t\right]$

Example #7. An object of mass 1.00 kg oscillates on a spring with an amplitude of 12.0 cm. Its maximum acceleration is 5.00 m/s^2. Find the total energy of the mass-spring system.

Picture the Problem. The total energy of the mass-spring system can be found from the spring constant and the amplitude. To find the spring constant, use Newton's 2nd Law.

1. Write out an expression for the total energy as a guide for the problem.	$E = \frac{1}{2}kA^2$

2. Find the spring constant from Newton's 2nd Law. To find k, we need to know both a and x at the same time. When a is at a maximum, the displacement is also at a maximum, but negative.	$F = ma$ $-kx = ma$ $k = -\dfrac{ma}{x}$ $k = \dfrac{ma_{max}}{A}$
3. Substituting from step 2 into step 1, we can find the total energy.	$E = \dfrac{1}{2}\dfrac{ma_{max}}{A}A^2 = \dfrac{ma_{max}A}{2}$ $= \dfrac{(1\,kg)(5\,m/s^2)(0.12\,m)}{2} = 0.300\,J$

Example #8—Interactive. A 400-g mass oscillates on the end of a spring with an amplitude of 2.50 cm. Its total energy is 2.00 J. What is the frequency of the oscillation?

Picture the Problem. Since you know the total energy and the amplitude, you can find the force constant of the spring. From the force constant and the mass, you can find the frequency of oscillation. **Try it yourself.** Work the problem on your own, in the spaces provided, to get the final answer.

1. Find the force constant of the spring from the energy of the oscillation.	
2. From the force constant and the mass, find the frequency of oscillation.	$f = 20.1\,Hz$

Example #9. Two simple pendula, each of length 1.80 m, with masses of 1.40 kg and 2.00 kg, hang side by side such that, at rest, their round steel bobs just touch (see Figure 14-3). The lighter pendulum on the left is pulled back 6 cm and released. It swings down and strikes the other, which is initially at rest. If the collision is elastic, to what distance does each pendulum rebound after the collision?

Picture the Problem. This is a three-part problem. The first pendulum falls, increasing its kinetic energy, and hence speed. There is an elastic collision in which both energy and momentum are conserved, and the two pendula then rise to new maximum heights. Rather than relating a decrease in potential energy to an increase in kinetic energy, we can instead relate the velocity of a pendulum to the amplitude of its motion and the angular frequency. The angular frequency can, in turn, be expressed in terms of the length of the pendulum and the acceleration due to gravity g.

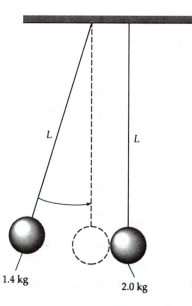

Figure 14-3

1. Find a relationship for the maximum velocity in terms of A, L, and g.	$T = 2\pi \sqrt{\dfrac{L}{g}} = \dfrac{2\pi}{\omega}$ $v = -\omega A \sin(\omega t + d)$ $v_{max} = \omega A = A\sqrt{\dfrac{g}{L}}$
2. For a one-dimensional elastic collision, the relative speed of recession after the collision equals the relative speed of approach before the collision. The initial speed of the 2-kg pendulum is zero.	$v_{2f} - v_{1f} = -(v_{2i} - v_{1i})$ $v_{2f} - v_{1f} = v_{1i}$
3. Now apply conservation of momentum to the collision.	$m_1 v_{1i} + m_2 v_{2i} = m_1 v_{1f} + m_2 v_{2f}$ $m_1 v_{1i} + m_2 (0) = m_1 v_{1f} + m_2 v_{2f}$ $v_{1i} = v_{1f} + \dfrac{m_2}{m_1} v_{2f}$
4. Adding the final expressions of steps 2 and 3, we can solve for v_{2f}.	$v_{2f} = \dfrac{2m_1}{m_1 + m_2} v_{1i}$
5. We can take this result and substitute it into the result from step 3 to solve for v_{1f}.	$v_{1f} = \dfrac{m_1 - m_2}{m_1 + m_2} v_{1i}$

6. Now that we know the final velocities, we can substitute them into our result from step 1 and solve for the maximum amplitudes after the collision.	$A_{1f} = v_{1f}\sqrt{\dfrac{L}{g}} = \left\lvert\dfrac{m_1 - m_2}{m_1 + m_2}\right\rvert v_{1i}\sqrt{\dfrac{L}{g}}$
	$= \left\lvert\dfrac{m_1 - m_2}{m_1 + m_2}\right\rvert A_{1i} = \dfrac{(1.4\,\text{kg}) - (2.0\,\text{kg})}{(1.4\,\text{kg}) + (2.0\,\text{kg})}(0.06\,\text{m})$
	$= 0.0106\,\text{m} = 1.06\,\text{cm}$
	$A_{2f} = v_{2f}\sqrt{\dfrac{L}{g}} = \dfrac{2m_1}{m_1 + m_2} v_{1i}\sqrt{\dfrac{L}{g}}$
	$= \dfrac{2m_1}{m_1 + m_2} A_{1i} = \dfrac{2(1.4\,\text{kg})}{(1.4\,\text{kg}) + (2.0\,\text{kg})}(0.06\,\text{m})$
	$= 0.0494\,\text{m} = 4.94\,\text{cm}$

Example #10—Interactive. Estimate the natural frequency of a man of mass 75.0 kg swinging from the end of a 5.00-m-long rope.

Picture the Problem. Assume the system can be approximated as a simple pendulum. In this case, the man's mass is simply window dressing because the period (and hence frequency) of a pendulum is determined only by its length. **Try it yourself.** Work the problem on your own, in the spaces provided, to get the final answer.

1. Find the frequency from the equation for the period of a simple pendulum.	
	$f = 0.223\,\text{Hz}$

Example #11. (a) How long is a simple pendulum whose period is 2.00 s? (Pendula with a period of 2 s are common in tall clocks.) (b) If this pendulum has a mass of 0.200 kg and a total energy of 0.0200 J, what is the amplitude of its oscillation?

Picture the Problem. The length of a pendulum can be determined only by the period. The amplitude of oscillation can be determined by its kinetic energy.

1. Find the length of the pendulum.	$T = 2\pi\sqrt{L/g}$
	$L = \left(\dfrac{T}{2\pi}\right)^2 g = \left(\dfrac{2.00\,\text{s}}{2\pi}\right)^2 (9.81\,\text{m/s}^2)$
	$= 0.994\,\text{m}$
2. Write an expression for the kinetic energy of the pendulum.	$K = \tfrac{1}{2}mv^2 = \tfrac{1}{2}m\left(-\omega A\sin(\omega t)\right)^2$

3. The total energy will be equal to the maximum kinetic energy, which will allow us to solve for the amplitude of oscillation.	$E = K_{max} = \frac{1}{2}mv_{max}^2 = \frac{1}{2}m(A\omega)^2 = \frac{1}{2}mA^2(2\pi/T)^2$ $A = \sqrt{\dfrac{2E}{m}}\dfrac{T}{2\pi} = \sqrt{\dfrac{2(.02\,\text{J})}{0.2\,\text{kg}}}\dfrac{2.00\,\text{s}}{2\pi} = 0.145\,\text{m}$

Example #12—Interactive. A damped harmonic oscillator of natural frequency 180 Hz loses 5 percent of its energy in each cycle. (*a*) What is the Q factor for this oscillator? (*b*) When this oscillator is driven, what is the width Δf of its resonance curve?

Picture the Problem. The Q value is defined in terms of the energy loss. The fractional width of the resonance of the oscillator is equal to $1/Q$. **Try it yourself.** Work the problem on your own, in the spaces provided, to get the final answer.

1. Find the Q factor of this oscillator from the fractional energy loss.	$Q = 126$
2. Find the width of the resonance curve from the natural frequency and Q.	$\Delta f = 1.40\,\text{Hz}$

Example #13. A child on a swing swings back and forth once every 3.75 s. The total mass of the child and the swing seat is 38.0 kg. At the bottom of the swing's arc, the child is moving at 1.82 m/s. (*a*) What is the total energy of the swing and the child? (*b*) If the Q factor is 25, what average power must be supplied to the swing to keep it moving with a constant amplitude? (Neglect the mass of the ropes.)

Picture the Problem. The child and swing can be modeled as a simple pendulum. The total energy is equal to the maximum kinetic energy of the child and swing. The Q factor is related to the fraction of energy dissipated in each cycle. The average power, energy dissipated per cycle, and period are also related. The average power supplied must equal the rate of energy dissipation in the swing.

1. Find the total energy of the swing and child. The child's maximum speed is at the bottom of the arc.	$E = K_{max} = \frac{1}{2}mv_{max}^2 = \frac{1}{2}(38\,\text{kg})(1.82\,\text{m/s}^2)^2 = 62.9\,\text{J}$
2. Because the swing has a Q factor, it must be losing energy each cycle.	$Q = 2\pi\dfrac{E}{|\Delta E|}$

3. The rate of energy loss per cycle is equivalent to the average power that must be delivered to the child and swing. We can get the power into the expression from step 2 by dividing by T/T.	$Q = 2\pi \dfrac{E/T}{\|\Delta E\|/T} = 2\pi \dfrac{E/T}{P_{av}}$ $P_{av} = \dfrac{2\pi E}{QT} = \dfrac{2\pi(62.9\,\text{J})}{25(3.75\,\text{s})} = 4.22\,\text{W}$

Example #14—Interactive. A certain damped harmonic oscillator loses 5 percent of its energy in each full cycle of oscillation. By what factor must the damping constant be increased in order to damp it *critically*? **Try it yourself.** Work the problem on your own, in the spaces provided, to get the final answer.

1. Relate the critical damping constant b_c with the mass and period.	
2. Obtain an equation relating the energy loss per cycle of a lightly damped oscillator with the energy, the damping constant, the mass, and the period.	
3. The requested factor is the ratio b_c/b.	251

Example #15. The rim of a 0.300-kg bicycle wheel 95.0 cm in diameter is hanging from a horizontal nail on the wall. If it is knocked slightly to one side and starts swinging, what is the period at which it oscillates?

Picture the Problem. This is a physical pendulum problem.

1. Write the expression for the period of a physical pendulum as a guide for the rest of the problem.	$T = 2\pi \sqrt{\dfrac{I}{mgd}}$
2. Find the moment of inertia for the wheel about the nail, using the parallel axis theorem.	$I = I_{cm} + MR^2 = MR^2 + MR^2 = 2MR^2$
3. The value of d is the distance of the center of mass to the rotation axis, in this case the radius of the wheel. Substitute everything in and solve for T.	$T = 2\pi \sqrt{\dfrac{2MR^2}{MgR}} = 2\pi \sqrt{\dfrac{2R}{g}}$ $= 2\pi \sqrt{\dfrac{0.95\,\text{cm}}{9.81\,\text{m/s}^2}} = 1.96\,\text{s}$

Example #16—Interactive. A nonuniform 400-g stick 120 cm long is found to have its center of mass 70 cm from one end. The stick is hung from a wall by a horizontal nail passing through a small hole 10 cm from the lighter end of the stick. When the stick is knocked gently to one side, it oscillates with a period of 1.8 s. (*a*) What is the moment of inertia of the stick about the nail? (*b*) What is the moment of inertia about an axis passing through the center of mass of the stick and parallel to the nail?

Picture the Problem. This is a physical pendulum problem. **Try it yourself.** Work the problem on your own, in the spaces provided, to get the final answer.

1. Find the moment of inertia of the stick from the equation for the period of a physical pendulum.	$I = 0.193\,\text{kg} \cdot \text{m}^2$
2. Use the parallel axis theorem to relate the moment of inertia about an axis along the top of the nail with the moment of inertia about the parallel axis through the center of mass.	$I = 0.049\,\text{kg} \cdot \text{m}^2$

Problem #17.   Oscillator number 1 has a peak resonance angular frequency of 19,500 s^{-1} and a Q factor of 20. Oscillator number 2 has a peak resonance angular frequency of 20,000 s^{-1} and a Q factor of 100. On the same graph sketch an estimate of the resonance curve for each oscillator showing the power delivered from a driving source versus the angular frequency of the source. On these sketches let the peak average power for each oscillator be the same.

Picture the Problem. The linewidths of each resonance peak can be found from the Q value and the resonance frequency of each oscillator.

1. Find the linewidth of each oscillator.	$\Delta\omega_1 = \dfrac{\omega_{01}}{Q_1} = \dfrac{19,500\,\text{s}^{-1}}{20} = 975\,\text{s}^{-1}$ $\Delta\omega_2 = \dfrac{\omega_{02}}{Q_2} = \dfrac{20,000\,\text{s}^{-1}}{100} = 200\,\text{s}^{-1}$

2. Plot the resonant peak power as a function of frequency.

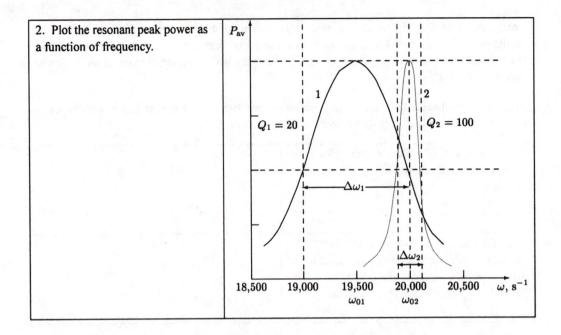

Example #18—Interactive. For the resonance curve shown in Figure 14-4 estimate the peak resonance frequency, the width of the resonance curve, and the Q factor.

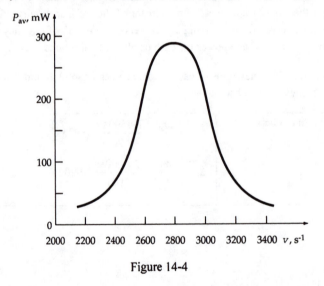

Figure 14-4

Picture the Problem. The peak frequency and the width of the resonance curve can be read from the graph. The Q factor can be determined from these two numbers. **Try it yourself.** Work the problem on your own, in the spaces provided, to get the final answer.

Answers: $\omega_0 = 2720\,\text{s}^{-1}$; $\Delta\omega = 400\,\text{s}^{-1}$; $Q = 6.8$

Chapter **15**

Traveling Waves

I.   Key Ideas

Wave motion is the transport of energy and momentum from one point to another without the transport of matter. Mechanical waves require a material medium through which to propagate (travel), whereas electromagnetic waves can propagate in a vacuum. For mechanical waves, a disturbance in a medium propagates because of the elastic properties of the medium.

Section 15-1. Simple Wave Motion. A wave in which the disturbance of the medium is perpendicular to the direction of propagation is called a **transverse wave.** Waves on a plucked string are examples of transverse waves. A wave in which the disturbance of the medium is parallel to the direction of propagation is called a **longitudinal wave.** Sound waves are longitudinal waves. Surface waves on water are a combination of the two types of waves.

Wave Pulses. If one end of a string under tension is given a flip, a wave pulse travels down the string. Energy and momentum are carried along the string, but the material of the string itself only twitches from side to side. The speed at which the pulse travels depends on the tension in the string and its mass density (mass per unit length). The size, shape, and location of a pulse is described by a **wave function** $y = f(x \pm vt)$, where y is the displacement of the string and v is the speed of the pulse. The plus sign is used to describe a pulse moving in the negative x direction, whereas the minus sign is for one moving in the positive x direction. The **wave speed** v is given by

$$v = \sqrt{\frac{F_T}{\mu}}$$ Speed of transverse waves on a taut string

where F_T is the tension in the string and μ is its mass per unit length. For a pressure wave in a fluid, like sound traveling through air, v is given by

$$v = \sqrt{\frac{B}{\rho}}$$ Speed of sound in a fluid

where B is the bulk modulus of the fluid and ρ is its mass per unit volume. (These formulas are derived by applying Newton's laws to a segment of the string or fluid.) For sound waves in a gas such as air this becomes

$$v = \sqrt{\frac{\gamma RT}{M}}$$ Speed of sound in a gas

where T is the absolute temperature in kelvins (K), $R = 8.314 \, \text{J}/(\text{mol} \cdot \text{K})$ is the universal gas constant, M is the molar mass of the gas (the number of kilograms per mole), and γ is a number that depends upon the molecular structure. For a monatomic gas $\gamma = 1.67$ and for a diatomic gas $\gamma = 1.40$.

The Wave Equation. The wave equation that expresses Newton's 2nd Law applied to an infinitesimal segment of a taut string is

$$\frac{\partial^2 y}{\partial x^2} = \frac{\mu}{F_T} \frac{\partial^2 y}{\partial t^2} = \frac{1}{v^2} \frac{\partial^2 y}{\partial t^2}$$ Wave equation

Section 15-2. Periodic Waves. If one end of a taut string is driven transversely in periodic motion (displacement a sinusoidal function of time), then a **periodic wave** is generated. If a periodic wave is traveling along a taut string or any other medium, each point along the medium oscillates with the same period.

Harmonic Waves. If a harmonic wave is traveling through a medium, each point of the medium oscillates in simple harmonic motion. When a harmonic wave travels along the string, the shape of the string at any particular instant in time is that of a sine function. The distance between two successive wave crests is called the wavelength λ. The wavelength is the distance after which the shape repeats itself. When a harmonic wave propagates along a string, each point on the string moves in simple harmonic motion with the same frequency f and period $T = 1/f$. During one period, the wave crest moves a distance equal to the wavelength. Then the speed of the wave is

$$v = \frac{\lambda}{T} = f\lambda = \frac{\omega}{k}$$ Wave speed

where we introduce the angular frequency ω, the frequency in radians per second, and the wave number k:

$$\omega = \frac{2\pi}{T} = 2\pi f$$ Angular frequency

$$k = \frac{2\pi}{\lambda}$$ Wave number

The wave function describing a harmonic wave traveling in the positive x direction with speed v is

$$y(x,t) = A\sin(kx - \omega t)$$ Displacement of harmonic transverse wave

where A is the **amplitude** of the wave. The argument of the sine function (θ is the argument of $\sin\theta$) is called the **phase.**

Next we consider the energy transported by waves on a string. The rate of energy transfer (power) at a point on a taut string is given by $P = \vec{F}_T \cdot \vec{v}_t = F_{Tx}v_x + F_{Ty}v_y = F_{Ty}v_y$, where $\vec{v}_t$ is the transverse velocity of the point. A taut string supports transverse waves where the point moves only in the transverse direction and $v_x = 0$. It follows that

$$P = F_{Ty}v_y = F_T \frac{\partial y}{\partial x}\frac{\partial y}{\partial t}$$

Power for waves on a taut string

where $v_y = \partial y/\partial t$ and $F_{Ty} = F_T \partial y/\partial t$. For a harmonic wave [$y = A\sin(kx - \omega t)$] this gives

$$P = \mu v \omega^2 A^2 \cos^2(kx - \omega t)$$

Power for harmonic waves on a taut string

where μv^2 has been substituted for the tension F_T. Since the average value of $\cos^2(kx - \omega t)$ is $\frac{1}{2}$, the average power at any point is

$$P_{av} = \tfrac{1}{2}\mu v \omega^2 A^2$$

Average power for harmonic waves on a taut string

Note that the average power is proportional to the square of the amplitude.

When considering the energy of a small segment of string, we know part of the energy is kinetic, due to its transverse motion, and the other part is potential, due to its stretching. This stretching is proportional to the magnitude of the slope $\partial y/\partial x$, which, for a harmonic wave, is greatest when the segment passes through the equilibrium position where its speed is also greatest.

In *harmonic sound waves* the molecules of the medium undergo simple harmonic motion. The wave function for such a wave is

$$s(x,t) = s_0 \sin(kx - \omega t)$$

Displacement for harmonic sound waves

where s is the displacement from equilibrium of the molecules and s_0 is the amplitude of their simple harmonic motion. These displacements cause variations (compressions and rarefactions) in the density of the medium, and therefore in the pressure of the medium. The associated pressure wave function is

$$p(x,t) = p_0 \sin(kx - \omega t - \pi/r)$$
$$= -p_0 \cos(kx - \omega t)$$

Pressure for harmonic sound waves

where p is the difference between the actual pressure and the equilibrium pressure and the pressure amplitude p_0 is the maximum value of p. The pressure is 90° ($\pi/2$ rad) out of phase with the displacement. The relation between the pressure amplitude and the displacement amplitude is

$$p_0 = \rho \omega v s_0$$

Pressure-amplitude/displacement-amplitude relation

Section 15-3. Waves in Three Dimensions. When a pebble is dropped onto the surface of a pond, a few surface waves ripple outwardly in a circular pattern. When a small, steadily vibrating object is placed on the surface of the pond, a steady train of surface waves will ripple outwardly in a circular pattern. When the vibrator is immersed deeply in the water, the steady train of compression maxima and minima produced will ripple outward in a spherical pattern. A spherical surface moving outward with one of these compression maxima is called a **wavefront** (a wavefront is a surface of constant phase). The outward motion of the wavefronts can be illustrated by **rays,** which are lines directed perpendicularly to the wavefronts. For the spherical wavefronts described here, the rays are straight lines directed away from the wave source.

Wave Intensity. The waves transport energy in the direction of the rays. If the source (the vibrator) generates waves steadily for a long time, the rate at which energy flows through any fixed spherical surface centered at the source equals the rate at which energy is generated by the source. The energy transferred during one cycle, divided by the duration of a cycle, is called the wave power P_{av}, and the wave intensity I is the wave power per unit area. I is measured in units of watts per square meter. The intensity a distance r from the point source is

$$I = \frac{P_{av}}{A} = \frac{P_{av}}{4\pi r^2} \qquad \text{Intensity of a point source}$$

where A is the surface area of a sphere of radius r. (In other words, for waves in three directions, the intensity varies inversely with the square of the distance from the source.)

If the incident wave is normal to the boundary, the power transferred through it is given by $P = \vec{F} \cdot \vec{v} = Fv$ where F is the force (pressure times area) exerted by the fluid and u is its velocity ($\partial s / \partial t$) at the boundary. It follows that the average power per unit area (intensity) is the pressure p times the velocity, averaged over one cycle. That is,

$$I = (pu)_{av} = \left(p\frac{\partial s}{\partial t}\right)_{av} = \tfrac{1}{2}\omega s_0 p_0 = \tfrac{1}{2}\rho\omega^2 s_0^2 v \qquad \text{Intensity of a harmonic sound wave}$$

The intensity must also equal the product of the average **energy density** η_{av} (energy per unit volume) and the wave speed v. It follows that

$$\eta_{av} = \frac{I}{v} = \tfrac{1}{2}\rho\omega^2 s_0^2 \qquad \text{Energy density of a harmonic sound wave}$$

One might expect that the sensation of loudness would vary directly with the intensity of sound. However, this is not the case. In fact, the sensation of loudness depends upon both intensity and frequency. However, the dependence of loudness on frequency will not be considered here. The sensation of loudness is found to vary more or less logarithmically with intensity, so the intensity level β is expressed in decibels (dB):

$$\beta = (10\,\text{dB}))\log_{10}\frac{I}{I_0} \qquad \text{Intensity level}$$

where I is the intensity of a sound and I_0 is a reference level which we take to be the threshold of hearing.

$$I_0 = 10^{-12} \text{ W/m}^2 \qquad \text{Threshold of hearing}$$

The threshold of pain occurs at an intensity of about 1 W/m^2. As the intensity varies from 10^{-12} W/m^2 to 1 W/m^2, the intensity level varies from 0 to 120 dB. A doubling of intensity results in a 3 dB increase in intensity level.

Section 15-4. Waves Encountering Barriers. When a wave is incident on a boundary that separates two regions of different wave speed, part of the wave is reflected and part is transmitted. A pulse on a string that is attached to a more dense string is reflected at the boundary as an inverted wave (an inversion corresponds to phase shift of 180°). However, if the second string is less dense than the first, the reflected pulse is not inverted. In either case, the transmitted pulse is not inverted. If the string is tied to a fixed point, the reflected pulse is inverted.

When a sound wave traveling in air strikes a solid or liquid surface, the reflected and incident rays make equal angles with the normal to the boundary, and the angle between the normal and the transmitted ray differs from that between the normal and the incident ray. This change of direction (bending) of the transmitted ray is called **refraction.** The angle that the refracted ray makes with the normal is greater or less than that of the incident ray, depending on whether the wave speed in the second region is less or greater than the wave speed in the incident medium.

Diffraction. In a homogeneous medium, such as air at constant density, sound waves travel in straight lines. In three dimensions a wavefront is a surface of constant phase, like a pressure maximum or minimum. At great distances from a compact source, a small part of a wavefront can be approximated by a plane (flat surface), and the wave propagates in a direction normal (perpendicular) to the wavefront. Such a wave is called a **plane wave.**

When a portion of a wavefront is blocked by an obstruction, the portions that are not blocked will spread out in all directions—including the region *behind* the obstruction. This spreading out, which in principle can be observed whenever a part of a wavefront is blocked, is called **diffraction.** Consider an obstruction consisting of a large screen that blocks waves incident upon it. If there is a hole (aperture) through the screen, and if the hole is just a fraction of a wavelength across, the portion of the wavefronts that pass through the hole will spread out—just as if the opening were a point source of waves. However, if the hole is many wavelengths across, the spreading (diffraction) of the wave into the region behind the screen will be hardly noticeable. The approximation that waves propagate along straight lines, that is, without diffracting, is known as the **ray approximation.** Wave motion for which the wavelengths are small compared to the size of holes can be adequately described via this approximation.

To experience the diffraction of sound, listen to your stereo from another room. Place yourself so the sound must pass through a doorway and spread out in order to reach your ears. The doorway will be narrow compared to the wavelength of the low (bass) frequencies and wide compared to the high (treble) frequencies. Low-frequency notes (like those of the bassoons) will therefore diffract as they pass through the doorway and you will be able to hear them. However, you will not be able to hear high-frequency notes (like those of the piccolos) because they will not diffract.

Section 15-5. The Doppler Effect. When a wave source and receiver are moving relative to each other, the rate at which the compressions leave the source is not the same as the rate at which they arrive at the receiver. Thus, the frequency observed by the receiver differs from the frequency emitted by the source. This phenomenon is called the **Doppler effect.** The relationship between the frequency emitted by the source, and the frequency measured by the receiver can be expressed as

$$f_r = \frac{v \pm u_r}{v \pm u_s} f_s$$

Doppler effect

where f_r is the frequency of the receiver, f_s is the frequency of the source, v is the speed of sound, u_r is the speed of the receiver, and u_s is the speed of the source. All speeds are relative to the propagating medium, which means Doppler shift problems are best done in a reference frame where there is no wind.

The correct choices for the signs are most easily obtained by remembering that the source moving toward the receiver produces an increase in frequency—as does the receiver moving toward the source. For example, if the receiver is moving toward the source and the source is moving toward the receiver, a plus sign is used in the numerator and a minus sign in the denominator. Notice that the observed frequency is increased when we have a smaller value for $v \pm u_s$ in the denominator. When the relative speed of approach u of the source and receiver is small compared to the speed of sound v, the **Doppler shift** in frequency Δf can be written

$$\frac{\Delta f}{f_s} \approx \pm \frac{u}{v} \quad (u \ll v)$$

Doppler shift

where the plus sign is used if the distance between the source and receiver is decreasing.

When a source moves at a speed greater than the speed of sound, a conical **shock wave** with the source at the apex forms behind the source at an angle θ with the path of the source. The angle θ between one side of the cone and the path of the source is related to the speed u of the source by the formula

$$\sin \theta = \frac{v}{u}$$

Mach angle

The ratio of the source speed u to the wave speed v is called the **Mach number.**

$$\text{Mach number} = \frac{u}{v}$$

Mach number

II. Physical Quantities and Key Equations

Physical Quantities

Universal gas constant	$R = 8.314 \, \text{J/(mol·K)}$
Gas constants	$\gamma = 1.67$ (monatomic gas)
	$\gamma = 1.40$ (diatomic gas)

Key Equations

Speed of transverse waves on a taut string

$$v = \sqrt{\frac{F_T}{\mu}}$$

Speed of sound in a fluid

$$v = \sqrt{\frac{B}{\rho}}$$

Speed of sound in a gas

$$v = \sqrt{\frac{\gamma RT}{M}}$$

Wave equation

$$\frac{\partial^2 y}{\partial x^2} = \frac{\mu}{F_T}\frac{\partial^2 y}{\partial t^2} = \frac{1}{v^2}\frac{\partial^2 y}{\partial t^2}$$

Wave speed

$$v = \frac{\lambda}{T} = f\lambda = \frac{\omega}{k}$$

Angular frequency

$$\omega = \frac{2\pi}{T} = 2\pi f$$

Wave number

$$k = \frac{2\pi}{\lambda}$$

Displacement of harmonic transverse wave

$$y(x,t) = A\sin(kx - \omega t)$$

Power for waves on a taut string

$$P = F_{Ty}v_y = F_T\frac{\partial y}{\partial x}\frac{\partial y}{\partial t}$$

Power for harmonic waves on a taut string

$$P = \mu v\omega^2 A^2 \cos^2(kx - \omega t)$$

Average power for harmonic waves on a taut string

$$P_{av} = \tfrac{1}{2}\mu v\omega^2 A^2$$

Displacement for harmonic sound waves

$$s(x,t) = s_0 \sin(kx - \omega t)$$

Pressure for harmonic sound waves

$$p(x,t) = p_0 \sin(kx - \omega t - \pi/r)$$
$$= -p_0 \cos(kx - \omega t)$$

Pressure-amplitude/displacement-amplitude relation

$$p_0 = \rho\omega v s_0$$

Intensity of a point source

$$I = \frac{P_{av}}{A} = \frac{P_{av}}{4\pi r^2}$$

Intensity of a harmonic sound wave

$$I = (pu)_{av} = \left(p\frac{\partial s}{\partial t}\right)_{av} = \tfrac{1}{2}\omega s_0 p_0 = \tfrac{1}{2}\rho\omega^2 s_0^2 v$$

Energy density of a harmonic sound wave	$\eta_{av} = \dfrac{I}{v} = \frac{1}{2}\rho\omega^2 s_0^2$
Intensity level	$\beta = (10\,\mathrm{dB}))\log_{10}\dfrac{I}{I_0}$
Threshold of hearing	$I_0 = 10^{-12}\ \mathrm{W/m^2}$
Doppler effect	$f_r = \dfrac{v \pm u_r}{v \pm u_s}f_s$
Doppler shift	$\dfrac{\Delta f}{f_s} \approx \pm\dfrac{u}{v}\ \ (u \ll v)$
Mach angle	$\sin\theta = \dfrac{v}{u}$
Mach number	$\text{Mach number} = \dfrac{u}{v}$

III. Potential Pitfalls

The term "amplitude" has the same meaning for a sinusoidal mechanical wave as it does for simple harmonic motion. It is the maximum displacement of a portion of the medium from its undisturbed, equilibrium position. It is not the distance from crest to valley.

Transverse waves propagating along a string disturb the string; that is, they displace it. Be careful not to confuse the transverse velocity of a segment of the string due to the wave with the velocity of the wave disturbance itself. The velocity of the wave is the rate at which the pattern of disturbance is moving along the string. The two velocities are not directly related and, in fact, are directed at right angles with each other. (In a longitudinal wave the two velocities are in opposite directions 50% of the time.)

The equation $v = \lambda/T = f\lambda$ is really just another way of stating that the speed equals the distance divided by the time. The wavelength is the distance and the frequency is the reciprocal of the period—the time it takes the disturbance to travel one wavelength. The wave speed v is determined by the properties of the medium that transports the wave and may or may not depend on the wave's frequency. The speed of sound in air, for example, is nearly independent of the frequency, whereas the speed of ripples on a pond depends strongly on the frequency.

It is really easy to make sign errors when doing Doppler-effect problems. Here is a case where simply memorizing formulas will not suffice. Remember, in the frequency formula given, u_r, u_s, and v are never negative because they are speeds, not velocities. A source moving in the direction of a receiver tends to produce an increase in the observed frequency, so the minus sign preceding u_s is used. A receiver moving in the direction of the source tends to produce an increase in frequency, so the plus sign preceding u_r is used. If both the source and receiver are moving, u_s and u_r are the speeds of the source and receiver relative to the medium, not to each other.

Although the terms intensity and intensity level are similar, they represent quite different quantities. The intensity I is the power per unit area, whereas the intensity level β equals $(10\,\text{dB})\log_{10}(I/I_0)$, where I_0 is a reference intensity. We suggest that you memorize that 3 dB refers to an intensity ratio of 2.

To simplify physics problems we usually assume that a sound source is isotropic—that is, the intensity is broadcast uniformly in all directions. As you know, actual sound sources, such as a loudspeaker or a human voice, generally have intensities that are greater in the forward direction.

IV. True or False Questions and Responses

True or False

_____ 1. A function of the form $y(x+vt)$ describes the shape or pattern represented by $y(x)$ moving with speed v in the $+x$ direction.

_____ 2. A harmonic wave is one in which the wave function has a sinusoidal shape.

_____ 3. The wavelength of a harmonic wave on a string is the distance along the string of one complete cycle of the wave.

_____ 4. The speed of a harmonic wave is the wavelength divided by the frequency.

_____ 5. Every element of a string on which a harmonic wave is propagating is undergoing simple harmonic motion.

_____ 6. If two elements of a string on which a harmonic wave is propagating move in phase, the two elements must be an integral number of wavelengths apart.

_____ 7. A wave motion in which the displacement of the propagating medium is parallel with the direction of propagation of the disturbance is called a longitudinal wave.

_____ 8. For a given pressure, the speed of sound in air is independent of temperature.

_____ 9. For a given temperature, the speed of sound in air is independent of pressure.

_____ 10. The speed at which sound waves propagate in a material medium depends upon the speed of the sound source relative to the medium.

_____ 11. In a medium transporting a harmonic sound wave, the pressure at a given point is 90° out of phase with the displacement at that point.

_____ 12. The ratio of the pressure amplitude to the displacement amplitude of a sound wave depends upon the wave frequency.

_____ 13. A wave pulse reflected at the end of a string is always inverted.

_____ 14. A wave passing through a very small hole in a barrier propagates outward from the hole in all directions on the far side of the barrier, as though from a point source.

___ 15. The received frequency of a sound wave is always equal to the frequency of the source, even when there is relative motion between the source and the receiver.

___ 16. The Doppler shift of a wave frequency depends only upon the relative speed between the receiver and the propagating medium.

___ 17. The intensity of a harmonic wave is the power per unit area.

___ 18. The intensity level of a harmonic wave is the logarithm of the intensity.

___ 19. The decibel is a unit of intensity.

___ 20. When the source and the receiver are moving such that they are getting closer together, the received frequency is higher than the frequency of the source.

Responses to True or False

1. False. It describes the shape or pattern represented by $y(x)$ moving with speed v in the $-x$ direction.

2. True.

3. True.

4. False. The speed is the wavelength divided by the period, that is, the wavelength multiplied by the frequency.

5. True.

6. True. The motions at distances separated by an integral number of wavelengths are identical.

7. True.

8. False. It is proportional to the square root of the absolute temperature.

9. True.

10. True.

11. True.

12. True.

13. False. Reflections at a free end are not inverted.

14. True.

15. False. That the frequency received and the frequency of the source differ when there is relative motion between the source and the receiver is known as the Doppler effect.

16. False. It depends on the speed of sound, the speed of the source, the speed of the receiver, on whether the source is moving toward the receiver or not, and on whether the receiver is moving toward the source or not.

17. True.

18. False. The intensity level is ten times the logarithm, to the base ten, of the ratio of the intensity to the reference intensity.

19. False. The decibel is a unit of intensity level.

20. True.

V.   Questions and Answers

Questions

1. For a function to be a wave function it must satisfy certain criteria. One necessary criterion is that it satisfy the condition

    $$y(x,t) = y_1(x - vt) + y_2(x + vt)$$

    A string must be continuous. Therefore only continuous functions can be wave functions for waves on a string.

    Which of the following functions can be wave functions for waves on a string?

    (a) $y(x, t) = Ae^{-(x-vt)^2/2a^2}$

    (b) $y(x, t) = Ae^{-x^2}e^{+bt}$

    (c) $y(x, t) = 0 \quad x \le vt - a$
         $y(x, t) = D \quad vt - a \le x \le vt + a$
         $y(x, t) = 0 \quad x \ge vt + a$

Pulse velocity

Figure 15-1

The function in (a) looks like the pulse shown in Figure 15-1 and the function in (c) is shown in Figure 15-2.

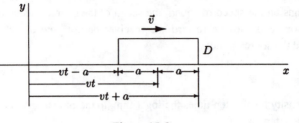

Figure 15-2

2. A Slinky is a children's toy that is just a long, loose-coiled spring. It is quite useful as a wave demonstrator. How would you generate longitudinal waves in a Slinky? What about transverse waves?

3. A tuning fork, which vibrates at a fixed frequency, is being waved back and forth along the direction from it to you. Describe what you hear. What will you hear if the fork is moved from side to side?

4. Consider a sound wave that propagates uniformly in all directions from a point source. How does the amplitude of the wave vary with distance from the source?

5. A nice, pleasant sound level at which to listen to music is 60 dB. Would 120 dB be twice as loud?

6. Imagine that you are listening to orchestral music being played on a stereo that is in another room down the hall. Will diffraction have any effect on the sound you hear?

7. Why is it that we can hear, but not see, around corners?

Answers

1. (*a*) The function $y(x, t) = Ae^{-(x-vt)^2/2a^2}$ is a possible wave function, as y is a function of $x - vt$.

   (*b*) The function $y(x, t) = Ae^{-x^2}e^{+bt}$ cannot be expressed as the sum of a function of $x + vt$ and a function of $x - vt$. Therefore it cannot be a wave function.

   (*c*) The function

   $$y(x, t) = 0 \quad x \le vt - a$$
   $$y(x, t) = D \quad vt - a \le x \le vt + a$$
   $$y(x, t) = 0 \quad x \ge vt + a$$

   is a function of $x - vt$. This can be seen by subtracting vt from each term in each of the inequalities giving

$y(x, t) = 0 \quad x - vt \le a$

$y(x, t) = D \quad -a \le x - vt \le a$

$y(x, t) = 0 \quad x - vt \ge a$

It would represent a rectangularly shaped pulse of height D and width $2a$ traveling in the positive x direction with speed v. However, this function is discontinuous at the points $x = vt - a$ and $x = vt + a$. A string cannot be discontinuous, so this function cannot be a wave function.

Discontinuous functions change abruptly and thus are not possible wave functions. In textbooks, however, discontinuous functions are sometimes used as wave functions. You should think of these as continuous functions that are close approximations to the discontinuous functions. Figure 15-3 shows a continuous function that is a close approximation to the discontinuous function in (c).

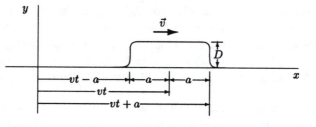

Figure 15-3

2. You can generate longitudinal waves in a Slinky very easily by tying one end of the spring to a wall and stretching it to a moderate degree. Moving the free end back and forth along the line of the spring will produce easily visible compression waves that travel at a few meters per second (see Figure 15-4a). Transverse waves are generated by moving the free end up and down (see Figure 15-4b.)

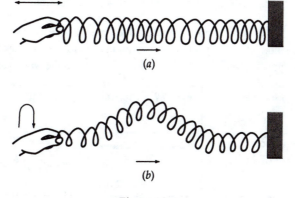

(a)

(b)

Figure 15-4

3. You will hear the pitch rise and fall as the tuning fork is moved toward or away from you, but you will hear only a steady pitch as it is moved from side to side.

4. For a wave propagating uniformly in all directions in three dimensions, the intensity decreases inversely with the square of the distance from the source. The intensity is also proportional to the square of the amplitude. Thus the amplitude must decrease inversely with the first power of the distance from the source.

5. No. In fact, just what "twice as loud" means isn't clear because loudness is a function of perception, but intensity level isn't. Typically, a tone sounds twice as loud when the intensity increases by a factor of ten, which corresponds to an intensity level increase of 10 dB. The 60 dB increase means the sound intensity is increased by a factor of 10^6 (or one million).

6. The diffraction of sound waves is the main reason you can hear "around corners" from the music room to the hall and then into your room. The smallest aperture involved is likely to be the width of a doorway, around 1 m. Sound waves with high frequencies and short wavelengths (much less than 1 m) diffract less than do waves with low frequencies and wavelengths much longer than 1 m. Thus, the high frequency sound waves reaching your ears are diminished more than are the low frequency sound waves. You will hear high notes, and particularly the high harmonics that give musical sounds their tone quality, less effectively, so the music will sound "flat" or "muffled" to you. Walls, drapes, and other surfaces selectively absorb some frequencies better than others. This will also affect what the listener hears.

7. The obstacles, apertures, and such that surround us in everyday circumstances—doorways, for instance—tend to be not very large compared to the wavelengths of audible sound, but all are very large compared to the wavelengths of visible light. Thus, sound waves are significantly diffracted, but light waves are not.

VI. Problems, Solutions, and Answers

Example #1. Far out at sea, your ship is traveling at a speed of 12.0 mi/hr with respect to the water in the same direction as the waves. You notice that one swell passes you every 9.00 s. If the waves are 500 ft apart, what is the speed of the waves relative to the water?

Picture the Problem. The wavelength divided by the observed period gives the velocity of the waves relative to the ship. Knowing your velocity with respect to the water and the velocity of the waves with respect to you, the velocity of the wave with respect to the water can be found.

1. Find the velocity of the waves relative to the ship.	$v = \lambda / T = (500\,\text{ft}) / (9\,\text{s}) = 55.6\,\text{ft/s}$
2. Using the relative velocity equation, find the speed of the waves relative to the water.	$v_{\text{waves,water}} = v_{\text{waves,ship}} + v_{\text{ship,water}}$ $= (55.6\,\text{ft/s}) + (12\,\text{mi/hr})(5280\,\text{ft/mi})(1\,\text{hr}/3600\,\text{s})$ $= 73.2\,\text{ft/s}$

Example #2—Interactive. Far out at sea, your sailboat is traveling at a speed of 8.00 mi/hr with respect to the water in the same direction as the waves. You notice that the boat moves so that it remains on top of a single wave crest. What is the speed of the waves relative to the water?

Picture the Problem. This is a relative velocity problem. **Try it yourself.** Work the problem on your own, in the spaces provided, to get the final answer.

1. Using the relative velocity equation, find the speed of the waves relative to the water.	$v_{\text{waves,water}} = 8\,\text{mi/hr}$

Example #3. The wave function for a harmonic wave on a string is $y(x,t) = 0.002\sin(31.4x - 628t + 1.57)$, where y and x are in meters and t is in seconds. (a) Find the amplitude, wavelength, frequency, period, and phase constant of this wave. (b) In what direction does this wave travel and what is its speed? (c) At $t = 0.02$ s find the phase at $x = 0.3$ m.

Picture the Problem. Use the formula of the harmonic wave to identify the terms in the wave function. The direction of travel can be determined from the relative sign of the position and time terms in the argument of the sine function. The phase is the argument of the sin function.

1. Compare the general formula of a harmonic wave with the function given to find the appropriate parameters.	$y(x,t) = A\sin(kx - \omega t + \delta)$ $A = 0.002\,\text{m}$ $\lambda = 2\pi/k = 2\pi/(31.4\,\text{rad/m}) = 0.200\,\text{m}$ $f = \omega/(2\pi) = (628\,\text{rad/s})/(2\pi) = 100\,\text{Hz}$ $T = 1/f = 0.01\,\text{s}$ $\delta = 1.57\,\text{rad}$
2. The direction of the wave is in the positive direction because the position and time terms have opposite signs. The speed can be found from the wavelength and frequency.	$v = \lambda f = (0.200\,\text{m})(100\,\text{Hz}) = 20.0\,\text{m/s}$
3. The phase is the argument of the sine function.	$\theta = kx - \omega t + \delta$ $= (31.4\,\text{rad/m})(0.3\,\text{m}) - (628\,\text{rad/s})(0.02\,\text{s}) + 1.57\,\text{rad}$ $= -1.57\,\text{rad}$

Example #4—Interactive. The wave function for a harmonic wave on a string is $y(x,t) = 0.001\sin(62.8x + 314t - 1.57)$ where y and x are in meters and t is in seconds. (a) Find the amplitude, wavelength, frequency, period, and phase constant of this wave. (b) In what direction does this wave travel and what is its speed? (c) Find the phase difference between the points at $x = 0.50$ m and at $x = 0.55$ m at any time t.

Picture the Problem. Use the formula of the harmonic wave to identify the terms in the wave function. The direction of travel can be determined from the relative sign of the position and time terms in the argument of the sine function. The phase is the argument of the sin function. **Try it yourself.** Work the problem on your own, in the spaces provided, to get the final answer.

1. Compare the general formula of a harmonic wave with the function given to find the appropriate parameters.	$A = 0.001\,\text{m}$ $\lambda = 0.100\,\text{m}$ $f = 50\,\text{Hz}$ $T = 0.02\,\text{s}$ $\delta = -1.57\,\text{rad}$
2. The direction of the wave can be found by the relative sign of the position and time terms. The speed can be found from the wavelength and frequency.	$v = 5.0\,\text{m/s}$ in the negative x direction
3. The phase is the argument of the sine function.	$\Delta\theta = 3.14\,\text{rad}$

Example #5. Show by direct substitution that the function $y = (x - vt)^2$ is a solution of the wave equation.

Picture the Problem. Substitute the function into the wave equation, and show you get an equality.

1. Substitute and go.	$$\frac{\partial^2 y}{\partial x^2} = \frac{1}{v^2}\frac{\partial^2 y}{\partial t^2}$$ $$\frac{\partial^2 (x - vt)^2}{\partial x^2} = \frac{1}{v^2}\frac{\partial^2 (x - vt)^2}{\partial t^2}$$ $$\frac{\partial\left[2(x - vt)\right]}{\partial x} = \frac{1}{v^2}\frac{\partial\left[-2v(x - vt)\right]}{\partial t}$$ $$2 = 2$$

Example #6—Interactive. Show by direct substitution that the functions $y = \cos(x + vt)$ and $y = -7(x - vt)^2 + 4\cos(x + vt)$ each satisfy the wave equation.

Picture the Problem. Substitute the function into the wave equation, and show you get an equality. **Try it yourself.** Work this proof on your own, in the space below.

Example #7. The temperature inside a house is maintained at 20.0°C, so the speed of sound in the air within the house is 340 m/s. On a cold day the speed of sound in the air outside the house is 334 m/s. A tuning fork in the house is struck, and starts vibrating at 440 Hz. A boy outside the house hears the sound from the tuning fork. What frequency does he hear? Determine the wavelength of the wave both in the air inside the house and in the air outside the house.

Picture the Problem. Determine the frequency with which the sound wave enters the cold air. Determine the frequency of the sound in the cold air. Wavelength can be found from the wave speed and frequency.

1. The frequency of the wave in the warm air is 440 Hz, the same as the tuning fork. The wave enters the cold air at the same frequency, so the boy outside still hears the same 440 Hz frequency.	
2. Find the wavelengths involved.	$\lambda_{cold} = v_{cold} / f = (334 \, \text{m/s}) / (440 \, \text{Hz}) = 0.759 \, \text{m}$ $\lambda_{house} = v_{house} / f = (330 \, \text{m/s}) / (440 \, \text{Hz}) = 0.773 \, \text{m}$

Example #8—Interactive. The extreme range of the human singing voice is roughly from 53 Hz (A flat below bass low C) to 1267 Hz (E flat above soprano high C). Find the range of wavelengths that corresponds to this frequency range.

Picture the Problem. The wavelengths can be found from the speed of sound in air and the frequencies. **Try it yourself.** Work the problem on your own, in the space provided, to get the final answer.

1. Find the wavelength range.	
	$6.40 \, \text{m} \geq \lambda \geq 26.8 \, \text{cm}$

Example #9. Air ($\gamma = 1.40$) has a molar mass of 28.9 g/mol and helium ($\gamma = 1.67$) has a molar mass of 4.00 g/mol. (*a*) At room temperature ($20°\text{C} = 293 \, \text{K}$), calculate the speed of sound in air and in helium. (*b*) In each medium calculate the frequency of a sound with a wavelength of 2.2 m.

Picture the Problem. Use the formula for the speed of sound waves in a gas in terms of temperature, molar mass, and γ. The frequency can be found from the speed of sound and the wavelength.

1. Find the speed of sound in air and helium at room temperature.	$v = \sqrt{\dfrac{\gamma RT}{M}}$ $v_{air} = \sqrt{\dfrac{1.40\left(8.31\,\text{J}/(\text{mol}\cdot\text{K})\right)(293\,\text{K})}{0.0289\,\text{kg/mol}}} = 343\,\text{m/s}$ $v_{helium} = \sqrt{\dfrac{1.67\left(8.31\,\text{J}/(\text{mol}\cdot\text{K})\right)(293\,\text{K})}{0.00400\,\text{kg/mol}}} = 1010\,\text{m/s}$
2. Calculate the frequency of a sound with a wavelength of 2.2 m in each medium.	$f = v/\lambda$ $f_{air} = v_{air}/\lambda = (343\,\text{m/s})/(2.20\,\text{m}) = 156\,\text{Hz}$ $f_{helium} = v_{helium}/\lambda = (1010\,\text{m/s})/(2.20\,\text{m}) = 459\,\text{Hz}$

Example #10—Interactive. You stand on the edge of a canyon 420 m wide and clap your hands. An echo comes back to you 2.56 s later. What is the temperature of the air? For air, $\gamma = 1.40$.

Picture the Problem. The speed of the wave can be found from the distance and time provided. Remember the sound has to go across the canyon and back. Use the formula for the speed of sound waves in a gas in terms of temperature, molar mass, and γ to find the temperature. **Try it yourself.** Work the problem on your own, in the spaces provided, to get the final answer.

1. Calculate the speed of the wave from the kinematic quantities given.	
2. Find the temperature.	$T = 267\,\text{K} = -5.7°\text{C}$

Example #11. Suppose each of the two loudspeakers of my stereo system is delivering 1 watt of average power in the form of audible sound in my office. Also suppose that this power is being delivered uniformly in all directions. (*a*) If I am sitting 2.5 m away from the speakers, what is the intensity of the sound I hear directly from the speakers (that is, neglecting the sound reflected from the walls of the room)? (*b*) To what intensity level does this correspond?

Picture the Problem. Calculate the direct sound intensity from the speakers using their power output and the receiver's distance from them. Convert the calculated intensity to an intensity level in dB.

1. Calculate the intensity from one of the speakers. Because the two loudspeakers are not coherent, the intensity for two speakers is twice the intensity for one speaker.	$I_1 = \dfrac{P_{av}}{4\pi r^2} = \dfrac{1\,\text{W}}{4\pi(2.5\,\text{m})^2} = 1.27 \times 10^{-2}\ \text{W/m}^2$ $I_{tot} = 2I_1 = 2.54 \times 10^{-2}\ \text{W/m}^2$
2. Find the intensity level. This is an extremely loud sound that would likely chase you out of the room. You don't buy amplifiers that put out lots of power because you need that much acoustical power; rather, it's because good-quality loudspeakers have low efficiency.	$\beta = 10\log_{10}\dfrac{I}{I_0}$ $= 10\log_{10}\dfrac{2.54 \times 10^{-2}\ \text{W/m}^2}{10^{-12}\ \text{W/m}^2} = 104\,\text{db}$

Example #12—Interactive. Assume that a barking dog can put out (on the average) 1 mW of sound power and that the sound propagates uniformly in all directions. What is the sound intensity level (in dB) 6 m away from a pen containing eight barking dogs?

Picture the Problem. Calculate the direct sound intensity from the dogs using their power output and the receiver's distance from them. Convert the calculated intensity to an intensity level in dB. **Try it yourself.** Work the problem on your own, in the spaces provided, to get the final answer.

1. Calculate the intensity from the dogs	
2. Find the intensity level.	$\beta = 72\,\text{dB}$

Example #13. A woman with perfect pitch, standing next to a railroad track on a still day, is amused to notice that the whistle of an approaching train is sounding a true concert A (440 Hz). After the train has passed her and is receding, the pitch has dropped to a true F natural (349 Hz). (*a*) How fast was the train going? (*b*) What was the actual frequency of the whistle?

Picture the Problem. The first frequency is heard when the train is coming straight toward the woman. Write a relationship for the received frequency in terms of the unknown source frequency, the source speed, the receiver's speed, and the speed of the sound. Repeat this for the second frequency, heard when the train is receding from the woman at the same source speed. Solve these two equations for the velocity of the train, which you can then use to find f_s.

1. Write a relationship for the received frequency in terms of the frequency of the source. The receiver is stationary. Initially, the source is moving toward the receiver. This motion results in an increase in the received frequency, so the minus sign in the denominator is used. When the train is receding, a lower frequency is heard, so we use the plus sign in the denominator.	$f_r = \dfrac{v \pm u_r}{v \pm u_s} f_s$ $f_{r1} = \dfrac{f_s}{1 - u_s/v}$ $f_{r2} = \dfrac{f_s}{1 + u_s/v}$
2. Taking the ratio of the two resulting frequencies we can solve for u_s.	$\dfrac{f_{r1}}{f_{r2}} = \dfrac{1 + u_s/v}{1 - u_s/v}$ $f_{r1} - f_{r1}u_s/v = f_{r2} + f_{r2}u_s/v$ $(f_{r1} + f_{r2})u_s/v = f_{r1} - f_{r2}$ $u_s = \dfrac{f_{r1} - f_{r2}}{f_{r1} + f_{r2}} v = \dfrac{(440\,\text{Hz}) - (349\,\text{Hz})}{(440\,\text{Hz}) + (349\,\text{Hz})}(340\,\text{m/s})$ $u_s = 39.2\,\text{m/s}$
3. Now substitute this value into the equation for f_{r1} and solve for f_s.	$f_s = f_{r1}\left(1 - \dfrac{u_s}{v}\right) = (440\,\text{Hz})\left(1 - \dfrac{39.2\,\text{m/s}}{340\,\text{m/s}}\right) = 389\,\text{Hz}$

Example #14—Interactive. My car's horn sounds a tone of frequency 250 Hz. If I am driving directly at you at 25.0 m/s, blowing my horn, on a hot, still day ($93.0°\,\text{F}$), (a) what is the wavelength of the sound that reaches you? (b) What frequency do you hear?

Picture the Problem. You will need to recalculate the speed of sound at the temperature given. The wavelength of the sound you hear will be reduced according to the Doppler equations. You can then find the frequency from the speed of sound and the wavelength. **Try it yourself.** Work the problem on your own, in the spaces provided, to get the final answer.

1. Calculate the speed of sound at the given temperature.	
2. Find the wavelength of the sound that reaches you.	$\lambda = 1.31\,\text{m}$
3. Find the frequency that you hear.	$f = 269\,\text{Hz}$

Example #15. On a still day, a car moves at a speed of 40.0 km/hr directly toward a stationary wall. Its horn emits a sound at a frequency of 300 Hz. (a) If the velocity of sound is 340 m/s, find the frequency at which the sound waves hit the wall. (*b*) The waves reflect off the wall and are received by the driver of the car. What frequency does she hear?

Picture the Problem. Work this problem in steps. Find the frequency of the sound "received" by the wall. Find the frequency of the sound the driver hears after the wave has bounced off the wall. Remember that in this part the "receiver" is also moving.

1. Convert the speed of the car to m/s.	$v_c = (40\,\text{km/hr})(1000\,\text{m/km})/(3600\,\text{s/hr}) = 11.1\,\text{m/s}$
2. Find the received frequency at the wall. Because the car is moving toward the wall, the frequency at the wall will be greater than the source frequency.	$f_r = \dfrac{v \pm u_r}{v \pm u_s} f_s$ $f_{r1} = \dfrac{f_s}{1 - v_c/v} = \dfrac{300\,\text{Hz}}{1 - (11.1\,\text{m/s})/(340\,\text{m/s})}$ $= 310\,\text{Hz}$
3. Find the frequency the driver hears. The frequency that hits the wall is reflected at that same frequency, so the wall acts like a source of a 310 Hz sound that is not moving. Because the car is moving toward the source, the frequency heard will again be greater than the source frequency.	$f_r = \dfrac{v \pm u_r}{v \pm u_s} f_r$ $f_{r2} = (1 + v_c/v) f_{r1}$ $= (1 + (11.1\,\text{m/s})/(340\,\text{m/s}))(310\,\text{Hz})$ $= 320\,\text{Hz}$

Example #16—Interactive. On a still day, a car traveling at 100 km/hr sounds its horn, which has a frequency of 250 Hz. What frequency is heard by a receiver who is proceeding at 60 km/h in the same direction (*a*) directly ahead of the car and (*b*) directly behind it?

Picture the Problem. Both the source and the receiver are moving. In part (*a*), the source is moving toward the receiver, which will cause an upward shift in the received frequency. Choose the correct sign in the denominator for this to happen. The receiver, however, is moving away from the source, which will cause a downward shift in the received frequency. Choose the correct sign in the numerator to provide this downward shift. In part (*b*), the source is moving away from the receiver and the receiver is moving toward the source. Use the signs that tend to provide the appropriate frequency shifts. **Try it yourself.** Work the problem on your own, in the spaces provided, to get the final answer.

1. Find the received frequency for the receiver ahead of the car.	
	259 Hz

2. Find the received frequency for the receiver behind the car.	
	242 Hz

Chapter 16

Superposition and Standing Waves

I.   Key Ideas

Section 16-1. Superposition of Waves. As a wave travels along a string, the transverse displacement y of the string varies with the position x along the string, and time t. The function $y = y(x, t)$ is called the **wave function.** The wave function $y_1(x - vt)$ gives the transverse displacement of the string for a wave moving in the positive x direction with speed v. The wave function $y_2(x + vt)$ gives the transverse displacement of the string for a wave moving in the negative x direction with speed v.

When two waves traveling along the same string superpose (overlap), the resultant displacement y of the string is the algebraic sum of the displacements y_1 and y_2 that would occur if only one or the other of the waves were present. The relation $y = y_1 + y_2$ is called the **principle of superposition.** Each wave propagates as if the other weren't there. Pulses traveling in opposite directions pass right through each other, emerging with their original size and shape.

The most general wave function $y(x, t)$ can always be expressed as the superposition of two wave functions describing waves traveling in opposite directions:

$$y(x, t) = y_1(x - vt) + y_2(x + vt) \qquad \text{Superposition principle}$$

The superposition of harmonic waves is called **interference.** Consider two harmonic waves of equal amplitude and frequency that are simultaneously traveling along the same string. If the crests of one of these waves exactly superpose the crests of the other, the difference between the phases of the two waves is zero. Such waves are said to be in phase and the resultant wave will have an amplitude equal to the sum of the amplitudes of the original waves. This phenomenon is called **constructive interference.** On the other hand, when the crests of one of these waves exactly superpose the troughs of the other, the difference between the phases of the two waves is $180°$ (π rad) and the resultant wave has an amplitude equal to zero—the difference between the amplitudes of the original waves. This is called **destructive interference.** For interference to occur, the waves do not have to be either in phase or $180°$ out of phase. They do, however, have to have a phase difference that remains fixed (is constant). For two waves y_1 and y_2 with an arbitrary phase difference δ, the two waves interfere, giving rise to the wave function y of the resultant wave

$$y_1 = A \sin(kx - \omega t)$$

$$y_2 = A \sin(kx - \omega t + \delta)$$

$$y = y_1 + y_2$$

Interference of harmonic waves

$$= A \sin(kx - \omega t) + A \sin(kx - \omega t + \delta)$$

$$= 2A \cos\left(\frac{\delta}{2}\right) \sin\left(kx - \omega t + \frac{\delta}{2}\right)$$

where $2A\cos(\delta/2)$ and $\delta/2$ are the amplitude and phase constant of the resultant harmonic wave. An analogous equation describes the resultant wave when two sound waves interfere. For a sound wave the y representing the transverse displacement of the medium is replaced either by s for the longitudinal displacement or by p for pressure.

Beats. When two harmonic waves with the same amplitude, but with different angular frequencies ω_1 and ω_2, overlap, the resulting pressure at a given point may be expressed in terms of the difference in angular frequency $\Delta\omega$ and the average angular frequency ω_{av}:

$$p = p_1 + p_2 = p_0 \sin(\omega_1 t) + p_0 \sin(\omega_2 t)$$

$$= 2p_0 \cos\left(\tfrac{1}{2}\Delta\omega t\right) \sin(\omega_{av} t)$$

Superposition of harmonic waves at a given point

where $\Delta\omega = \omega_1 - \omega_2$ and $\omega_{av} = (\omega_1 + \omega_2)/2$. The resultant wave can also be expressed in terms of the frequencies by substituting $2\pi \Delta f$ and $2\pi f_{av}$ for $\Delta\omega$ and ω_{av}. When the difference in frequency is small compared with the average frequency, the term $2\pi(1/2)\Delta f\, t$ varies slowly in comparison with the term $2\pi f_{av} t$. The resulting pressure at a fixed point varies with the frequency f_{av}. Therefore, the ear hears the intensity periodically getting louder and then quieter at a frequency Δf. This frequency is called the **beat frequency** f_{beat}:

$$f_{beat} = |\Delta f| = |f_1 - f_2|$$

Beat frequency

The interference of two waves of different but nearly equal frequencies produces a tone whose intensity varies alternately between loud and soft—a phenomenon known as **beats.**

Phase Difference due to Path Difference. A common cause of phase difference between two waves is the difference in path length between each source and the point of interest. The formula relating the phase difference δ with the associated path-length difference Δx is

$$\delta = (2\pi \,\text{rad})\frac{\Delta x}{\lambda} = (360°)\frac{\Delta x}{\lambda}$$

Phase difference–path-length difference relation

Coherent sources are sources that vibrate so that the phase difference between the two sources remains fixed. Coherent sound sources can be produced by two loudspeakers that are driven by the same electrical signal. Sources that vibrate so that the phase difference is not constant but varies randomly over time are called **incoherent sources.** Two violins playing the same music in an orchestra are incoherent sources.

Even when two coherent sources vibrate in phase, the waves are typically not in phase when they arrive at a given point of interest. Suppose there are two loudspeakers vibrating coherently and in phase. If you are half a wavelength farther from one of the speakers than the other, the sound from the two speakers will arrive at your ears 180° out of phase. At such a location you hear nothing because the interference is destructive. (We are considering only the sound that travels directly from the speaker to you, neglecting any that reaches you after reflecting off a wall or some other object.) Destructive interference occurs in locations where the difference in path lengths is an odd number of half wavelengths. Constructive interference occurs any time this difference is an even number of wavelengths. A region where destructive interference occurs is called a **node.** A region where constructive interference occurs is called an **antinode.**

Section 16-2. Standing Waves. Waves traveling along a taut string with one or both ends fixed reflect at the ends of the string. For harmonic waves of certain specific frequencies, the waves reflecting back and forth can superpose to form a stationary vibration pattern called a **standing wave.**

Standing Waves on Strings. When one end of the string is vibrated at one of these frequencies, it responds "enthusiastically." This enthusiastic response is called **resonance,** and the frequencies at which the string resonates (vibrates in a standing wave) are called the **resonant frequencies** of the string. On a string that is vibrating in a standing wave there are locations called **nodes** where the string remains stationary. Between each pair of adjacent nodes is a location of maximum amplitude called an **antinode.**

The standing-wave patterns of a string that is fixed at both ends are shown in Figure 16-1. The lowest resonance frequency for such a string is called its **fundamental frequency** f_1. When the string vibrates at its fundamental frequency, the wave pattern produced is referred to as the **fundamental** mode of vibration or the **first harmonic**. The wave pattern produced when the string vibrates at the second lowest resonance frequency f_2 is called the **second harmonic,** and the pattern when it vibrates at the nth lowest frequency f_n is the ***n*th harmonic.** For these standing waves, the resonance frequency f_n of the nth harmonic is related to the fundamental frequency f_1 by

$$f_n = nf_1 \quad n = 1, 2, 3, \ldots \qquad \text{Resonance frequencies, both ends fixed}$$

where

$$f_1 = \frac{v}{2L} = \frac{1}{2L}\sqrt{\frac{F_T}{\mu}} \qquad \text{Fundamental frequency}$$

For a flexible string fixed at both ends, the wavelength λ_n of the nth harmonic is related to the length L of the string by the equation

$$L = n\frac{\lambda_n}{2} \quad n = 1, 2, 3, \ldots \qquad \text{Standing-wave condition, both ends fixed}$$

The wavelength of a standing wave is the wavelength of the harmonic waves that interfere to form the resultant standing wave.

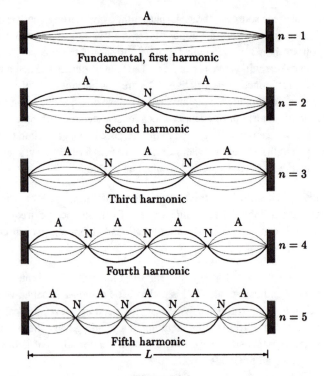

Figure 16-1

In the terminology often used in music, the frequencies of the second and higher harmonics are called **overtones.** Thus the frequency of the second harmonic is called the *first overtone,* that of the third harmonic is called the *second overtone,* and so forth. The resonant frequencies of the string are called its **natural frequencies,** and each vibrational pattern is called a **mode of vibration.** A sequence of natural frequencies that are integral multiples of a fundamental frequency is called a **harmonic series,** and the individual natural frequencies are called **harmonics.**

Standing waves on a taut string can also be produced on a string with one end fixed and the other end free. The standing-wave patterns for such a string are shown in Figure 16-2.

The pertinent relations are

$$f_n = nf_1 \quad n = 1, 3, 5, \ldots \qquad \text{Resonance frequencies, one end free}$$

$$f_1 = \frac{v}{4L} = \frac{1}{4L}\sqrt{\frac{F_T}{\mu}} \qquad \text{Fundamental frequency}$$

$$L = n\frac{\lambda_n}{4} \quad n = 1, 3, 5, \ldots \qquad \text{Standing-wave condition, one end free}$$

Note that a string with one end free resonates only at the odd multiples of the fundamental frequency. That is, only the odd harmonics occur.

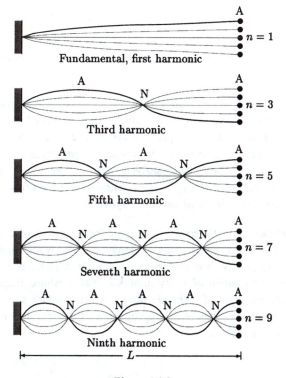

Figure 16-2

To realize the condition just described, we must provide a connection that maintains tension F_T in the string, yet allows the "free" end to oscillate up and down with negligible friction. This can only occur in an ideal world. The analogous arrangement (an open-ended organ pipe) for sound waves is more readily achieved.

Standing Sound Waves. Standing sound waves have much in common with the standing waves of a vibrating string. If air is confined in a tube of length L with both ends open, any standing waves in the air in the tube will have pressure nodes (points of atmospheric pressure), and displacement antinodes, at both ends of the tube. As a consequence, the standing wave condition and resonant frequencies are the same for an open tube as they are for a string fixed at both ends.

In a tube with only one end open to the atmosphere, the standing waves of the air in the tube have a displacement antinode at the open end. This displacement antinode is also a pressure node since the pressure at the open end is fixed at the pressure throughout the room, usually atmospheric pressure. Thus, for a standing wave in such a tube, the standing wave condition and resonant frequencies are the same as for a string fixed at one end.

Section 16-3. The Superposition of Standing Waves. For a vibrating string of finite length, the motion of the string can always be treated as a superposition of standing waves. The wave function is a sum of the standing wave functions:

$$y(x, t) = \sum_n A_n \sin(k_n x) \cos(\omega_n t + \delta_n) \qquad \text{Superposition of standing wave functions}$$

where A_n, ω_n, δ_n, and k_n are the amplitude, angular frequency, phase constant, and wave number of the nth harmonic.

Section 16-4. Harmonic Analysis and Synthesis. When an air column, such as the air in a clarinet or trombone, vibrates, it usually vibrates not as a single standing wave but as the superposition of two or more standing waves. A plot of the pressure versus time for the sound produced is called a **waveform.** Waveforms can be analyzed to determine the amplitudes, frequencies, and phase constants of the harmonics present. This analysis is called **harmonic analysis** or **Fourier analysis.** The inverse of harmonic analysis—the construction of the original periodic waveform from its harmonic components—is called **harmonic synthesis.**

Section 16-5. Wave Packets and Dispersion. A periodic (repetitive) waveform, like the sound of a sustained note on a clarinet, is a superposition of harmonic waves whose frequency distribution is discrete (not continuous). A waveform that is not periodic, such as the sound from a firecracker, can be considered a superposition of many harmonic waves whose frequency distribution is continuous. Such a superposition of harmonic waves, which produce a pulse, is called a **wave packet.** For a wave packet, the range of angular frequencies $\Delta\omega$ and the time interval Δt for the packet to pass any point are such that

$$\Delta\omega \, \Delta t \sim 1 \qquad \text{Temporal wave packet relation}$$

The analogous relation for the spatial parameters is

$$\Delta k \, \Delta x \sim 1 \qquad \text{Spatial wave packet relation}$$

where Δk is the range of wave numbers and Δx is the spatial length of the packet. If a wave packet is to sustain its size and shape, the speed of all the harmonic waves that make up the packet must be the same. This will occur only in a nondispersive medium, that is, a medium in which the speed of a harmonic wave is independent of its wavelength or frequency. In a **dispersive medium,** the speed of harmonic waves varies with frequency. The average speed of the individual harmonic waves that make up a wave packet is called the **phase velocity,** and the velocity of the packet itself is called the **group velocity.**

II. Physical Quantities and Key Equations

Physical Quantities

There are no new physical quantities for this chapter.

Key Equations

Superposition principle $\qquad\qquad\qquad y(x, t) = y_1(x - vt) + y_2(x + vt)$

Interference of harmonic waves

$$y = y_1 + y_2$$
$$= A \sin(kx - \omega t) + A \sin(kx - \omega t + \delta)$$
$$= 2A \cos\left(\frac{\delta}{2}\right) \sin\left(kx - \omega t + \frac{\delta}{2}\right)$$

Superposition of harmonic waves at a given point

$$p = p_1 + p_2 = p_0 \sin(\omega_1 t) + p_0 \sin(\omega_2 t)$$
$$= 2p_0 \cos\left(\tfrac{1}{2}\Delta\omega t\right) \sin(\omega_{av} t)$$

Beat frequency

$$f_{\text{beat}} = |\Delta f| = |f_1 - f_2|$$

Phase difference–path-length difference relation

$$\delta = (2\pi\,\text{rad})\frac{\Delta x}{\lambda} = (360°)\frac{\Delta x}{\lambda}$$

Standing-wave condition, both ends fixed (closed) or both ends free (open)

$$\lambda_n = \frac{2L}{n} \quad n = 1, 2, 3, \ldots$$

Resonance frequencies, both ends fixed (closed) or both ends free (open)

$$f_n = \frac{v}{\lambda_n} = n\frac{v}{2L} = nf_1 \quad n = 1, 2, 3, \ldots$$

Standing-wave patterns for string, both ends fixed

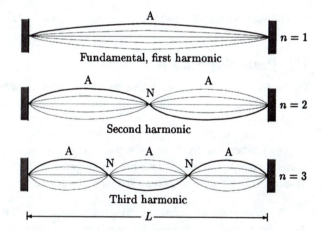

Standing-wave condition, one end fixed (closed) and one end free (open)

$$\lambda_n = n\frac{4L}{n} \quad n = 1, 3, 5, \ldots$$

Resonant frequencies, one end fixed (closed) and one end free (open)

$$f_n = \frac{v}{\lambda_n} = n\frac{v}{4L} = nf_1 \quad n = 1, 3, 5, \ldots$$

Standing-wave patterns for string, one end free

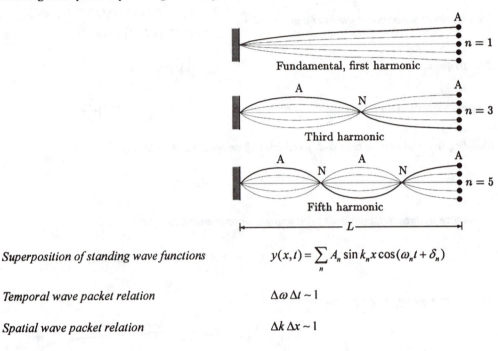

Superposition of standing wave functions $y(x,t) = \sum_n A_n \sin k_n x \cos(\omega_n t + \delta_n)$

Temporal wave packet relation $\Delta\omega\,\Delta t \sim 1$

Spatial wave packet relation $\Delta k\,\Delta x \sim 1$

III. Potential Pitfalls

Destructive interference means that the two interfering waves are 180° out of phase with one another and so will tend to cancel. The cancellation is total, however, only if the two waves have equal amplitudes.

The fundamental frequency of a taut string is the lowest frequency at which it can resonate. Correspondingly, the fundamental has the longest wavelength of any of the resonance frequencies. That is, the higher harmonics have shorter wavelengths.

The frequencies of a vibrating string are calculated as if there were nodes at both ends. Clearly this doesn't accurately describe the usual physics laboratory experiment in which one of the ends is being driven in simple harmonic motion. However, the driven end is very nearly a node if the amplitude of the standing wave is large compared to that of the driven end.

Two wave sources need not be vibrating in phase in order for interference to occur, but they must be coherent—that is, the phase difference between their motions must be constant.

A vibrating string, such as a guitar string or a piano string, can act as a source of sound waves of the same frequency in air as on the string. But notice that the wavelength of the sound waves and the wavelength of the standing waves on the string are not the same. This is because the wave speeds for transverse waves on the string and for sound waves in air are different.

IV. True or False Questions and Responses

True or False

_____ 1. If two wave pulses traveling in the opposite directions along the same string meet, they reflect off each other.

_____ 2. When one end of a taut string is shaken transversely at one of the string's natural frequencies, there is an antinode at the point where the string is attached to the shaker.

_____ 3. Standing waves on a taut string that is fixed at each end have wavelengths that are integral multiples of the length of the string.

_____ 4. When a string is vibrating in a standing wave pattern, every element of the string, except for those at nodes, is undergoing simple harmonic motion.

_____ 5. When a string is vibrating in a standing wave pattern, the motion at one antinode is in phase with the motion at all other antinodes.

_____ 6. When a standing wave exists in the air within a tube that has an open end, a pressure node can be found just beyond the open end of the tube.

_____ 7. The fundamental frequency of standing waves in a tube with one end open, or in a tube approximately twice as long with both ends open, is the same.

_____ 8. The fundamental frequency of standing waves in a tube with both ends open, or a tube of approximately equal length with both ends closed, is the same.

_____ 9. The complete destructive interference of two harmonic waves requires that they have the same amplitude.

_____ 10. For the destructive interference of two harmonic waves from two point sources that vibrate in phase to occur, the path-length difference must be an integral number of wavelengths.

_____ 11. At a maximum in the interference pattern of the waves from two identical point sources, the intensity is double the intensity due to either source alone.

_____ 12. Two sources are coherent if the phase difference between them is fixed.

_____ 13. If piano strings of frequencies 257 Hz and 261 Hz are sounded together, you will hear 2 beats per second.

_____ 14. A periodic waveform can be synthesized by the superposition of its harmonic components.

_____ 15. In a nondispersive medium the speed of a harmonic wave depends upon its frequency.

Responses to True or False

1. False. The waves "pass through" each other.

2. False. The amplitude of vibration of the shaker is much less than the amplitude of the string's motion at an antinode.

3. False. The wavelengths are given by $2L/n$, where L is the length of the string.

4. True.

5. False. At adjacent antinodes the motions are 180° out of phase. That is, when the displacement of the string at one antinode is $+A$, the displacement at the next antinode is $-A$.

6. True.

7. True.

8. True.

9. True.

10. False. The path lengths must differ by $\left(n-\frac{1}{2}\right)\lambda$, where n is a positive integer and λ is the wavelength.

11. False. It is the amplitude that is twice the value of the amplitude due to the wave from one source alone. However, because the intensity is proportional to the square of the amplitude, the intensity is four times greater than the intensity due to the wave from one source alone.

12. True.

13. False. The beat frequency equals the difference in the frequencies, which is 4 beats per second.

14. True.

15. False. In a nondispersive medium all waves travel at the same speed.

V.   Questions and Answers

Questions

1. Two long strings, one heavy, the other light, are joined end to end and then stretched under tension between two trees. A harmonic wave travels in the direction from the light string to

the heavy string. Are the wave speed, the frequency, and the wavelength the same in the heavy string as they are in the light string?

2. The wave pulse shown in Figure 16-3 travels along a stretched string. What is the velocity of the element of string that is momentarily at the top of the pulse? What is the direction of the velocities of the elements of string in the leading half of the pulse? The trailing half? Sketch both the displacement y and the velocity v_y of the string elements as a function of position x along the string.

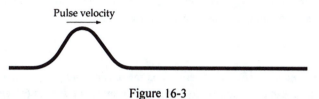

Figure 16-3

3. In Figure 16-4, two wave pulses (one erect and one inverted) of the same size and shape approach one another along a string under tension. A short time later, the pulses superpose (overlap). When the pulses completely superpose, destructive interference occurs and the string is momentarily flat. At the moment when the string is flat, where is the energy that the pulses were carrying? Sketch both the displacements y and velocities v_y of the string elements as a function of position x along the string at the moment when the string is flat.

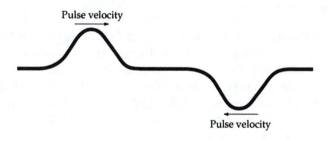

Figure 16-4

4. Most orchestral instruments are based on the resonant frequencies of either a stretched string or air in a pipe. What happens to these frequencies as the temperature increases?

5. The fundamental frequency of an air column in a pipe closed at both ends is 240 Hz; the fundamental frequency is 117.5 Hz if the pipe is open at one end. Why is the ratio of the frequencies approximately 2:1? Why is it not exactly 2:1? What would you expect to be the fundamental frequency of the pipe if it is open at both ends?

6. What happens to the frequency of a clarinet if the clarinetist breathes out pure helium?

7. Which of the properties that determine the fundamental frequency of a vibrating system (a string, air in a pipe, or the like) is normally varied in playing a musical instrument at different frequencies?

8. How can the phenomenon of beats be used to tune musical instruments?

9. When we play chords on the piano, several notes (frequencies) are sounded at once. Why don't we hear beats?

10. It's hard to set up a really convincing demonstration of two-source interference using sound waves in a lecture room. You can set up two loudspeakers vibrating in phase, and as you move from one to another part of the room, you can hear variations in the loudness of the sound, but they aren't terribly striking. Why not?

11. One end of a stretched string is vibrated to produce a standing wave on the string. What is it that determines the amplitude of the standing wave?

Answers

1. The wave speed equals $\sqrt{F_T / \mu}$, where F_T is the tension and μ is the mass per unit length. The tension is the same on either side of the knot joining the strings (or else the knot would accelerate toward the side with the highest tension). Therefore the wave speed is greatest in the light string.

    The two string elements adjacent to the knot, one on either side, move up and down at the same frequency as the knot. On each string the wavelength is related to the wave speed by the equation $v = \lambda f$. Since the frequency is the same on either side, the wavelength is directly proportional to the wave speed. Therefore the wavelength is greater in the light string, which has the greater wave speed.

2. The velocity of the string element at the top has just finished rising to the top and is about to start descending to its equilibrium position. Its velocity is momentarily zero. The string elements in the leading half of the pulse are rising, so their velocities are directed upward. The string elements in the trailing half of the pulse are returning to their equilibrium positions, so their velocities are directed downward. Figure 16-5(a) shows the displacement y of the string as a function of position x along the string; Figure 16-5(b) shows the velocity v_y of the string as a function of position x.

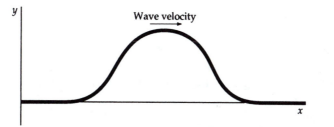

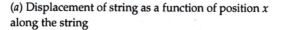

(a) Displacement of string as a function of position x along the string

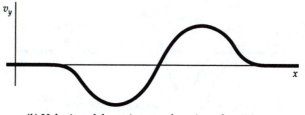

(b) Velocity of the string as a function of position x along the string

Figure 16-5

3. Figure 16-6(a) shows the displacement y of the string as a function of position x along the string; as the pulses approach each other, Figure 16-6(b) shows the velocity v_y of the string as a function of position x.

(a) Displacement of string as a function of position x along the string

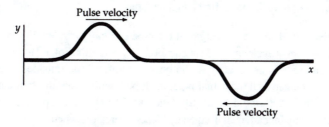

(b) Velocity of the string as a function of position x along the string

Figure 16-6

When the pulses superpose, the resultant displacement and velocity of the string are the algebraic sums of the displacements and velocities of the individual pulses. At the moment the pulses completely superpose, Figure 16-7 shows the displacement y and velocity v_y of the string as a function of the position x along the string. At the moment the pulses exactly superpose, the energy being carried by the two pulses is in the kinetic energy of the string.

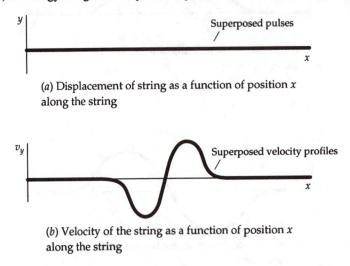

(a) Displacement of string as a function of position x along the string

(b) Velocity of the string as a function of position x along the string

Figure 16-7

4. The frequencies of the wind instruments increase because the speed of sound in air increases with increasing temperature. Since strings expand a little as the temperature goes up, the tension in a string decreases somewhat; the wave speed and fundamental frequency therefore decrease. Thus, wind instruments go sharp and string instruments go flat as they warm up. This is why an orchestra warms up first and then tunes.

5. The ratio is about 2:1 because the length of the pipe is a half wavelength when both ends are closed and a quarter-wavelength when one end is open. It's not exactly 2:1 because the displacement antinode is not exactly at the open end but a little outside, so the effective air column is a little longer than the actual pipe. With both ends open, the fundamental is around 240 Hz again. (In fact it would be a little less than 240 Hz—perhaps 232 Hz—because the ends are open.) The fundamental frequency depends not on whether the ends are open or closed, but more on whether the two ends are the same or different.

6. All the frequencies that the clarinet produces are higher by the ratio of the speed of sound in helium to that in air, which is about 2.5 times. Thus all its tones sound more than an octave higher—provided that the clarinetist tightens his lips.

7. Most often, it is the length of the system that is varied. You finger the violin string to change its vibrating length, and you open or close holes or insert extra lengths of pipe on a wind instrument to "end" the air column at different places. Length modification is secondary, however, in playing some brass instruments. With them, you select along the harmonic frequencies of a pipe of fixed length by how you hold your mouth.

8. The phenomenon of beats is a clearly audible indicator of a small frequency difference. For example, when you simultaneously sound two strings that are supposed to have the same frequency, any difference in frequency produces beats that you can hear. Then you adjust the tension in one string, decreasing the beat frequency until no beats are heard.

9. In a way we do here them. In this case the beat or difference frequencies are tens to hundreds of hertz, and we can't hear them as intensity variations. However, due to the way the ear responds to sound waves, "tones" at the different frequencies are present in the combination of sounds we hear and, in some circumstances, are easily distinguished.

10. In almost any room, a large part of the sound you are hearing comes not directly from the sources but has been reflected (one or more times) from the walls. It is only the part coming directly from the sources in which you perceive strong interference maxima and minima, so these are usually not very noticeable.

11. The amplitude of the standing wave grows until the power being dissipated in the string plus the acoustic power being transferred from the string to the air equals the power being delivered to the string by the signal source.

VI. Problems, Solutions, and Answers

Example #1. A string has a tension of 10.0 N and a mass per unit length of 5.00 g/m. Two harmonic waves traveling in the same direction on this string, each with a 20.0-cm wavelength and a 1.00-cm amplitude, differ in phase by 90°. Determine the amplitude and the power of the wave that results when these waves interfere.

Picture the Problem. Because both waves are on the same string and have the same wavelength, they will have the same frequency. The amplitude of the wave resulting from interference of the two harmonic waves can be expressed in terms of the phase difference between the two waves. The power depends on the angular frequency, the amplitude, the wave speed, and the linear mass density.

1. Determine the wave function of the resultant wave by adding the two harmonic waves.	$y = A\sin(kx - \omega t) + A\sin(kx - \omega t + \delta)$ $$= 2A\cos\left(\frac{\delta}{2}\right)\sin\left(kx - \omega t + \frac{\delta}{2}\right)$$
2. The amplitude of the resultant wave is the time invariant piece of the wave function.	$A_r = 2A\cos\left(\frac{\delta}{2}\right)$ $$= 2(1.00\,\text{cm})\cos\left(\frac{90°}{2}\right) = \sqrt{2}\,\text{cm} = 1.41\,\text{cm}$$

3. The resultant wave has a single, fixed frequency, so it is also a harmonic wave. We learned how to calculate the power of a harmonic wave on a string in Chapter 15.	$P = \dfrac{1}{2}\mu v \omega^2 A_r^2$ $v = \sqrt{F_T / \mu} \qquad \omega = kv = 2\pi v / \lambda$ $P = \dfrac{1}{2}\mu v \left(\dfrac{2\pi v}{\lambda}\right)^2 A_r^2 = \dfrac{1}{2}\mu \left(\dfrac{2\pi A_r}{\lambda}\right)^2 \left(\dfrac{F_T}{\mu}\right)^3$ $= \dfrac{1}{2}(0.005\,\text{kg/m}) \left(\dfrac{2\pi\left(\sqrt{2}\,\text{cm}\right)}{20.0\,\text{cm}}\right)^2 \left(\dfrac{10.0\,\text{N}}{0.005\,\text{kg/m}}\right)^{3/2}$ $= 44.1\,\text{W}$

Example #2—Interactive. A string has a tension of 40.0 N and a mass per unit length of 5.00 g/m. Two harmonic waves of equal amplitude, each with a wavelength of 15.0 cm, that are traveling in the same direction differ in phase by 120°. When these waves interfere, the amplitude of the resultant wave is 1.00 cm. Determine the amplitude of each of the original waves and the power of the resultant wave. **Try it yourself.** Work the problem on your own, in the spaces provided, to get the final answer.

1. Determine the wave function of the resultant wave by adding the two harmonic waves.	
2. The amplitude of the resultant wave is the time invariant piece of the wave function. Solve for the amplitude of the original waves using the amplitude of the resultant wave and the phase. $A_r = 1.00\,\text{cm}$	
3. The resultant wave has a single, fixed frequency, so it is also a harmonic wave. We learned how to calculate the power of a harmonic wave on a string in Chapter 15. $P = 314\,\text{W}$	

Example #3. A string is stretched between fixed supports 0.70 m apart, and the tension is adjusted until the fundamental frequency of the string is 264 Hz. If the string has a mass of 4.2 g, what is the tension?

1. The wave speed is related to the tension and mass density of the string. It is also equal to the product of frequency and wavelength.	$v = \sqrt{F_T/\mu} = \lambda_1 f_1$
2. The wavelength of the fundamental is twice the length of the string. Substitute, and solve for the tension.	$\lambda_1 = 2L$ $\sqrt{F_T/\mu} = 2Lf_1$ $F_T = (2Lf_1)^2 \dfrac{m}{L} = 4mLf_1^2$ $= 4(0.0042\,\text{kg})(0.70\,\text{m})(264\,\text{Hz})^2 = 820\,\text{N}$

Example #4—Interactive. The lowest pitch on the standard piano is A (27.5 Hz). On my piano (a small console) the 85.0-cm-long "string" that sounds this pitch has a linear mass density of 0.235 g/cm. Assuming the string is vibrating in its fundamental mode, what is the tension in the string? **Try it yourself.** Work the problem on your own, in the spaces provided, to get the final answer.

1. The wave speed is related to the tension and mass density of the string. It is also equal to the product of frequency and wavelength.	
2. The wavelength of the fundamental is twice the length of the string. Substitute in, and solve for the tension.	$F_T = 51.4\,\text{N}$

Example #5. Three successive resonant frequencies of a taut string are 273, 364, and 455 Hz. Determine the fundamental frequency of the string.

Picture the Problem. We do not know if the string is fixed at both ends, or simply one end. If it is fixed at both ends, the fundamental frequency is simply the difference between successive resonant frequencies. If it is fixed at one end, the fundamental frequency is one-half the difference between successive resonant frequencies. Take the ratios of the resonant frequencies to determine which case we have.

1. Determine the ratio of two successive resonant frequencies if the string is fixed at both ends.	$f_n = nf_1 \qquad n = 1,2,3,\ldots$ $\dfrac{f_{n+1}}{f_n} = \dfrac{(n+1)f_1}{nf_1} = \dfrac{n+1}{n} \quad n = 1,2,3,\ldots$

2. Determine the ratio of two successive resonant frequencies if the string is fixed at one end.	$f_n = nf_1 \qquad n = 1,3,5,\ldots$ $\dfrac{f_{n+2}}{f_n} = \dfrac{(n+2)f_1}{nf_1} = \dfrac{n+2}{n} \qquad n = 1,3,5,\ldots$
3. Take the ratio of the given resonant frequencies, and determine which of the above patterns match the ratios of the provided frequencies.	$\dfrac{f_H}{f_M} = \dfrac{455}{364} = \dfrac{5}{4}$ $\dfrac{f_M}{f_L} = \dfrac{364}{273} = \dfrac{4}{3}$
4. The pattern matches that for a string fixed at both ends. Find the fundamental frequency.	$f_1 = f_{n+1} - f_n = 364\,\text{Hz} - 273\,\text{Hz} = 91\,\text{Hz}$

Example #6—Interactive. Three successive resonance frequencies of a taut string are 363, 605, and 847 Hz. Determine the fundamental frequency of the string.

Picture the Problem. We do not know if the string is fixed at both ends, or simply one end. If it is fixed at both ends, the fundamental frequency is simply the difference between successive resonant frequencies. If it is fixed at one end, the fundamental frequency is one-half the difference between successive resonant frequencies. Take the ratios of the resonant frequencies to determine which case we have. **Try it yourself.** Work the problem on your own, in the spaces provided, to get the final answer.

1. Determine the ratio of two successive resonant frequencies if the string is fixed at both ends.	
2. Determine the ratio of two successive resonant frequencies if the string is fixed at one end.	
3. Take the ratio of the given resonant frequencies.	
4. Determine which pattern matches, and find the fundamental frequency.	$f_1 = 121\,\text{Hz}$

Example #7. (*a*) A man is sitting in a room directly between two loudspeakers 5.00 m apart that are vibrating in phase. He is 1.80 m from the nearer speaker, at point A in Figure 16-8. If the lowest frequency at which he observes maximum destructive interference is 122 Hz, what is the speed of sound in air? (*b*) If, instead, he listens from point B, what is the lowest frequency at which he will observe destructive interference?

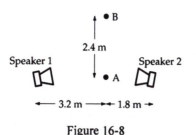

Figure 16-8

Picture the Problem. When the man observes maximal destructive interference, the waves reaching him are 1/2 cycle out of phase. We know it's one-half, and not 3/2 or 5/2 cycles because we are interested in the lowest frequency.

1. Find the path-length difference of the two sound waves from the speakers to point A.	$\|\Delta s\| = s_1 - s_2 = 3.20\,\text{m} - 1.80\,\text{m} = 1.40\,\text{m}$
2. Apply the condition for complete destructive interference to find the wavelength of the sound.	$\Delta s = \lambda/2$ $\lambda = 2\,\Delta s = 2(1.40\,\text{m}) = 2.80\,\text{m}$
3. Find the speed of sound in air.	$v = \lambda f = (2.80\,\text{m})(122\,\text{Hz}) = 342\,\text{m/s}$
4. Find the path-length difference of the two sound waves from the speakers to point B.	$\|\Delta s\| = s_1 - s_2 = 4.00\,\text{m} - 3.00\,\text{m} = 1.00\,\text{m}$
5. Apply the condition for complete destructive interference to find the wavelength of the sound.	$\Delta s = \lambda/2$ $\lambda = 2\,\Delta s = 2(1.00\,\text{m}) = 2.00\,\text{m}$
6. Find the frequency of this sound.	$f = v/\lambda = (342\,\text{m/s})/(2.00\,\text{m}) = 171\,\text{Hz}$

Example #8—Interactive. Two loudspeakers radiate sound waves in phase at a frequency of 100 Hz. A listener is 8.5 m from one speaker and 13.6 m from the other. Either speaker alone would produce a sound intensity level of 75 dB at the position of the listener. (*a*) What is the sound intensity level at the listener due to both speakers? (*b*) What is the intensity level at the listener if the same two speakers are 180° out of phase?

Picture the Problem. Determining the intensity level requires knowledge of the intensity, which requires knowing the net amplitude of the sound where the listener is. You will first have to determine how the sound from the two speakers is interfering, which will allow you to determine the sound amplitude, and hence the intensity level. **Try it yourself.** Work the problem on your own, in the spaces provided, to get the final answer.

1. Find the path-length difference of the two sound waves from the speakers to the listener.	

2. Find the wavelength of the sound.	
3. Determine whether or not the waves interfere constructively or destructively.	
4. Find the intensity of the resulting sound.	Because the waves interfere destructively, the intensity is zero, which cannot be expressed on the decibel scale.
5. Determine whether or not the waves interfere constructively or destructively if one speaker is 180° out of phase from the other.	
6. Find the intensity of the resulting sound.	Because the waves interfere constructively, the intensity is four times the intensity of an individual wave, or 81 dB.

Example #9. Two loudspeakers 3.00 m apart and a few meters away from you are producing sound of the same frequency. At about what frequency can you tell by the sound that there are two sources rather than one?

Picture the Problem. A wave cannot be used to observe or locate details much smaller than one wavelength. Since the speakers are 3 m apart, the wavelength must be less than 3 m. This corresponds to a frequency greater than $f = v/\lambda = (342\,\text{m/s})/(3.00\,\text{m}) = 114\,\text{m}$.

Example #10—Interactive. A loudspeaker that is producing sound with a wavelength much longer than the diameter of the speaker acts as a point source; that is, the sound radiates more or less uniformly in all directions. If the wavelength is very small compared to the diameter of the speaker, on the other hand, the sound travels straight ahead from the speaker in approximately a straight line. If each speaker in a portable stereo is 18.0 cm in diameter, for what frequency is the sound wavelength (*a*) ten times and (*b*) one-tenth of the loudspeaker diameter? **Try it yourself.** Work the problem on your own, in the spaces provided, to get the final answer.

1. Find the frequency for part (*a*).	$f = v/\lambda = (342\,\text{m/s})/(1.8\,\text{m}) = 189\,\text{Hz}$
2, Find the frequency for part (*b*).	$f = v/\lambda = (342\,\text{m/s})/(.018\,\text{m}) = 18.9\,\text{kHz}$

Discussion. Long wavelength (low frequency) sounds correspond to the bass sounds. If someone has their stereo volume turned way up, the low frequency (bass) sounds radiate uniformly, but you can only hear the high frequency (treble) sounds if you are directly in front of a speaker.

Example #11. Two identical strings on a piano are tuned to concert A (440 Hz). Each is under a tension of 1300 N. Over the course of time, one string loosens to the point that, when you strike the two strings simultaneously, you hear beats every 1.10 s. By how much has the tension in the loose string decreased?

Picture the Problem. The beat frequency is the same as the difference in the frequency of the two strings. This decrease in frequency comes from a reduction in the speed of the wave on the string, from which you can find the reduction of the tension of the loose string.

1. Find the new frequency. Because the tension has decreased, the frequency of the loose string will decrease, not increase.	$f' = 440\,\text{Hz} \pm \dfrac{1}{1.10\,\text{s}} = 439.09\,\text{Hz}$
2. To find the tension in the loose string, take a ratio of the two frequencies.	$\dfrac{f'}{f} = \dfrac{v'/2L}{v/2L} = \dfrac{v'}{v} = \sqrt{\dfrac{F'/\mu}{F/\mu}} = \sqrt{\dfrac{F_\text{T}'}{F_\text{T}}}$ $F_\text{T}' = F_\text{T}\left(\dfrac{f'}{f}\right)^2 = (1300\,\text{N})\left(\dfrac{439.09\,\text{Hz}}{440\,\text{Hz}}\right)^2$ $= 1294.6\,\text{N}$
3. The change in tension can now be calculated. Note that we carried extra significant figures throughout this calculation. This was required to get a precise result in the *change* of the tension.	$\Delta F_\text{T} = F_\text{T} - F_\text{T}' = 1300\,\text{N} - 1294.6\,\text{N} = 5.4\,\text{N}$

Example #12—Interactive. When a violin string is played simultaneously with a tuning fork of frequency 264.0 Hz, beats occurring once every 0.65 s are heard in the sound. Tightening the string slightly causes the beats to disappear. What was the initial frequency of the violin string?

Picture the Problem. The beat frequency is the same as the difference in the frequency of the two waves. When the string is tightened, the frequency increases. **Try it yourself.** Work the problem on your own, in the spaces provided, to get the final answer.

1. Calculate the beat frequency.	
2. Find the frequency of the violin string.	
	262.5 Hz

Example #13. The fundamental frequency of a violin string 30.0 cm long is sounded next to the open end of a closed organ pipe 41.0 cm long. The strongest standing sound wave in the pipe occurs when the string is vibrating at the pipe's fundamental frequency. This happens when the tension in the string is 220.0 N. What is the mass of the string?

Picture the Problem. Find the fundamental frequency for an organ pipe open at one end. Knowing the string has the same frequency, and the length of the string, the wave speed on the string can be determined. From the wave speed and tension, the mass can be found.

1. Find the resonant frequency for an organ pipe open at one end.	$\lambda_1 = 4L = 4(0.410\,\text{m}) = 1.64\,\text{m}$ $f_1 = v/\lambda_1 = (342\,\text{m/s})/(1.64\,\text{m}) = 209\,\text{Hz}$
2. The string has this same fundamental frequency. Find the wave speed on the string.	$v = f_1\lambda_1 = f_1(2L) = (209\,\text{Hz})(2)(0.300\,\text{m})$ $= 125\,\text{m/s}$
3. Find the mass of the string.	$v = \sqrt{F_T/\mu} = \sqrt{F_T/(m/L)} = \sqrt{F_T L/m}$ $m = F_T L/v^2 = (220\,\text{N})(0.300\,\text{m})/(125\,\text{m/s})^2$ $= 4.22\times10^{-3}\,\text{kg}$

Example #14—Interactive. The maximum range of human hearing is about 20 Hz to 20 kHz. What is the longest organ pipe that could have a fundamental frequency within this range? **Try it yourself.** Work the problem on your own, in the spaces provided, to get the final answer.

1. Find the wavelengths corresponding to the frequency range.	
2. Find the length of pipe for which these are the fundamental frequencies. Consider pipes open at both ends, and pipes open at only one end.	
3. Come up with your answer.	It is 8.5 m long, open at both ends.

Chapter 17

Temperature and the Kinetic Theory of Gases

I. Key Ideas

Section 17-1. Thermal Equilibrium and Temperature. **Thermodynamics** is the study of temperature and energy exchange. When two objects are in **thermal contact,** energy in the form of heat flows from the warmer object to the cooler object. This normally results in the warmer object cooling down and the cooler object warming up. As this process continues, both the rate of cooling of the warmer object and the rate of warming of the cooler object become less and less. When each of the objects is neither warming up nor cooling down, the objects have reached a steady-state situation called **thermal equilibrium.** The **zeroth law of thermodynamics** states that

> If two objects are each in thermal equilibrium with a third, then they are in thermal equilibrium with each other.

Two objects are defined to have the same *temperature* if they are in thermal equilibrium with each other. The zeroth law, as we will see, enables us to define a temperature scale.

Temperature is the measure of hotness or coldness of a substance. Warmer objects are at a higher temperature than cooler objects, and objects that are in thermal equilibrium are at the same temperature. When objects at different temperatures are in thermal contact, the natural direction of heat flow is always from the warmer object to the cooler object.

Section 17-2. The Celsius and Fahrenheit Temperature Scales. A physical property that changes with temperature is called a **thermometric property.** Any thermometric property can be used as the basis of a thermometer. A common example is the thermal expansion of a liquid sealed in a glass envelope. For such a thermometer, a temperature scale is established by assigning values at which the thermometer is in thermal equilibrium with reproducible states of some system, such as water at its ice (freezing) and steam (boiling) points. The interval between these points is then divided up into equal increments called degrees.

For the **Celsius scale,** a value of zero degrees Celsius (0°C) is assigned to the ice point of water, and a value of 100 degrees Celsius (100°C) is assigned to its steam point at atmospheric pressure. A Celsius thermometer is constructed by selecting a convenient thermometric property, measuring that property at the ice and steam points, and dividing the difference between these measured values into 100 equally spaced increments. The length L_t of a metal bar is a thermometric property that can be used to construct a Celsius scale. The Celsius temperature t_C is then given by

$$t_C = \frac{L_t - L_0}{L_{100} - L_0} \times 100° \qquad \text{Celsius scale}$$

where L_0 and L_{100} are the lengths of the bar when it is in thermal equilibrium with an ice bath and a steam bath, respectively.

In the **Fahrenheit scale,** still commonly used in the United States, a value of 32 degrees Fahrenheit (32°F) is assigned to the ice point of water, and 212 degrees Fahrenheit (212°F) is assigned to the steam point at atmospheric pressure. The relationship between a Fahrenheit temperature t_F and a Celsius temperature t_C is

$$t_C = \tfrac{5}{9}(t_F - 32°) \qquad \text{Fahrenheit–Celsius conversion}$$

Section 17-3. Gas Thermometers and the Absolute Temperature Scale. A drawback to most thermometers is that the temperature readings of different kinds of thermometers do not agree, except at the defining temperatures, because different thermometric properties do not change with temperature in the same way. One class of thermometers that do agree over a wide range of temperatures is **constant-volume gas thermometers,** for which pressure is the thermometric property. If any gas at low pressure is confined to a constant volume, its pressure will increase with increasing temperature over a wide range of temperatures. The Celsius temperature t_C of such a thermometer is given by

$$t_C = \frac{P_t - P_0}{P_{100} - P_0} \times 100° \qquad \text{Celsius scale, gas thermometer}$$

where P_0 and P_{100} are the pressures of the gas when it is in thermal equilibrium with an ice bath and a steam bath, respectively. For gases at very low densities, all constant-volume gas thermometers give the same value of the temperature, independent of the particular gas used. When the pressure in a constant-volume gas thermometer is extrapolated to zero, it approaches zero as the temperature approaches −273.15°C. This temperature is known as absolute zero.

The ice and steam points of water are less easy to reproduce precisely than the **triple point of water,** which is the single state at which ice, water, and steam can coexist in equilibrium. It occurs at a pressure of 4.58 mmHg and a temperature of 0.01°C. The **ideal-gas temperature scale** is defined by making the triple-point temperature 273.16 kelvins (K). Using a constant-volume gas thermometer, this gives

$$T = \frac{273.15 \text{ K}}{P_3} P \qquad \text{Ideal-gas temperature scale}$$

where P_3 is the pressure when the gas is at the temperature of the triple point of water and the **kelvin (K)** is a degree unit that is the same size as the Celsius degree. The ideal-gas temperature scale is identical with another scale, the absolute temperature scale (also called the **Kelvin scale),** which is defined independently of the properties of any substance. The symbol T is used when referring to the absolute temperature. To convert from degrees Celsius to kelvins, we simply add 273.15:

$$T = t_C + 273.15$$

Celsius–kelvin conversion

Section 17-4. The Ideal-Gas Law. Experimentally, when a confined gas kept at low pressure P and at a constant temperature T is either compressed or allowed to expand, the product of the pressure of the gas and its volume remains constant. This observation is known as **Boyle's law.** Also, at low pressures the absolute temperature of a gas kept at a constant volume V is proportional to its pressure. These relations are contained in the **ideal-gas law,** the equation of state for gases at low pressures:

$$PV = NkT$$

Ideal-gas law

where N is the number of molecules and k is **Boltzmann's constant.**

$$k = 1.381 \times 10^{-23} \text{ J/K} = 8.617 \times 10^{-5} \text{ eV/K}$$

Boltzmann's constant

An amount of gas is often expressed in moles. A **mole** (mol) of any substance is the amount of that substance that contains **Avogadro's number** N_A of atoms or molecules, defined as the number of carbon atoms in 12 grams of ^{12}C:

$$N_A = 6.022 \times 10^{23} \text{ molecules/mol}$$

Avogadro's number

If we have n moles of a substance, the number of molecules is then $N = n N_A$. The ideal-gas law is then

$$PV = n N_A kT = nRT$$

where $R = N_A k$ is called the **universal gas constant.** Its value, which is the same for all gases, is

$$R = N_A k = 8.314 \text{ J/} (\text{mol} \cdot \text{K}) = 0.08206 \text{ L} \cdot \text{atm/} (\text{mol} \cdot \text{K})$$

Universal gas constant

The mass of one mole of a chemical element or compound is called its **molar mass M,** for which the customary unit is g/mol. The molar mass of ^{12}C is, by definition, 12 g/mol or 12×10^{-3} kg/mol. Approximate molar masses of some elements and compounds are

Species	H	H_2	He	C	N	O	N_2	O_2	CO_2
Molar mass (g/mol)	1	2	4	12	14	16	28	32	44

A given amount of a substance, such as 1 kg of water, is placed in a closed, otherwise empty container where the substance is maintained at a constant pressure and volume, and the container is maintained at a constant temperature. Eventually the substance and the container reach thermal equilibrium, at which time the temperature of the substance and the container are equal. After thermal equilibrium is established, it is found that no further changes in the sample—such as changes from liquid to solid or from liquid to gas—can be observed. However, scientists understand that even after thermal equilibrium is reached, and the temperature, pressure, and volume of the sample remain constant, changes continue to occur within the sample. These changes occur at the microscopic level, a level that is *not* directly observable. At the microscopic level, matter consists of numerous molecules that are in constant, random motion. Consequently, the positions of the individual molecules vary with time, and so, at least at the microscopic level, the sample is constantly changing.

The state of the sample is determined only by its observable (macroscopic) properties. Because microscopic variations of the sample do not always result in observable changes, they do not necessarily result in a change in its state. Using instruments such as a pressure gauge, a graduated cylinder, a thermometer, and your senses, the pressure, volume, temperature, and phase (solid, liquid, or gaseous) of the sample can be observed. Thus, any changes in pressure, volume, temperature, or phase represent a change in the state of the sample.

The equation $PV = nRT$, which relates macroscopic variables P, V, and T for a given amount of a gas, is called an **equation of state**. It is the equation of state for a sample of gas at low density. The state of the sample is specified if any two of the three variables P, V, and T are specified.

Section 17-5. The Kinetic Theory of Gases. An ideal gas is one in which the molecules are separated, on the average, by distances that are large compared with their diameters, and they exert no forces on each other except when they collide. The properties of an ideal gas can be understood in terms of a simple kinetic model. We can picture the gas as consisting of noninteracting molecules in motion and the pressure as the result of the collisions of the gas molecules with the walls of their container. Applying the impulse–momentum theorem to calculate the pressure resulting from these collisions gives

$$PV = 2N\left(\tfrac{1}{2}mv_x^2\right)_{av} \qquad \text{Pressure of an ideal gas}$$

where P is the pressure, V is the volume containing N molecules, m is mass of a gas molecule, and v_x is the x component of its velocity. The ideal-gas law $\left(PV = NkT\right)$ together with this formula yields $kT = m(v_x^2)_{av}$. Furthermore, combining the relations $v^2 = v_x^2 + v_y^2 + v_z^2$ and $(v_x^2)_{av} = (v_y^2)_{av} = (v_z^2)_{av}$ (the first obtained via the Pythagorean theorem and the second via symmetry) we get $(v^2)_{av} = 3(v_x^2)_{av}$. It follows that $kT = m(v_x^2)_{av} = \tfrac{1}{3}m(v^2)_{av}$ or

$$K_{av} = \left(\tfrac{1}{2}mv^2\right)_{av} = \tfrac{3}{2}kT \qquad \text{Average translational kinetic energy}$$

where K_{av} is the average translational kinetic energy of a gas molecule of mass m.

Thus the absolute temperature of the gas is a measure of the average translational kinetic energy of its molecules. The total translational kinetic energy K of n moles of a gas containing N molecules is

$$K = N\left(\tfrac{1}{2}mv^2\right)_{av} = \tfrac{3}{2}NkT = \tfrac{3}{2}nN_A kT = \tfrac{3}{2}nRT \qquad \text{Total translational kinetic energy}$$

We can use these results to estimate the speeds of the molecules in a gas by solving for the **root mean square** (rms) speed, which is the square root of $\left(v^2\right)_{av}$:

$$v_{rms} = \sqrt{\left(v^2\right)_{av}} = \sqrt{\frac{3kT}{m}} = \sqrt{\frac{3RT}{N_A m}} = \sqrt{\frac{3RT}{M}}$$ Root mean square speed

where $M = N_A m$ is the molar mass.

The translational kinetic energy of a molecule of a gas can be expressed as the sum of three terms, that is, $\frac{1}{2}mv^2 = \frac{1}{2}mv_x^2 + \frac{1}{2}mv_y^2 + \frac{1}{2}mv_z^2$. If the gas is in equilibrium, these three terms are, on average, equal. This sharing of the energy equally between the three terms in the translational kinetic energy is a special case of the **equipartition theorem,** a result that follows from classical statistical mechanics. Each component of position and momentum (including angular position and angular momentum) that appears as a squared term in the expression for the energy of the system is called a **degree of freedom.** More generally, the equipartition theorem states that

When a substance is in equilibrium, there is an average energy of $\frac{1}{2}kT$ per molecule or $\frac{1}{2}RT$ per mole associated with each degree of freedom.

The average distance λ a gas molecule travels between collisions is called its **mean free path.** For an ideal gas, the expression for the mean free path is

$$\lambda = \frac{1}{\sqrt{2}n_v \pi d^2}$$ Mean free path

Where n_v is the number density (the number of molecules per unit volume) and d is the molecular diameter. If v_{av} is the average speed, the average distance traveled between collisions is $\lambda = v_{av}\tau$, where τ is the collision time (the mean time between collisions).

It is desirable to statistically connect the details of molecular motions to the macroscopic thermodynamic parameters. To do this involves making use of the Maxwell-Boltzmann distribution functions. The **Maxwell–Boltzmann speed distribution function** contains information concerning the distribution of molecular speeds. In a gas of N molecules, the number that have speeds in the range between v and $v+dv$ is dN, given by

$$dN = N f\left(v\right) dv$$

where $f\left(v\right)dv$ is the fraction dN/N that have speeds between v and $v+dv$. The Maxwell–Boltzmann speed distribution function $f\left(v\right)$ can be derived using statistical mechanics. The result is

$$f(v) = \frac{4}{\sqrt{\pi}}\left(\frac{m}{2kT}\right)^{3/2} v^2 e^{-mv^2/2kT}$$ Maxwell–Boltzmann speed distribution function

The distribution of the translational kinetic energies of the molecules in a gas is described by the **Maxwell–Boltzmann energy distribution function.** This function is

$$F(E) = \frac{2}{\sqrt{\pi}} \left(\frac{1}{kT}\right)^{3/2} E^{1/2} e^{-E/kT} \qquad \text{Maxwell–Boltzmann energy distribution function}$$

where $E = \frac{1}{2}mv^2$. In the language of statistical mechanics, the energy distribution is considered to be the product of two factors: one, called the **density of states,** is proportional to $E^{1/2}$; the other is the probability of a state being occupied, which is $e^{-E/kT}$ and is called the **Boltzmann factor.**

The Maxwell–Boltzmann speed and energy distribution functions are continuous fractional distribution functions. They are valid in the classical approximation, that is, they are valid if the quantum nature of the energy distribution can be ignored. One property of such functions, called the **normalization condition,** is that the sum of the fractions equals 1. That is, $\int f(x)\, dx = 1$, where f is any continuous fractional distribution function. Furthermore, the average value of any function $g(x)$ is given by

$$\left[g(x)\right]_{\text{av}} = \int g(x)\, f(x)\, dx$$

II. Physical Quantities and Key Equations

Physical Quantities

Boltzmann's constant $\qquad k = 1.381 \times 10^{-23} \text{ J/K} = 8.617 \times 10^{-5} \text{ eV/K}$

Avogadro's number $\qquad N_A = 6.022 \times 10^{23} \text{ molecules/mol}$

Universal gas constant $\qquad R = N_A k = 8.314 \text{ J/(mol·K)} = 0.08206 \text{ L·atm/(mol·K)}$

Key Equations

Celsius scale $\qquad t_C = \dfrac{L_t - L_0}{L_{100} - L_0} \times 100°$

Fahrenheit–Celsius conversion $\qquad t_C = \frac{5}{9}(t_F - 32°)$

Celsius scale, gas thermometer $\qquad t_C = \dfrac{P_t - P_0}{P_{100} - P_0} \times 100°$

Ideal-gas temperature scale $\qquad T = \dfrac{273.15 \text{ K}}{P_3} P$

Celsius–kelvin conversion $\qquad T = t_C + 273.15$

Ideal-gas law $\qquad PV = NkT = nRT$

Average translational kinetic energy $\qquad K_{\text{av}} = \left(\frac{1}{2}mv^2\right)_{\text{av}} = \frac{3}{2}kT$

| Total translational kinetic energy | $K = N\left(\tfrac{1}{2}mv^2\right)_{av} = \tfrac{3}{2}NkT = \tfrac{3}{2}nN_A kT = \tfrac{3}{2}nRT$ |

| Root mean square speed | $v_{rms} = \sqrt{\left(v^2\right)_{av}} = \sqrt{\dfrac{3kT}{m}} = \sqrt{\dfrac{3RT}{N_A m}} = \sqrt{\dfrac{3RT}{M}}$ |

| Mean free path | $\lambda = \dfrac{1}{\sqrt{2}n_v \pi\, d^2}$ |

| Maxwell–Boltzmann speed distribution function | $f(v) = \dfrac{4}{\sqrt{\pi}}\left(\dfrac{m}{2kT}\right)^{3/2} v^2 e^{\,mv^2/2kT}$ |

| Maxwell–Boltzmann energy distribution function | $F(e) = \dfrac{2}{\sqrt{\pi}}\left(\dfrac{1}{kT}\right)^{3/2} E^{1/2} e^{\,E/kT}$ |

III. Potential Pitfalls

Be careful not to confuse a specific value of temperature with a change in temperature or a temperature range. When dealing with Celsius or Fahrenheit degrees, °C or °F should be used to denote a specific temperature and C° or F° should be used to denote either a change in temperature or a temperature range; no such distinction is possible with the kelvin.

Many of the equations in physics that include a temperature factor are valid only when temperature is expressed in kelvins. Be sure you know for which equations this is true. Of course, if you are dealing with a temperature difference, a kelvin and a Celsius degree are equal.

At a given temperature, molecules of all gases have the same average translational kinetic energy. Their average and root mean square speeds will not be the same, however, because the molecular mass of different gases are not the same.

In applying the ideal-gas law, be sure that all quantities, including the gas constant R, are in consistent units. Trying to work with pressure in torr, volume in cubic meters, and R in L•atm/(mol•K) will only give you a headache. A good rule is to use only SI units.

If a gas law is used for proportional calculations, such as Boyle's law ($P_1 V_1 = P_2 V_2$), you can use any units you like, but be sure to use absolute pressure, not gauge pressure. Of course, the units have to be the same on both sides of the equation.

IV. True or False Questions and Responses

True or False

_____ 1. If both object A and object B are in thermal equilibrium with object C, then object A must be in thermal equilibrium with object B.

_____ 2. All Celsius thermometers must agree that the steam point of water is 100°C.

____ 3. All Celsius thermometers must agree that normal body temperature is about 37°C.

____ 4. All low-pressure, constant-volume gas thermometers agree that normal body temperature is about 37°C.

____ 5. The molar mass of a sample of a gas is the mass of one mole of the gas.

Responses to True or False

1. True.

2. True.

3. False. Celsius thermometers must agree that the ice point and boiling point of water are 0 and 100°C, respectively, but they do not have to be in agreement for any other temperatures because not all thermometric properties vary with temperature in the same way.

4. True.

5. True.

V. Questions and Answers

Questions

1. In what way is the Celsius scale more convenient than the Kelvin scale for everyday use? In what way is the Kelvin scale more suitable than the Celsius scale for scientific use?

2. What change must be made in the temperature of an ideal gas to halve the rms speed of its molecules?

3. Two different ideal gases are at the same temperature. How do the rms speeds of their molecules compare? What is the average translational kinetic energy of their molecules?

4. What properties should a physical system possess in order to serve as a good thermometer?

5. What is the mass of a mole of carbon monoxide gas (molecular formula CO)? How many molecules are there in one mole?

Answers

1. On the Celsius scale, the temperatures that we ordinarily deal with tend to be numbers between −15 and +40. These are more convenient to use and remember than temperature values of several hundred kelvins would be. The Kelvin scale is more suitable for scientific use in that many formulas are less complex when the Kelvin scale is used. Also, the Kelvin scale is independent of the thermometric properties of any particular substance.

2. The average molecular translational kinetic energy, which is proportional to the rms speed squared, must be reduced to $\frac{1}{4}$ its original value in order to halve the rms speed. Because this energy is proportional to the absolute temperature for an ideal gas, the absolute temperature must be reduced to $\frac{1}{4}$ its original value.

3. If two ideal gases are at the same temperature, they have the same average translational kinetic energy per molecule. If they do not happen to have the same molecular mass, however, then the molecules of the lighter gas will, on average, be moving at higher speeds.

4. It depends a lot on what the thermometer is supposed to be good for, but there are some general requirements that apply to most cases. The thermometric property used—length, electrical resistance, pressure, or whatever—should be easily measurable itself and should vary linearly, or nearly so, with absolute temperature. Otherwise, the thermometer will be usable only for a limited number of applications. In most cases, the thermometer will need to be physically small so that it will quickly come to thermal equilibrium with the system to be measured, and also so that it will affect that system as little as possible. Other considerations will arise depending on the particular situation.

5. The molar mass of CO is $12 + 16 = 28$, so a mole of CO has a mass of 28 g. There are 6.02×10^{23} molecules (Avogadro's number) in a mole of anything.

VI. Problems, Solutions, and Answers

In some of these problems you will need the molar mass of one or more gases. These can be determined from the atomic mass values given in Appendix C of the text.

Example #1. A certain constant-volume gas thermometer reads a pressure of 88 torr at the temperature of the triple point of water. (*a*) What pressure will the thermometer read at a temperature of 310 K? (*b*) What is the temperature when the thermometer reads a pressure of 70 torr?

1. Temperature is proportional to pressure in a constant-volume gas thermometer.	$\dfrac{P_3}{T_3} = \dfrac{P}{T}$ $P = \dfrac{TP_3}{T_3} = \dfrac{(310\,\text{K})(88\,\text{torr})}{(273\,\text{K})} = 99.9\,\text{torr}$
2. The same concepts as for part (*a*) apply here.	$\dfrac{P_3}{T_3} = \dfrac{P}{T}$ $T = \dfrac{PT_3}{P_3} = \dfrac{(70\,\text{torr})(273\,\text{K})}{(88\,\text{torr})} = 217\,\text{K}$

Example #2—Interactive. On a morning when the thermometer reads 52°F, you check the pressure in your bicycle tires using your pressure gauge, and the gauge reading is 75 lb/in². Later in the day, after riding several miles on hot pavement, the temperature of the air in the tires reaches 125°F. Assuming that the volume of the tires hasn't changed, what is the pressure in them now?

Picture the Problem. This is a constant volume problem, so the absolute pressure is proportional to the absolute temperature. Because you will be dealing with ratios of pressure, you will not have to convert pressure to units of torr. **Try it yourself.** Work the problem on your own, in the spaces provided, to get the final answer.

1. Convert the temperatures to kelvins. You may have to convert to degrees Celsius as an intermediate stage.	
2. Convert the pressure readings to absolute pressure. Your tire gauge measures the pressure relative to atmospheric pressure, so you should add $14.7 \, \text{lb/in}^2$, the value of atmospheric pressure, to each reading.	
3. Apply the ratio of pressure and temperature as in the previous problem to find the final pressure in the tire.	$P = 103 \, \text{lb/in}^2$ absolute $= 88 \, \text{lb/in}^2$ gauge

Example #3—Interactive. Mercury freezes at $-39°C$. Express this temperature on both the Fahrenheit and Kelvin scales. **Try it yourself.** Work the problem on your own, in the spaces provided, to get the final answer.

1. Apply the conversion formula to convert the given temperature to Fahrenheit.	$T = -38.2°\, \text{F}$
2. Apply the conversion formula to convert the given temperature to the Kelvin scale.	$T = 234 \, \text{K}$

Example #4. The gas in intergalactic space is mostly atomic hydrogen at a temperature of 3 K and a pressure of the order of 10^{-21} atm. How many hydrogen atoms are there per cubic centimeter? (*Note:* It is interesting to compare this result with that of the next example, which deals with a very good laboratory vacuum.)

1. Apply the ideal-gas law to solve for the molar volume.	$PV = nRT$ $\dfrac{n}{V} = \dfrac{P}{RT} = \dfrac{10^{-21} \, \text{atm}}{\left(0.0821 \, \text{atm} \cdot \text{L}/\left(\text{mol} \cdot \text{K}\right)\right)\left(3 \, \text{K}\right)}$ $= 4.1 \times 10^{-21} \, \text{mol/L}$

2. Use the fact that there are Avogadro's number of atoms in 1 mol of a gas to find the total number of atoms per cubic centimeter.	$N = nN_A$ $= \left(4.1 \times 10^{-21}\ \text{mol/L}\right)\left(6.02 \times 10^{23}\ \text{atoms/mol}\right)$ $= \left(2500\ \text{atoms/L}\right)\left(10^{-3}\ \text{L/cm}^3\right)$ $= 25\ \text{atoms/cm}^3$

Example #5—Interactive. A high-vacuum pump reduces the pressure in a container to 10^{-8} torr. If the temperature is 20°C, how many gas molecules per cubic centimeter are there in the container?

Picture the Problem. To use the ideal-gas law, you will have to convert the values for *P, V, T,* and the gas constant *R* to compatible units. Units conversions play an important role in this sort of problem. **Try it yourself.** Work the problem on your own, in the spaces provided, to get the final answer.

1. Convert *P, V, T,* and the gas constant *R* to compatible units.	
2. Apply the ideal-gas law to solve for the molar volume.	
3. Use the fact that there are Avogadro's number of molecules in 1 mol of a gas to find the total number of molecules per cubic centimeter.	$N = 3.3 \times 10^8$ molecules/cm^3

Example #6. The mass of a certain gas sample that occupies 3.00 L at 20°C and 10 atm pressure is found to be 55 g. What is this gas?

Picture the Problem. Apply the ideal-gas law to find the molar mass of the gas. Use what you remember from chemistry to determine the kind of gas.

1. Apply the ideal-gas law.	$PV = nRT$ $n = \dfrac{PV}{RT}$

2. Use this to get the molar mass.	$M = \dfrac{m}{n} = \dfrac{mRT}{PV}$ $= \dfrac{(55\,g)(0.0821\,atm\cdot L/(mol\cdot K))(293\,K)}{(10\,atm)(3\,L)}$ $= 44.0\,g/mol$
3. This may be any gas with a molar mass of 44.0. Convince yourself that carbon dioxide (CO_2), nitrous oxide (NO_2), and propane (C_3H_8) all satisfy this condition.	

Example #7—Interactive. A rigid, high-pressure gas cylinder has a mass of 21.22 kg when empty. Its interior volume is 1.33 L. Its mass is 21.61 kg after it is filled with nitrogen gas (N_2) at room temperature (20°C). What is the pressure in the filled cylinder?

Picture the Problem. From the mass of the gas in the cylinder, and the molar mass of nitrogen, you can find the number of moles of nitrogen. With this result, you can use the ideal-gas law to solve for the pressure. **Try it yourself.** Work the problem on your own, in the spaces provided, to get the final answer.

1. Find the mass of the nitrogen in the cylinder.	
2. Find the number of moles of nitrogen in the cylinder from the total mass of nitrogen and the molar mass of nitrogen.	
3. Solve the ideal gas law for the pressure of the gas in the cylinder.	$P = 252\,atm = 2.55 \times 10^7\,Pa$

Example #8. Two gases present in the atmosphere are water vapor (H_2O) and argon. What is the ratio of their rms speeds?

Picture the Problem. Both gases should have the same average translational kinetic energies because they are at the same temperature. Use Appendix C in the text to find the atomic masses of hydrogen, oxygen, and argon.

1. Equate the average translational kinetic energies of the two gases, and solve for the ratio of their rms speeds.	$\frac{1}{2}m_{H_2O}v_{rms,H_2O}^2 = \frac{1}{2}m_{Ar}v_{rms,Ar}^2$ $\dfrac{v_{rms,H_2O}}{v_{rms,Ar}} = \left(\dfrac{m_{Ar}}{m_{H_2O}}\right)^{1/2} = \sqrt{\dfrac{39.9}{18}} = 1.49$

Example #9—Interactive. Naturally occurring uranium contains a rare isotope with an atomic mass of 235 and a common isotope with an atomic mass of 238. Uranium reacts with fluorine (atomic weight 19) to form the gas uranium hexafluoride (UF_6). What is the ratio of the rms speed of the UF_6 gas molecules containing atoms of uranium 238 to the rms speed of the gas molecules containing atoms of uranium 235? **Try it yourself.** Work the problem on your own, in the spaces provided, to get the final answer.

1. Use Appendix C of the text to find the masses of uranium 235, uranium 238, and fluorine, as well as the mass of uranium hexafluoride with either species of uranium.	
2. Equate the average translational kinetic energies of both species, and solve for the ratio of the rms speeds.	0.996

Example #10. The surface of the sun consists primarily of monatomic hydrogen gas at a temperature of 6000 K. Compare the rms speed of a hydrogen atom at the surface of the sun with the escape speed from the sun's surface. The sun's radius is 6.96×10^8 m and its mass is 1.99×10^{30} kg.

1. Find the escape velocity of the sun. You may want to refer to Chapter 11.	$v_e = \sqrt{\dfrac{2GM_s}{R_s}}$
2. Find an expression for the rms speed of monatomic hydrogen in terms of temperature, the universal gas constant, and the molar mass of hydrogen.	$v_{rms} = \sqrt{\dfrac{3RT}{M}}$
3. Find the ratio of these two speeds.	$\left(\dfrac{v_{rms}}{v_e}\right)^2 = \dfrac{3RTR_s}{2MGM_s}$ $= \dfrac{3\left(8.314\,\text{J/}\left(\text{mol}\cdot\text{K}\right)\right)\left(6\times10^3\ \text{K}\right)\left(6.96\times10^8\ \text{m}\right)}{2\left(0.0001\,\text{kg/mol}\right)\left(6.67\times10^{-11}\ \text{N}\cdot\text{m}^2/\text{kg}^2\right)\left(1.99\times10^{30}\ \text{kg}\right)}$ $= 3.92\times10^{-4}$ $\dfrac{v_{rms}}{v_e} = 0.02$

322   Chapter Seventeen: Temperature and the Kinetic Theory of Gases

Example #11—Interactive. Like the molecules of all monatomic gases, the molecules of neon have no rotational or vibrational kinetic energy, only translational. (*a*) What is the total kinetic energy of the molecules in 1 L of neon gas (atomic mass of 22) at 1 atm pressure? (*b*) If the gas is expanded at constant temperature to a volume of 2 K, by how much does the total kinetic energy change?

Picture the Problem. The average kinetic energy of a gas molecule is $\frac{3}{2}kT$. The ideal-gas law can be used to find an expression for the number of neon molecules present. You can then find the total kinetic energy of all molecules present. Do all the algebra first, and see what cancels. **Try it yourself.** Work the problem on your own, in the spaces provided, to get the final answer.

1. Use the ideal-gas law to find an expression for the number of neon molecules present.	
2. Multiply this by the average kinetic energy of each molecule to find the total energy of the gas.	$K = 152\,\text{J}$
3. Ask yourself how the calculations will change if the volume changes at constant temperature.	No change.

Example #12. One way to compare the root mean square speed v_{rms} with the mean speed v_{av} of the molecules of a gas is to find the ratio of the two speeds. Find this ratio using the result

$$\int_0^\infty x^3 e^{-a^2 x^2}\, dx = \frac{1}{2a^4}$$

Picture the Problem. You know the rms speed of a gas molecule is given by $\sqrt{3kT/m}$. The average speed of a gas molecule can be found from the Maxwell-Boltzmann speed distribution. Finally, take the ratio of the two speeds.

1. Write out the Maxwell-Boltzmann speed distribution function.	$f(v) = \dfrac{4}{\sqrt{\pi}}\left(\dfrac{m}{2kT}\right)^{3/2} v^2 e^{-mv^2/2kT}$

2. Find the average value of this function. Don't forget to use the equation given in the problem to help solve the integral.	$v_{av} = \int_0^\infty v f(v)\,dv$ $= \dfrac{4}{\sqrt{\pi}}\left(\dfrac{m}{2kT}\right)^{3/2}\int_0^\infty v^3 e^{-mv^2/2kT}\,dv$ $= \dfrac{4}{\sqrt{\pi}}\left(\dfrac{m}{2kT}\right)^{3/2}\dfrac{1}{2\left(\frac{m}{2kT}\right)^2}$ $= \dfrac{2}{\sqrt{\pi}}\left(\dfrac{m}{2kT}\right)^{-1/2} = \left(\dfrac{8kT}{\pi m}\right)^{1/2}$
3. Take the ratio of the two speeds.	$\dfrac{v_{rms}}{v_{av}} = \dfrac{\left(\dfrac{3kT}{m}\right)^{1/2}}{\left(\dfrac{8kT}{\pi m}\right)^{1/2}} = \sqrt{\dfrac{3\pi}{8}} = 1.09$

Example #13—Interactive. Compare the root mean square speed of a gas molecule with the most probable speed by finding their ratio. **Try it yourself.** Work the problem on your own, in the spaces provided, to get the final answer.

1. Write an expression for the rms speed of a gas molecule.	
2. Find the most probable speed of a gas molecule. This will be the speed at which the Maxwell-Boltzmann speed distribution function is a maximum. Take a derivative of this function to find this speed.	
3. Take the ratio of the two speeds.	$\dfrac{v_{rms}}{v_{most\ probable}} = \sqrt{3/2} = 1.22$

Chapter 18

Heat and the First Law of Thermodynamics

I. Key Ideas

Heat is energy in transit from a warmer object to a colder object because of the temperature difference between them. When energy is transferred to or from an object as heat, a change in the internal energy (thermal energy) of the object occurs. Neither thermal energy nor mechanical energy is necessarily conserved. What is always conserved in an isolated system is the sum of the mechanical energy and the thermal energy.

Section 18-1. Heat Capacity and Specific Heat. The transfer of energy via heat is often accompanied by changes in temperature. The amount of heat Q needed to raise the temperature of an object by one degree is called its **heat capacity** C:

$$Q = C \, \Delta T = mc \, \Delta T \qquad\qquad\qquad \text{Heat capacity}$$

where m is the mass of the substance. The **specific heat** c is the heat capacity per unit mass of a substance. The **molar specific heat** c' of a substance is the heat capacity per mole:

$$C = mc = nc' \qquad\qquad\qquad \text{Specific heat and molar heat capacity}$$

where n is the number of moles.

The specific heat of water is quite large compared to that of other ordinary materials. The specific heat of liquid water is nearly constant over a wide range of temperature, and the specific heats of other materials are often measured by comparison with water. This is done by heating a sample of a material to some known temperature, placing the sample in a water bath of known mass and temperature, and then measuring the final equilibrium temperature. If this system is isolated from its surroundings, the heat leaving the object equals the heat entering the water and its container. This process is called *calorimetry* and the insulated water container is called a *calorimeter.*

The traditional unit of heat is the **calorie** (cal). It is approximately the amount of heat required to raise the temperature of 1 gram of water by 1 K. It is defined in terms of the joule.

$$1 \text{ cal} = 4.184 \text{ J} \qquad\qquad\qquad \text{Definition of calorie}$$

The "calorie" used in measuring the nutritional value of food is actually a kilocalorie. The U.S. customary unit of heat is the **Btu** (British thermal unit). It was originally defined as the amount of energy needed to raise the temperature of one pound of water by one degree Fahrenheit. The Btu is related to the calorie and the joule by

$$1\,\text{Btu} = 252\,\text{cal} = 1.054\,\text{kJ}$$ Definition of Btu

Section 18-2. Change of Phase and Latent Heat. Under certain conditions, a substance will remain at the same temperature while energy is being transferred to or from it in the form of heat. This happens when the substance undergoes a **phase change** such as **fusion** (a change from the liquid phase to the solid phase). The other common phase changes are **vaporization** (a change from a liquid to a vapor or gas) and **sublimation** (a change from a solid directly into a gas). The heat required to melt a substance of mass m with no change in its temperature is

$$Q_f = mL_f$$ Latent heat of fusion

where L_f is the **latent heat of fusion** of the substance. For the melting of ice to water at a pressure of 1 atm, the latent heat of fusion is 333.5 kJ/kg = 79.7 kcal/kg. When the phase change is from liquid to gas, the heat required is

$$Q_v = mL_v$$ Latent heat of vaporization

where L_v is the **latent heat of vaporization**. The latent heat of vaporization for water at 1 atm is 2.26 MJ/kg = 540 kcal/kg .

Section 18-3. Joule's Experiment and the First Law of Thermodynamics. When Joule measured the amount of mechanical work required to produce a given amount of thermal energy, he found it took 4.184 joules of work (or mechanical energy) to raise the temperature of 1 g of water 1 C°. This quantity (≈ 4.184 J/cal) is known as the **mechanical equivalent of heat**. This lead to the first law of thermodynamics, which states that energy is conserved. Conservation of energy means that the sum of all types of energy in a system at an instant in time equals the sum at an earlier time, but that the distribution of the different types of energy, such as potential, kinetic, and thermal, may vary. If the sum varies, then energy has been added to or taken away from the system.

$$\Delta E_{int} = Q_{in} + W_{on}$$ First law of thermodynamics

The change in the internal energy of the system equals the heat transferred into the system plus the work done on the system.

In the first law of thermodynamics, ΔE_{int} is the change in the internal energy of the system, Q_{in} is the heat transferred into the system, and W_{on} is the work done by the surroundings *on* the system. Q_{in} is positive if heat is transferred into the system by the surroundings, and negative if heat is transferred out of the system into its surroundings. W_{on} is positive if, for example, a piston compresses the gas in a system, and negative if that gas would expand against the piston.

For very small amounts of heat added, work done, or changes in internal energy, it is customary to express the first law of thermodynamics in differential form:

$$dE_{int} = dQ_{in} + dW_{on}$$ Differential form of the first law of thermodynamics

In this equation, dE_{int} is the differential of the internal-energy function. However, neither dQ_{in} nor dW_{on} is a differential of any function. Rather, dQ_{in} merely represents a small amount of heat added to the system, and dW_{on} represents a small amount of work done on the system.

Section 18-4. The Internal Energy of an Ideal Gas. At low densities, the total volume of the molecules of a gas, compared with the volume occupied by the gas, is negligible, and the forces the molecules exert on each other during the intervals between collisions are negligible. Gases at densities low enough to fulfill these conditions are called ideal gases. The average translational kinetic energy of the molecules of an ideal gas is proportional to the absolute temperature of the gas. If the internal energy E_{int} of a gas consists only of this translational kinetic energy, then

$$E_{int} = K = \tfrac{3}{2}nRT$$ Internal energy of a monatomic ideal gas

where n is the number of moles, R is the universal gas constant, and T is the absolute temperature. If the molecules have other types of energy in addition to translational energy, such as rotational energy, the internal energy will be greater than $\tfrac{3}{2}nRT$. According to the equipartition theorem, the average energy associated with any degree of freedom will be $\tfrac{1}{2}kT$ per molecule ($\tfrac{1}{2}RT$ per mole). *Note:* The internal energy of an ideal gas depends only on the temperature and not on the volume or pressure. (In a real gas the forces between molecules cannot always be neglected. The internal energy of such a gas includes the potential energy associated with these forces. This potential energy depends upon the intermolecular separation, which does depend on the pressure and volume of such a gas.)

To test whether or not the particles of real gases are noninteracting, Joule performed an experiment in which a gas confined in a rigid, thermally insulated container was allowed to expand into a second rigid, thermally insulated container that, prior to the expansion, was evacuated. This process is called a **free expansion**. In a free expansion the gas does no work on its surroundings and no heat is transferred to or from the gas, so the internal energy of the gas does not change. Joule found that if a gas initially at low density (with the molecules well separated) underwent a free expansion, any changes in its temperature were too small to observe. However, when a gas initially at high density (with the molecules in close proximity) underwent a free expansion, he observed a slight decrease in its temperature. These results demonstrate that for a gas at low density, the interactions between molecules are negligible; however, for a gas that is at a higher density, the attractive forces between molecules have a small effect. When a gas expands, the potential energy associated with the attractive forces between the molecules must increase. Since the internal energy of the gas is constant, this increase in potential energy is accompanied by a corresponding decrease in translational kinetic energy—a decrease that is evidenced by the observed slight temperature drop.

Section 18-5. Work and the PV Diagram for a Gas. In the following discussion, the first law of thermodynamics is applied to gases. For simplicity this discussion will be restricted to gases at low densities that satisfy the ideal-gas law.

Processes such as expansions and compressions may occur so slowly that during the process the gas is never far from an equilibrium state. In this kind of process, called a **quasi-static**

process, the gas moves through a series of equilibrium states. In practice it is possible to execute processes slowly enough to approximate quasi-static processes fairly well. If a confined gas is quasi-statically compressed, during the compression the pressure is the same throughout the gas. However, if the compression occurs very rapidly (violently), there will be pressure shock waves in the gas and the pressure of the confined gas will vary with location in the gas at any moment in time.

Consider a gas confined to a cylinder with a movable piston. When the piston slowly (quasi-statically) recedes a small distance dx, the gas expands. The work dW done by the gas on the piston is

$$dW_{\text{by gas}} = F_x \, dx = PA \, dx = P \, dV$$

where A is the area of the piston and $dV = A \, dx$ is the increase in volume of the gas. When the piston moves through a finite distance, from x_1 to x_2, the volume of the gas increases from V_1 to V_2.

The equation of state of a given sample of gas relates the three variables: pressure P, temperature T, and volume V. Thus, the state of the gas can be specified by any two of these variables, say P and V. Therefore, each point on a PV diagram specifies a particular state of the gas. Because specifying P and V specifies the temperature, a temperature T is associated with each point on the graph. The variables P, V, and T are called functions of state (or state functions) because they only depend upon the state of the gas. The internal energy E_{int} is also a state function.

During an expansion or compression, the total work done *on* the gas is simply the negative of the work done by the gas

$$W_{\text{on gas}} = -W_{\text{by gas}} = -\int_{V_i}^{V_f} P \, dV = \text{negative area under } PV \text{ curve} \qquad \text{Work done on a gas}$$

It is particularly easy to calculate the work done by a gas during an **isobaric expansion,** which is one that takes place at constant pressure. It is simply

$$W_{\text{on gas, isobaric}} = -P \, \Delta V \qquad \text{Isobaric work done on a gas}$$

For an **isothermal expansion**—one that takes place at constant temperature—the work is

$$W_{\text{on gas, isothermal}} = -\int_{V_i}^{V_f} P \, dV = -\int_{V_i}^{V_f} \frac{nRT}{V} \, dV$$

$$= -nRT \int_{V_i}^{V_f} \frac{1}{V} \, dV = -nRT \, \ln\frac{V_f}{V_i} = nRT \ln\frac{V_i}{V_f} \qquad \text{Isothermal work done on a gas}$$

Because pressures are often given in atmospheres and volumes are often given in liters, it is convenient to have a conversion factor between liter-atmospheres and joules:

$$1 \text{ L·atm} = (10^{-3} \text{ m}^3)(1.013 \times 10^5 \text{ Pa}) = 101.3 \text{ J} \qquad \text{Liter-atmospheres}$$

If a gas undergoes a **cyclic process,** by definition the final state of the gas is the same as its initial state. Thus, for a cyclic process the path on a PV curve is closed. The total work done by the gas during a cycle equals the area enclosed by the PV curve during the cycle.

Section 18-6. Heat Capacities of Gases. If a sample of a material is allowed to expand as it is heated at a constant pressure (as nearly all materials tend to do), more heat input is required for a given temperature increase than would be required for the same temperature increase if the volume of the material were held constant. This is because in expanding, the sample does work by pushing back on its surroundings. This loss of energy to the surroundings is made up for by an increased input of heat.

For solids and liquids, which expand only slightly, the additional heat required for a given temperature increase is very small. Thus, the specific heat at constant pressure is approximately equal to the specific heat at constant volume. The specific heat at constant pressure is normally specified because solids and liquids generate enormous pressure if they are not allowed to expand, making it difficult to measure the specific heat at constant volume.

Unlike solids and liquids, the heat capacity at constant pressure for gases is substantially greater than the heat capacity at constant volume. At constant pressure, gases undergo significant fractional increases in volume, so an appreciable increase of heat is required for a given temperature change.

When heat Q_v is added to a gas confined to a constant volume, no work is done by the gas so all of the heat goes into increasing the internal energy ΔE_{int} of the gas. That is,

$$dQ_v = dE_{int} = C_v\,dT$$

so

$$C_v = \frac{dE_{int}}{dT} \qquad\qquad \text{Heat capacity at constant volume}$$

For an ideal gas E_{int} depends only on T. Thus $dE_{int} = C_v\,dT$ even if V and/or P are not constant. For an ideal gas $V = nRT/P$. Thus, for an expansion at constant pressure, $dV = (nR/P)\,dT$, so

$$dQ = dU + dW = dU + P\,dV = C_v\,dT + nR\,dT$$
$$= C_p\,dT$$

Hence,

$$C_p = C_v + nR \qquad\qquad \text{Heat capacity at constant pressure}$$

The heat capacity at constant pressure is greater than that at constant volume because of the work done by the expanding gas. The heat capacity per mole c_p' of a gas at constant pressure exceeds the heat capacity per mole c_v' at constant volume by R.

If the translational kinetic energy of the molecules of a gas is the only form of internal energy E_{int}, then

$$E_{\text{int}} = \tfrac{3}{2} RT$$

and the heat capacity at constant volume C_v would be just $\tfrac{3}{2} nR$ (and the heat capacity per mole c_v' at constant volume is $\tfrac{3}{2} R$). Measurements show that this holds for monatomic gases but not for more complex molecules.

Heat Capacities and the Equipartition Theorem. The average translational kinetic energy per molecule of a monatomic gas is $\tfrac{3}{2} kT$ (recall that $R = N_A k$). There are three equivalent, independent directions of motion, so a kinetic energy $\tfrac{1}{2} kT$ associated with each direction contributes to this total. This division of the kinetic energy into three terms in the translational kinetic energy equation ($K = \tfrac{1}{2} mv_x^2 + \tfrac{1}{2} mv_y^2 + \tfrac{1}{2} mv_z^2$) is a special case of the **equipartition theorem**. Each coordinate, velocity component, angular velocity component, and so forth that appears in the expression for the energy of a molecule is called a **degree of freedom**. The equipartition theorem states that there is an average energy of $\tfrac{1}{2} kT$ per molecule associated with each degree of freedom. For gases with complex molecules, the rotational and vibration degrees of freedom must also be considered. Diatomic molecules are found to have molar heat capacities at constant volume equal to $\tfrac{5}{2} R$. This is understood by recognizing that in addition to the three degrees associated with its translational kinetic energy, a diatomic molecule has two degrees of freedom associated with its rotational kinetic energy. It follows that the energy per molecule, energy per mole, and heat capacity at constant volume are, respectively, $\tfrac{5}{2} kT$, $\tfrac{5}{2} RT$, and $\tfrac{5}{2} nR$.

Section 18-7. Heat Capacities of Solids. Experimental measurements show the molar specific heats of most solids to be approximately $3R$. This result is called the **Dulong–Petit law**. The molecules in a solid are fixed in place, but each can vibrate in three dimensions. Solids thus have a total of six degrees of freedom (three for vibrational potential energy and three for vibrational kinetic energy), and the equipartition theorem predicts a heat capacity per mole of $3R$.

Section 18-8. Failure of the Equipartition Theorem. Although the equipartition theorem has been highly successful, it also has shortcomings. Ultimately it fails because at the molecular level Newtonian mechanics fails. To predict observed results successfully, the theory of quantum mechanics must be used.

Section 18-9. The Quasi-Static Adiabatic Expansion of a Gas. An **adiabatic process** is one in which there is no net flow of heat into or out of the system. A **quasi-static process** is one that happens in infinitesimal steps with delays between steps allowing the system to reach equilibrium. If a gas undergoes a quasi-static adiabatic expansion (or compression), it can be shown that

$$PV^{\gamma} = \text{constant} \quad \text{and} \quad TV^{\gamma-1} = \text{constant} \qquad \text{Quasi-static adiabatic process}$$

where $\gamma = C_p / C_v$ is the ratio of the constant pressure and constant volume heat capacities. The work done on a gas undergoing a quasi-static adiabatic compression is

$$W_{\text{adiabatic}} = C_v \, \Delta T = \frac{P_f V_f - P_i V_i}{\gamma - 1} \qquad \text{Quasi-static adiabatic work done on a gas}$$

The speed of sound is given by $v = \sqrt{B/\rho}$ where ρ is the mass density and B the bulk modulus. For an isothermal process the bulk modulus equals the pressure P, but for an adiabatic process it equals γP. Using the adiabatic bulk modulus it can be shown that

$$v = \sqrt{\frac{B_{\text{adiab}}}{\rho}} = \sqrt{\frac{\gamma RT}{M}}$$

Speed of sound

where M is the molar mass. This result is in close agreement with measured values for the speed of sound.

II.   Physical Quantities and Key Equations

Physical Quantities

Definition of calorie	1 cal = 4.184 J
Definition of Btu	1 Btu = 1.054 kJ
Liter-atmospheres	1 L·atm = $(10^{-3} \text{ m}^3)(1.013 \times 10^5 \text{ Pa}) = 101.3$ J
Latent heat of fusion for water	L_f =333.5 kJ/kg = 79.7 kcal/kg
Latent heat of vaporization for water	L_v =2.26 MJ/kg = 540 kcal/kg

Key Equations

Heat capacity	$Q = C\,\Delta T = mc\,\Delta T$
Specific heat and molar heat capacity	$C = mc = nc'$
Latent heat of fusion	$Q_f = mL_f$
Latent heat of vaporization	$Q_v = mL_v$
First law of thermodynamics	$\Delta E_{\text{int}} = Q_{\text{in}} + W_{\text{on}}$
Work done on a gas	$W_{\text{on gas}} = -\int_{V_i}^{V_f} P\,dV$ = neagative area under PV curve
Heat capacity at constant volume	$C_v = \dfrac{dE_{\text{int}}}{dT}$

Except for the Dulong–Petit law, the following formulas are restricted to ideal gases:

Translational kinetic energy of a gas	$K = \frac{3}{2}nRT$
Internal energy of a monatomic ideal gas	$U = \frac{3}{2}nRT$

Internal energy of a diatomic gas	$U = \frac{5}{2}nRT$
Heat capacity per mole for a monatomic gas	$c'_v = \frac{3}{2}R$
Heat capacity per mole for a diatomic gas	$c'_v = \frac{5}{2}R$
Heat capacity per mole at constant pressure	$c'_p = c'_v + R$
Dulong–Petit law (for solids)	$c' = 3R$
Ratio of specific heats	$\gamma = \dfrac{C_p}{C_v} = \dfrac{c'_p}{c'_v}$
Quasi-static adiabatic ($Q = 0$) process	$PV^\gamma = \text{constant} \text{ and } TV^{\gamma-1} = \text{constant}$
Isobaric ($P = \text{constant}$) work	$W_{\text{on gas, isobaric}} = -P\,\Delta V$
Isothermal ($T = \text{constant}$) work	$W_{\text{on gas, isothermal}} = -\int_{V_i}^{V_f} P\,dV = nRT\ln\dfrac{V_i}{V_f}$
Quasi-static adiabatic ($Q = 0$) work	$W_{\text{on gas, adiabatic}} = C_v\,\Delta T = \dfrac{P_f V_f - P_i V_i}{\gamma - 1}$

III. Potential Pitfalls

Be very careful with signs in first-law problems, heat exchange problems, and such. It's easy to get them mixed up. The convention is that Q_{in} is positive when heat flows into the system and W_{on} is positive when work is done by the surroundings on the system.

In a free expansion the gas does no work. For any other type of expansion, negative work is done on the gas. As a gas is compressed, positive work is done on the gas.

For a solid or a liquid, the heat capacity specified in a table is almost always the heat capacity at a pressure of 1 atmosphere. The heat capacity at constant volume is slightly less than this, but the difference is so small it is usually negligible. This is because, for a given temperature increase, the work done by the solid or liquid in pushing back the atmosphere is small compared to the increase in its internal energy.

In doing first-law-of-thermodynamics problems, don't assume that a heat transfer and a temperature change always go together. A quasi-static adiabatic expansion is an example of a temperature change without heat transfer. An isothermal expansion is an example of a heat transfer without a temperature change. There can be a temperature change without any heat transfer, and there can be a heat transfer without any temperature change.

Remember that the heat capacities C_p and C_v refer to a specific sample of gas and are proportional to the quantity of gas. By contrast the molar heat capacities (heat capacities per mole) c'_p and c'_v are independent of the total amount of gas.

The internal energy of an ideal gas depends only on its temperature. For other systems the internal energy depends on additional conditions, such as the density of the material.

When an ideal gas undergoes some process and thus changes state, the ideal-gas law may be used to relate initial and final states. However, this equation does not specify whether heat flowed into or from the gas in the process. The amount of heat exchanged by the gas depends on how the process was carried out.

IV. True or False Questions and Responses

True or False

_____ 1. The only way in which the internal energy of a system can increase is for heat to be added to the system.

_____ 2. The heat capacity of a material is its specific heat per unit mass.

_____ 3. A quasi-static process is one in which no work is done.

_____ 4. The specific heat commonly given for solids and liquids is the specific heat at constant pressure.

_____ 5. The internal energy of an ideal gas depends only on the temperature of the gas.

_____ 6. In an isothermal process, the temperature of the system remains constant.

_____ 7. In an adiabatic process, the internal energy of the system remains constant.

Responses to True or False

1. False. Doing work on a system can also cause an increase in its internal energy.

2. False. It is the other way around. The specific heat is the heat capacity per unit mass.

3. False. A quasi-static process is one that occurs sufficiently slowly for the system to progress through a sequence of equilibrium states.

4. True. For solids and liquids the difference between the specific heat at constant pressure and the specific heat at constant volume is so small that it is usually considered negligible.

5. True.

6. True.

7. False. In an adiabatic process, no heat is added to or removed from the system.

V.   Questions and Answers

Questions

1. Can a system absorb heat without its temperature increasing? Can a system absorb heat without its internal energy increasing?

2. Can the temperature of 2 L of water be increased from 20 to 30°C without any work being done by the water? Explain.

3. For solids and liquids, we usually don't distinguish between specific heats at constant volume and specific heats at constant pressure. Why not? If we did want to make this distinction, which one would be larger? Which specific heat are you most likely to find tabulated?

4. An ideal gas expands at constant temperature and does work on a piston. Does the internal energy of the gas decrease? If not, where does the energy to do the work come from?

5. Give an example of a process in which the temperature of a system is increased without any heat input or output.

6. An ideal gas expands slowly to twice its initial volume (*a*) at constant pressure and (*b*) at constant temperature. In which case does it do more work on its surroundings? Why?

7. Consider nitrogen (N_2) and helium (He) gases. For which gas is the internal energy per mole greater at a given temperature? For which gas is the internal energy per gram greater at a given temperature?

8. The ratio of molar heat capacities ($\gamma = c_p' / c_v'$) for a monatomic ideal gas is $\frac{5}{3} = 1.67$; for a diatomic ideal gas it is $\frac{7}{5} = 1.4$. Would you expect the ratio to be higher or lower than 1.4 for an ideal gas having three or more atoms per molecule?

Answers

1. Consider a process in which heat is added to a system and the system does work on its surroundings. According to the first law of thermodynamics, the internal energy of the system will increase if the heat added to the system is greater than the work done by the system; the internal energy will remain the same if the heat added to the system is equal to the work done by the system, and the internal energy will decrease if the heat added to the system is less than the work done by the system. Thus, it is possible for a system to absorb heat without its temperature and its internal energy increasing, if the work done by the system equals or is greater than the heat it absorbs.

    Even without doing work, a system can absorb energy without its temperature increasing—for example, when it is melting or vaporizing.

2. If the volume of the water is kept the same during the temperature increase (which is **very** difficult to do, by the way), then the water does no work.

3. For any material that expands when its temperature is increased, c_p is greater than c_v, just as for a gas. That is because, in an expansion at constant pressure, the material does work in pushing back its surroundings. We normally pay no attention to the distinction in solids and liquids because the change in volume is very small. The value ordinarily measured and tabulated is c_p.

4. The internal energy of an ideal gas depends only upon its temperature. Thus in any isothermal process the internal energy of the gas remains constant. When a gas expands, it does work on its surroundings. According to the first law of thermodynamics, when the gas does work on its surroundings, its internal energy will decrease by an amount equal to the work done, unless energy is transferred to the gas via heat. To keep the temperature—and thus the internal energy—constant, the heat added to the gas during expansion must equal the work done by the gas.

5. Consider an ideal gas, confined in a cylinder with insulated walls, being compressed by a piston. The piston does work on the gas, which results in an increase in the internal energy of the gas. The internal energy of the gas is proportional to its temperature, so the increase in internal energy will result in an increase in temperature.

6. An ideal gas is initially at pressure P_0 and volume V_0. If the gas expands to a volume of $2V_0$ at constant pressure, the work it does on its surroundings is $P_0 \Delta V = P_0 V_0$. When a gas expands at constant temperature, the pressure varies inversely with the volume, so the pressure steadily drops as it expands. Thus, during the isothermal expansion the average pressure is always less than P_0, so the work done by the gas is less than $P_0 V_0$.

7. The internal energies per mole of all monatomic and diatomic gases are $1.5RT$ and $2.5RT$, respectively. Helium is a monatomic gas and nitrogen is diatomic. Thus at a given temperature the internal energy per mole for nitrogen is greater than it is for helium. The internal energy per gram of a substance equals the internal energy per mole divided by the molar mass, and the molar masses are 28 g/mol for nitrogen and 4 g/mol for helium. Thus, the internal energy per gram is $(2.5/28)RT = 0.089RT$ for nitrogen and $(1.5/4)RT = 0.375RT$ for helium. The internal energy per unit mass for helium is greater than it is for nitrogen.

8. The ratio of molar heat capacities is

$$\gamma = \frac{c_p'}{c_v'} = \frac{c_v' + R}{c_v'} = 1 + \frac{R}{c_v'}$$

It is smaller when the number of degrees of freedom is larger because c_v' is proportional to the number of degrees of freedom. A gas whose molecules have three or more atoms can be expected to have more degrees of freedom than a diatomic gas. Thus, for a gas with three or more atoms per molecule, the ratio of molar heat capacities is expected to be lower than it is for either monatomic or diatomic gases. Measurements show that many polyatomic gases have a ratio around $\gamma = 1.35$.

VI. Problems, Solutions, and Answers

Example #1. A piece of iron (specific heat $0.431 \text{ kJ}/(\text{kg} \cdot \text{K})$) of mass 80.0 g at a temperature of 98.0°C is dropped into an insulated vessel containing 120 g of water at 20.0°C. At what final temperature does the system come to equilibrium?

Picture the Problem. Assume the heat losses to the container and the surroundings are negligible. Then the heat lost by the iron is equal to that gained by the water. The heat loss (or gain) is equal to the mass times the specific heat times the temperature change.

1. The heat that leaves the iron object is equal to the heat received by the water.	$Q_{in, \, water} = Q_{out, \, object}$ $m_{water} c_{water} \Delta T_{water} = m_{object} c_{object} \left\| \Delta T_{object} \right\|$ $m_{water} c_{water} \left(T_f - T_{i,water} \right) = m_{object} c_{object} \left(T_{i,object} - T_f \right)$
2. Solve for the final temperature.	$T_f = \dfrac{m_{object} c_{object} T_{i,object} + m_{water} c_{water} T_{i,water}}{m_{object} c_{object} + m_{water} c_{water}}$ $= \dfrac{(0.0345 \text{ kJ/K})(371 \text{ K}) + (0.502 \text{ kJ/K})(293 \text{ K})}{(0.0345 \text{ kJ/K}) + (0.502 \text{ kJ/K})}$ $= 298 \text{ K} = 25°\text{C}$

Example #2—Interactive. The specific heat of copper is $0.386 \text{ kJ}/(\text{kg} \cdot \text{K})$. If 180 g of copper at 200°C is dropped into an insulated container containing 280 g of water at 20°C, what is the final equilibrium temperature of the copper and water?

Picture the Problem. Assume the heat losses to the container and the surroundings are negligible. Then the heat lost by the copper is equal to that gained by the water. The heat loss (or gain) is equal to the mass times the specific heat times the temperature change. This approach neglects any boiling of the water. When this experiment is actually done some water boils when the 200°C surface of the copper comes in contact with it. Boiling continues until the surface temperature of the copper falls below 100°C. Under these circumstances the equilibrium temperature of the copper and the remaining water would be somewhat lower than the temperature calculated without considering the boiling. **Try it yourself.** Work the problem on your own, in the spaces provided, to get the final answer.

1. The heat that leaves the copper object is equal to the heat received by the water.	
2. Solve for the final temperature.	$T_f = 30.1°\text{C}$

Example #3. A 160-g mass of a certain metal at an initial temperature of 88°C is dropped into 140 g of water in an insulated container. The water is initially at 10°C. The system finally comes to equilibrium at 18.4°C. (*a*) If heat losses to the container and the surroundings are negligible, what is the specific heat of the metal? (*b*) If the Dulong–Petit law holds, what is the molar mass of the metal?

Picture the Problem. If the heat losses to the container and the surroundings are negligible, the heat lost by the metal object is equal to that gained by the water. The heat lost (or gained) is equal to the mass times the specific heat times the temperature change. Find the heat capacity per mole of the metal, and by using the Dulong-Petit law, determine the molar mass of the metal, and hence the composition of the metal.

1. The heat lost by the metal object is equal to the heat gained by the water. Use this to solve for the specific heat of the metal object.	$$Q_{in,\,water} = Q_{out,\,object}$$ $$m_{water} c_{water}\, \Delta T_{water} = m_{object} c_{object} \left\lvert \Delta T_{object} \right\rvert$$ $$c_{object} = \frac{m_{water} c_{water} \left(T_f - T_{i,water}\right)}{m_{object}\left(T_{i,object} - T_f\right)}$$ $$= \frac{(0.0140\,\text{kg})(4.18\,\text{kJ}/(\text{kg}\cdot\text{K}))(18.4°\text{C} - 10°\text{C})}{(0.160\,\text{kg})(88°\text{C} - 18.4°\text{C})}$$ $$= 0.441\,\text{kJ}/(\text{kg}\cdot\text{K})$$
2. The heat capacity per mole is the product of the specific heat and the molar mass M. The heat capacity per mole for a solid is approximately $3R$ according to the Dulong-Petit law.	$$C_{mo} = c_0 M = 3R$$ $$M = \frac{3R}{c_0} = \frac{3(8.31\,\text{J}/(\text{mol}\cdot\text{K}))}{0.441\,\text{kJ}/(\text{kg}\cdot\text{K})} = 56.5\,\text{g/mol}$$
3. Based on this molar mass, predict the composition of the metal.	It is likely iron.

Example #4—Interactive. Imagine that you want to take a warm bath, but there's no hot water. You draw 40.0 kg of tap water at 18°C in the bathtub and heat water on your stove to warm the bath water up. If you heat the water to 100°C in a 2-L saucepan, how many panfuls must you add to the bath to raise its temperature to 40°C?

Picture the Problem. If the heat losses to the bathtub and the surroundings are negligible, the heat lost by the heated water is equal to that gained by the cold bath water. The heat lost (or gained) is equal to the mass times the specific heat times the temperature change. **Try it yourself.** Work the problem on your own, in the spaces provided, to get the final answer.

1. The heat lost by the heated water is equal to that gained by the cold bath water. Use this to solve for the required mass of hot water.	

2. Use the density of water to determine how many liters of water are required.	
3. Convert the number of liters to the number of panfuls.	7.33 panfuls

Example #5. A copper bar of mass 2.50 kg at an initial temperature of 66.0°C is dropped into an insulated vessel containing 400 g of water and 70.0 g of ice at 0.00°C. The specific heat of copper can be found in Table 18-1 on page 560 of the text. At what final temperature does the system come to thermal equilibrium?

Picture the Problem. Assume the heat losses to the container and the surroundings are negligible. Unlike in Example #2, here we will also consider the effects of the melting ice. The ice must melt before the water temperature increases. As a result, the heat lost by the copper is equal to the latent heat of fusion required to melt the ice plus the heat gained by the resulting total amount of water.

1. Determine the amount of heat required to melt the ice.	$Q_f = mL_f = (0.700\,\text{kg})(333.5\,\text{kJ/kg}) = 23.3\,\text{kJ}$
2. Determine how much the temperature of the copper will drop if it gives up this much heat.	$Q = mc\,\Delta T$ $\Delta T = \dfrac{Q}{mc} = \dfrac{-23.3\,\text{kJ}}{(2.50\,\text{kg})(0.386\,\text{kJ/}(\text{kg}\cdot\text{K}))}$ $= -24.1\,\text{K} = -24.1\,°\text{C}$
3. Find the final temperature of the copper after the ice has melted. Because the copper is still warmer than the water, it will increase the temperature of the water. What would it mean if the temperature of the copper at this point was less than 0°C?	$66.0\,°\text{C} - 24.1\,°\text{C} = 41.9\,°\text{C}$
4. The remaining heat lost by the copper is equal to the heat gained by the total mass of water, which must include the mass of the newly melted water. Use this to solve for the final temperature of the water.	$Q_{\text{in, water}} = Q_{\text{out, object}}$ $m_{\text{water}} c_{\text{water}} \Delta T_{\text{water}} = m_{\text{object}} c_{\text{object}} \left| \Delta T_{\text{object}} \right|$ $T_f = \dfrac{m_{\text{object}} c_{\text{object}} T_{i,\text{object}} + m_{\text{water}} c_{\text{water}} T_{i,\text{water}}}{m_{\text{object}} c_{\text{object}} + m_{\text{water}} c_{\text{water}}}$ $= \dfrac{(40.4\,\text{kJ}\cdot°\text{C/K})(371\,\text{K}) + (0\,\text{kJ}\cdot°\text{C/K})}{(0.965\,\text{kJ/K}) + (1.96\,\text{kJ/K})}$ $= 13.8\,°\text{C}$

Example #6—Interactive. A cup contains 240 g of fresh-brewed coffee at a temperature of 97°C. If heat losses to the surroundings are negligible, to what final temperature is the coffee cooled if you drop an 18-g ice cube into it?

Picture the Problem. The heat lost by the coffee will first melt the ice cube, and then, if there is enough heat left, increase the temperature of the melted water, which starts out at 0.00°C. **Try it yourself.** Work the problem on your own, in the spaces provided, to get the final answer.

1. Determine the amount of heat required to melt the ice.	
2. Determine how much the temperature of the coffee will drop if it gives up this much heat. Coffee is essentially hot water.	
3. Find the final temperature of the coffee.	
4. If the coffee is still hotter than the melted water, the remaining heat lost by the coffee is equal to the heat gained by the water. Use this to solve for the final temperature of the weakened coffee.	$T_f = 84.7°C$

Example #7. A piece of aluminum (specific heat $0.9 \text{ kJ}/(\text{kg}\cdot\text{K})$) of mass 135 g initially at 20°C is placed in a large container of liquid nitrogen at 77 K (its normal boiling point). If the latent heat of vaporization of nitrogen is 199 kJ/kg, what mass of nitrogen is vaporized in cooling the aluminum to 77 K?

Picture the Problem. Since a large amount of liquid nitrogen is available, the aluminum will be cooled all the way to 77 K. The heat extracted from the aluminum is equal to that absorbed by the nitrogen while it boils.

1. The heat gained by the nitrogen goes into boiling it off: the latent heat of vaporization. This must be equal to the heat lost by the aluminum.	$Q_{gained, N} = Q_{lost, Al}$ $m_N L_{v,N} = m_{Al} c_{Al} \lvert \Delta T_{Al} \rvert$ $m_N = \dfrac{m_{Al} c_{Al} \lvert \Delta T_{Al} \rvert}{L_{v,N}}$ $= \dfrac{(0.135\,\text{kg})(0.900\,\text{kJ}/(\text{kg}\cdot\text{K}))(293\,\text{K} - 77\,\text{K})}{199\,\text{kJ}/(\text{kg}\cdot\text{K})}$ $= 0.132\,\text{kg}$

Example #8—Interactive. An abundance of steam at 100°C is passed into an insulated flask containing 200 g of ice at −25°C. Neglecting any heat loss from the flask, what mass of liquid water at 100°C will finally be present in the flask? The specific heats of ice and liquid water can be found in Table 18-1 on page 560 of the text. The latent heats of fusion and vaporization of water can be found in Table 18-2 on page 563 of the text.

Picture the Problem. The temperature of the ice will first be raised to 0°C, then the ice will melt, and finally the temperature of the melted water will be raised to 100°C. The heat gained by the ice in this process will be equal to the heat lost by some of the steam as it condenses. The final mass of water at 100°C will be equal to the initial mass of the ice, plus the mass of the condensed steam. **Try it yourself.** Work the problem on your own, in the spaces provided, to get the final answer.

1. Determine how much heat is required to raise the temperature of the ice to 0°C.	
2. Determine how much heat is required to melt the ice.	
3. Determine how much heat is required to raise the temperature of the water from 0°C to 100°C.	
4. The total heat gained by the ice in steps 1 through 3 is equal to the heat lost by the steam that condenses. Use this to solve for the mass of the condensed steam.	
5. The total mass of liquid water at 100°C is the sum of the mass of the initial ice and the condensed steam.	$m = 271\,\text{g of water}$

Example #9. Consider a 1.0-mol block of ice as it slowly warms from a temperature of −10°C to −3°C. As it warms, it expands 21 mm^3. The molar mass of ice is 18 g/mole. (*a*) Calculate the work done by the ice as it expands the 21.0 mm^3 at atmospheric pressure. What does it do this work on? (*b*) How much heat is transferred to the block during this process? (*c*) What is the increase in the block's internal energy?

Picture the Problem. The work at constant pressure is the product of pressure and the change in volume. Use the values for specific heats and molar heat capacities of solids and liquids in Table 18-1 on page 560 the text. Use the first law of thermodynamics to find the change in internal energy.

1. The work done by the ice on the atmosphere is the opposite of the work done by the atmosphere on the ice. This work is done at constant pressure.	$W_{\text{by ice}} = \int P\,dV = P\int dV = P\Delta V$ $= (21\,\text{atm})(21\,\text{mm}^3)\dfrac{1.01\times10^5\ \text{Pa/atm}}{\left(10^3\ \text{mm/m}\right)^3}$ $= 2.12\times10^{-3}\ \text{J}$
2. Heat is transferred to the ice under constant pressure.	$Q = nc_p'\,\Delta T = (1.00\,\text{mol})\big(36.9\,\text{J/}(\text{K}\cdot\text{mol})\big)(+7\,\text{K})$ $= 258\,\text{J}$
3. The total change in the internal energy can be found from the first law of thermodynamics. Because the ice does positive work on the atmosphere, the atmosphere does *negative* work on the ice.	$\Delta E_{\text{int}} = Q_{\text{in}} + W_{\text{on}}$ $= 258\,\text{J} - 2.12\times10^{-3}\ \text{J} = 258\,\text{J}$

Example #10—Interactive. A sample of gas initially occupies a volume of 15 L at 20°C and a pressure of 240 kPa. It is compressed at constant pressure to a volume of 6 K. How much work is done by the gas in the process?

Picture the Problem. In an isobaric (constant pressure) process, the work that the gas does is the product of the pressure and the change in volume. **Try it yourself.** Work the problem on your own, in the space provided, and check your answer.

$$W = -2160\,\text{J}$$

Example #11. In the *PV* diagram for a monatomic ideal gas, shown in Figure 18-1, path A is an isothermal expansion and path B is an expansion at a constant pressure followed by cooling at constant volume. For each process, calculate the heat input, the work done, and the change in internal energy for 1 mol of the gas.

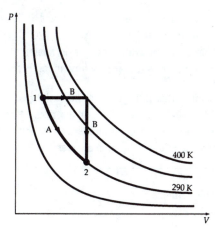

Figure 18-1

1. For path A, use the expression for a quasi-static isothermal expansion to find the work done on the gas. The ratio V_i/V_f is found by realizing that the ratio of the volumes on the constant pressure segment of path B is also V_i/V_f, and that for an expansion at constant pressure, the volume is directly proportional to the temperature.	$W_{\text{A, on gas}} = nRT \ln \dfrac{V_i}{V_f}$ $\dfrac{V_i}{V_f} = \dfrac{T_{\text{low}}}{T_{\text{high}}}$ $W_{\text{A, on gas}} = nRT \ln \dfrac{T_{\text{low}}}{T_{\text{high}}}$ $= (1\,\text{mol})(8.31\,\text{J/}(\text{mol}\cdot\text{K}))(290\,\text{K}) \ln \dfrac{290\,\text{K}}{400\,\text{K}}$ $= -775\,\text{J}$
2. The change in the internal energy of the gas is zero because the temperature remains constant. The first law of thermodynamics can then be used to calculate the heat added to the gas.	$\Delta E_{\text{int}} = Q_{\text{in}} + W_{\text{on}}$ $0 = Q_{\text{in}} + W_{\text{on}}$ $Q_{\text{in}} = -W_{\text{on}} = 775\,\text{J}$
3. For path B, no work is done on the gas during the constant-volume cooling, so the only work done on the gas occurs during the isobaric expansion. The ideal gas law provides relationships between temperature, pressure, and volume at the two points.	$W_{\text{on gas}} = -P\Delta V = P_1(V_1 - V_2) = P_1V_1 - P_1V_2$ $P_1V_1 = nRT_{\text{low}}$ $P_1V_2 = nRT_{\text{high}}$ $W_{\text{on gas}} = nR(T_{\text{low}} - T_{\text{high}})$ $= (1\,\text{mol})(8.31\,\text{J/}(\text{mol}\cdot\text{K}))(290\,\text{K} - 400\,\text{K})$ $= -914\,\text{J}$
4. The change in the internal energy of the gas is zero because the temperature remains constant. The first law of thermodynamics can then be used to calculate the heat added to the gas.	$\Delta E_{\text{int}} = Q_{\text{in}} + W_{\text{on}}$ $0 = Q_{\text{in}} + W_{\text{on}}$ $Q_{\text{in}} = -W_{\text{on}} = 914\,\text{J}$

Example #12—Interactive. One-half mole of nitrogen gas is heated from room temperature (20°C) and a pressure of 1 atm to a final temperature of 120°C. (*a*) How much heat must be supplied if the volume is kept constant while the gas is heated? (*b*) How much heat must be supplied if the heating is at constant pressure? (*c*) By how much is the internal energy changed?

Picture the Problem. Remember that nitrogen is a diatomic gas. The molar heat capacities of nitrogen can be found in Table 18-3 on page 574 of the text. For an ideal gas, the change in internal energy depends only upon the change in temperature. **Try it yourself.** Work the problem on your own, in the spaces provided, to get the final answer.

1. For part (*a*), use the first law of thermodynamics. If the volume remains constant, no work is done on the gas.	
	$Q = 1040\,\text{J}$

2. For part (b), the surroundings will do work on the gas. The internal energy of the system should remain constant if the temperature is constant.	
	$Q = 1450\,J$
3. The change in internal energy is due to the change in temperature.	
	$Q = 1040\,J$

Example #13. For 195 g of a certain ideal gas, the heat capacity at constant volume is 145 J/K and the heat capacity at constant pressure is 203 J/K. (a) How many moles of the gas are there? (b) What is its molar mass?

1. For 1 mole of any ideal gas, the difference in the molar heat capacities at constant pressure and constant volume is R. Use this to solve for the number of moles.	$C_p - C_v = nR$ $n = \dfrac{C_p - C_v}{R} = \dfrac{(203\,J/K) - (145\,K/K)}{8.31\,J/(mol\cdot K)} = 6.98\,mol$
2. The molar mass of the gas is simply its total mass divided by the number of moles.	$M = \dfrac{m}{n} = \dfrac{195\,g}{6.98\,mol} = 27.9\,g/mol$
3. The gas might be CO, or N_2, or C_2H_4.	

Example #14. Two moles of nitrogen, initially at a temperature of 293 K, undergo a quasi-static, adiabatic expansion from a pressure of 5 atm to a pressure of 1 atm. Find the work done on the gas.

Picture the Problem. For any process involving an ideal gas, the change in internal energy depends only upon the change in temperature of the gas. For an adiabatic process, the change in internal energy plus the work done by the gas equals zero. For a quasi-static adiabatic expansion of an ideal gas, PV^γ is constant. Use this together with the ideal-gas law to find the temperature of the gas.

1. Use the relationship for an adiabatic process to solve for the work done on the gas.	$W_{\text{on gas, adiabatic}} = C_v\,\Delta T$
2. We must first find the final temperature, using the fact that this is a quasi-static adiabatic expansion, and the ideal-gas law.	$P_1 V_1^\gamma = P_2 V_2^\gamma$ $P_1\left(\dfrac{nRT_1}{P_1}\right)^\gamma = P_2\left(\dfrac{nRT_2}{P_2}\right)^\gamma$ $P_1^{(1-\gamma)/\gamma} T_1 = P_2^{(1-\gamma)/\gamma} T_2$ $T_2 = T_1\left(\dfrac{P_1}{P_2}\right)^{(1-\gamma)/\gamma}$

3. Solve for γ before substituting back in for the work done on the gas.	$\gamma = \dfrac{C_p}{C_v} = \dfrac{C_v + nR}{C_v} = \dfrac{(7/2)nR}{(5/2)nR} = \dfrac{7}{5}$ $\dfrac{1-\gamma}{\gamma} = \dfrac{1-(7/5)}{7/5} = -\dfrac{2}{7}$
4. Finally, substitute everything back into the equation of step 1, and solve.	$W = C_v \, \Delta T = C_v \left(T_2 - T_1\right)$ $= \dfrac{5}{2}nR\left[T_1 \left(\dfrac{P_1}{P_2}\right)^{(1-\gamma)/\gamma} - T_1 \right]$ $= \dfrac{5}{2}nRT_1 \left[\left(\dfrac{P_1}{P_2}\right)^{-2/7} - 1 \right]$ $= \dfrac{5}{2}(2\,\text{mol})(8.31\,\text{J/(mol·K)})(293\,\text{K})\left[\left(\dfrac{5\,\text{atm}}{1\,\text{atm}}\right)^{-2/7} - 1 \right]$ $= -4490\,\text{J}$

Example #15—Interactive. Ten grams of argon at a pressure of 1 atm are placed in a thermally insulated, flexible 4-L container. The container and the argon are slowly lowered into the ocean until the volume occupied by the argon is 1 L. (*a*) What is the pressure of the gas after being compressed? (*b*) How much work is done on the gas as it is compressed from 4 L to 1 L? (*c*) What is the change in internal energy of the gas?

Picture the Problem. Argon is a monatomic gas, so $C_v = \frac{3}{2}nR$, $C_p = C_v + nR$, and $\gamma = C_p / C_v$. The number of moles of argon can be obtained from the given mass and the information in Appendix C of the text. For a quasi-static adiabatic compression of an ideal gas, PV^γ is constant. Use this to find the final pressure of the gas. There is a formula for the work done by the gas in a quasi-static adiabatic expansion or compression. For any adiabatic process, the change in the gas's internal energy plus the work done by the gas equals zero. **Try it yourself.** Work the problem on your own, in the spaces provided, to get the final answer.

1. Find the pressure of the gas after being compressed by using the fact that PV^γ is constant.	$P = 10.1\,\text{atm}$
2. Use the equation for the work done on a gas during adiabatic expansion to find the work done.	$W_{\text{on gas}} = 9.12\,\text{L·atm}$
3. Use the fact that the change in internal energy plus the work done on the gas is equal to zero to find the change in internal energy.	$\Delta E_{\text{int}} = -9.12\,\text{L·atm}$

Chapter 19

The Second Law
of Thermodynamics

I.   Key Ideas

Recall from Chapter 18 that energy is transferred as either heat or work. Heat is energy in transit due to a temperature difference, whereas work is energy in transit due to something other than a temperature difference (the displacement of a piston, for example).

By the first law of thermodynamics, total energy is conserved. In a given case, however, not all the energy is available for use. Consider a block sliding to rest along a horizontal table top where friction is not negligible. In this process, the block and table become warmer as mechanical energy, the initial kinetic energy of the block, is transformed into internal energy of the block and table. This is an example of an irreversible process; that is, the reverse process never occurs. Internal energy of the block and table is never spontaneously converted into kinetic energy to send the block sliding along the table while the table and block cool. Another example of an irreversible process occurs when a hot object is placed in thermal contact with a cold object. Heat will flow from the hot object to the cold object until they are at the same temperature. However, the reverse never occurs. Two bodies at the same temperature that are in thermal contact with each other remain at the same temperature. That is, heat does not flow from one to the other, making one colder and the other warmer.

The second law of thermodynamics states that processes of this type do not occur. One statement of the **second law of thermodynamics** is

> It is not possible to remove a given amount of energy from a system as heat and use it all to do work without some other change.

The conversion of work and mechanical energy into heat is an irreversible process, and the second law is fundamentally a statement about the direction of irreversible processes.

As a matter of convenience, several alternative statements of the second law of thermodynamics are in common use. These statements are completely equivalent. The Kelvin-Planck or **heat-engine statement of the second law of thermodynamics** is

It is impossible for a heat engine working in a cycle to produce no other effect than that of extracting energy as heat from a reservoir and performing an equivalent amount of work.

Section 19-1. Heat Engines and the Second Law of Thermodynamics. Historically, the second law of thermodynamics was first formulated in terms of the efficiency of heat engines. A **heat engine** is a device that operates in a cyclic process. It extracts heat from a high-temperature heat reservoir, converts some of the heat into mechanical energy, and transfers the mechanical energy, as work, to some external agent. In accordance with the first law of thermodynamics, the heat that is not converted into work is rejected to a low-temperature reservoir. A **heat reservoir** is a system with a heat capacity so large that it can absorb or give off heat with no appreciable change in its temperature. The surrounding atmosphere, lakes, or the ocean often act as practical heat reservoirs. A working substance in the engine absorbs heat Q_h from a high-temperature reservoir, converts some of the heat into work W done by the engine on its surroundings, and rejects the remaining heat Q_c to a low-temperature reservoir. *Note:* Our **sign convention** is that W, Q_h, and Q_c represent magnitudes, and as such are always positive. Thus, if heat leaves the system, it will always be represented as $-Q$. Applying the first law of thermodynamics to the system gives the work done *by* the heat engine during one cycle:

$$W = Q_h - Q_c \qquad \qquad \text{Work done by a heat engine in one cycle}$$

The **efficiency** ε of the engine is the ratio of the work done by the heat engine to the heat absorbed by the engine from the high-temperature reservoir.

$$\varepsilon = \frac{W}{Q_h} = \frac{Q_h - Q_c}{Q_h} = 1 - \frac{Q_c}{Q_h} \qquad \qquad \text{Efficiency of a heat engine}$$

In practical engines, the efficiency may be as high as 50% or more, but it is impossible to make a heat engine with an efficiency of 100%, according to the heat engine statement of the second law of thermodynamics.

Section 19-2. Refrigerators and the Second Law of Thermodynamics. A **refrigerator** is simply a heat engine run backward. It operates in a cyclic process, in which work W is done by an external agent on the refrigerator, heat Q_c is absorbed from a low-temperature reservoir by the working substance, and heat Q_h is rejected to a high-temperature reservoir. The coefficient of performance (COP) of a refrigerator is the ratio of the heat extracted by the refrigerator from the low-temperature reservoir to the work done on the refrigerator by the external agent. That is,

$$\text{COP} = \frac{Q_c}{W} \qquad \qquad \text{Coefficient of performance for a refrigerator}$$

where both Q_c and W, the work done by the external agent on the refrigerator, are positive quantities. Practical refrigerators have COPs as high as 5 or 6 or more. The larger the COP, the better.

The Clausius, or **refrigerator statement of the second law of thermodynamics** is

It is impossible for a refrigerator working in a cycle to produce no other effect than the transfer of energy as heat from a cold object to a hot object.

An ideal refrigerator would require zero work input and would transfer heat from a lower- to a higher-temperature reservoir without any other change to the surroundings. This is impossible according to the second law.

Section 19-3.   Equivalence of the Heat-Engine and Refrigerator Statements.   Although appearing to be very different, the heat-engine and refrigerator statements of the second law are equivalent in that, if we could make a device that violated one, it would also violate the other. This can be seen from simple heat flow diagrams and is independent of any specific engine design features. See the text for more information.

Section 19-4.   The Carnot Engine. We have seen that, according to the second law of thermodynamics, it is impossible for a heat engine working between two heat reservoirs to be 100% efficient. What, then, is the maximum possible efficiency for such an engine? This question was answered by Sadi Carnot. Carnot found that all reversible engines working between two heat reservoirs have the same efficiency and that no engine could have a greater efficiency than that of a reversible engine. This is known as **Carnot's theorem:**

No engine working between two given heat reservoirs can be more efficient than a reversible engine working between those reservoirs.

Any reversible engine working between two heat reservoirs is called a **Carnot engine.** The conditions necessary for a process to be reversible are as follows:

1. No mechanical energy can be transformed into thermal energy by friction, viscous forces, or other dissipative forces.

2. Energy transfer as heat can only occur between objects at the same temperature (or infinitesimally near the same temperature).

3. The process must be quasi-static so that the system is always in an equilibrium state (or infinitesimally near an equilibrium state).

Any process that violates any of the above conditions is irreversible. Most processes in nature are irreversible. Nevertheless, one can come very close to a reversible process, and the concept is very important.

A Carnot cycle consists of a sequence of four stages:

(1)   a reversible isothermal expansion,

(2)   a reversible adiabatic expansion,

(3)   a reversible isothermal compression, and

(4)   a reversible adiabatic compression.

Prior to state (1) the working substance is brought to thermal equilibrium with a high-temperature reservoir of temperature T_h. During stage (1) the working substance, in thermal contact with a high-temperature reservoir, expands isothermally (at constant temperature). During this expansion it does positive work on an external agent and heat is transferred to it from the high-temperature reservoir. Stage (1) ends and stage (2) begins when the working substance is thermally isolated from the high-temperature reservoir. During stage (2) it continues to expand, this time adiabatically; that is, no heat is transferred to or from the material. During this expansion it does positive work on the external agent and the material cools as its internal energy is converted to work. Its temperature continues to drop until it reaches the temperature of the low-temperature reservoir T_c. At this point the working substance is placed in thermal contact with the low-temperature reservoir; stage (2) ends, and stage (3) begins. During stage (3) the working substance, now in thermal contact with the low-temperature reservoir, is isothermally compressed. During the compression the external agent does positive work on it and the working substance rejects heat to the low-temperature reservoir. When the working substance is thermally isolated from the cold-temperature reservoir, stage (3) ends. During stage (4) the compression continues, this time adiabatically. The external agent does additional positive work on the working substance, which results in its internal energy increasing along with its temperature. The compression continues until its temperature again equals T_h, at which point the engine has completed stage (4), the final stage of the cycle. At this moment the working substance is in the exact state it was in when the cycle began.

The efficiency of a reversible engine (a Carnot engine) can be calculated if an ideal gas is used for the working substance. By Carnot's theorem, this must be the efficiency of every reversible engine and the upper limit possible for any real heat engine operating between the same two temperatures. The efficiency for a reversible heat engine using an ideal gas is calculated to be

$$\varepsilon_c = 1 - \frac{T_c}{T_h}$$ Efficiency of a Carnot engine

where T_c and T_h are absolute temperatures.

Since the efficiency is the same for all reversible heat engines operating in a cycle, it can be used to define the temperature of the two reservoirs. This allows the definition of a temperature scale that is independent of any particular material or thermometer. To use this definition to construct a temperature scale we choose one fixed point, say the triple point of water, and assign it a value. Then the temperature of a reservoir is measured by means of a reversible heat engine operating between the reservoir and a second reservoir maintained at the triple point of water. By measuring the efficiency of this engine we determine the temperature of the reservoir. If the temperature of the triple point of water is given the value of 273.16 K, then the absolute temperature and the ideal-gas temperature will agree over the range for which gas thermometers are able to be used.

Section 19-5. Heat Pumps.   A heat pump, which is essentially a refrigerator, is used to pump heat from a colder region (for example, outdoors) to a warmer region (for example, the interior of a building). For a heat pump we are interested in the heat $|Q_h|$ delivered into the high-temperature reservoir, the house. Consequently the parameter of interest is usually not the COP (Q_c / W) but the ratio $|Q_h| / W$, where the work W is the amount of energy that will appear on our electric bill.

The parameter of interest for a heat pump can be shown to equal 1 plus the coefficient of performance:

$$\text{COP}_{HP} = \frac{|Q_h|}{W} = 1 + \text{COP}$$

Coefficient of performance for heat pump

The maximum COP theoretically attainable is that of a reversible cycle. It can be shown that

$$\text{COP}_{HP,\,max} = \frac{T_h}{\Delta T}$$

Maximum coefficient of performance

where ΔT is the difference $T_h - T_c$. However, because perfectly adiabatic and isothermal processes cannot be carried out and because of frictional losses and the like, $\text{COP}_{HP,\,max}$ cannot be attained in practice.

Section 19-6. Irreversibility and Disorder. The second law is related to the fact that physical processes go only in one direction. In all irreversible processes, the direction is such that the system and its surroundings, taken together, tend to a less ordered state.

Section 19-7. Entropy. Entropy S is a thermodynamic quantity that is a measure of the "disorder" of a system. It is defined as

$$dS = \frac{dQ_{rev}}{T}$$

Definition – Change in entropy

where dQ_{rev} is the energy that must be transferred to the system as heat in a reversible process. Reversible heat transfer is not required, however, for there to be a change in entropy. Changes in entropy, like changes in internal energy, depend only on the initial and final states of the system, and not on the process taking the system from the initial state to the final state. Thus entropy, like internal energy, is a state function.

The change in entropy of an ideal gas that undergoes a reversible expansion from volume V_1 and temperature T_1 to volume V_2 and temperature T_2 can be derived to be

$$\Delta S = C_v \ln\frac{T_2}{T_1} + nR \ln\frac{V_2}{V_1}$$

Change in entropy for a reversible expansion of an ideal gas

where C_v is the heat capacity at constant volume, n is the number of moles of the gas, and R is the universal gas constant.

The total entropy of the universe never decreases. In an irreversible process it always increases, and in a reversible process it remains constant.

Section 19-8. Entropy and the Availability of Energy. In an irreversible process, some energy, an amount equal to the product of the entropy change of the universe ΔS and the temperature T of the coldest available reservoir, becomes unavailable for doing work. This energy, or the "work lost," is

$$W_{lost} = T\Delta S_u$$

Work lost in an irreversible process

Section 19-9. Entropy and Probability. If two or more states of a system have the same energy but different amounts of order, there is a greater probability that the system will be found in the state with the greater disorder. If for some reason the system is in one of the more ordered states, it will tend to spontaneously change states to a state with greater disorder. In an irreversible process the entropy of the universe must increase.

II. Physical Quantities and Key Equations

Physical Quantities

There are no new physical quantities introduced in this chapter.

Key Equations

Work done by a heat engine in one cycle $W = Q_h - Q_c$

Efficiency of a heat engine $\varepsilon = \dfrac{W}{Q_h} = 1 - \dfrac{Q_c}{Q_h}$

Efficiency of a Carnot engine $\varepsilon_c = 1 - \dfrac{T_c}{T_h}$

Coefficient of performance, refrigerator $COP = \dfrac{Q_c}{W}$

Coefficient of performance, heat pump $COP_{HP} = \dfrac{|Q_h|}{W} = 1 + COP$

Maximum coefficient of performance $COP_{HP,\ max} = \dfrac{T_h}{\Delta T}$

Entropy change $dS = \dfrac{dQ_{rev}}{T}$

Work lost in an irreversible process $W_{lost} = T \Delta S_u$

III. Potential Pitfalls

The work done by a heat engine during one cycle must be calculated as the work done by the working substance during its expansion plus the work done by it during its compression, not just the work during the expansion; it is the total for the entire cycle. (The work done by the engine during the expansion is positive, whereas during the compression the work is negative.)

Changes in entropy may be calculated only for reversible processes. Entropy is a function of state and a given entropy change is associated with a given change in state, but not with any specific process bringing the system from the initial state to the final state. To calculate the entropy change, we must calculate the entropy change for any reversible process bringing the system from the initial state to the final state.

Note that $\int dQ_{rev}/T$ does not equal the entropy of the initial or final state in a process but only the difference in the entropy between the two states. It is rather like potential energy in that only differences, or changes, are defined.

The second law does not say that the entropy of some object or system cannot decrease; in fact, it can decrease. But the decrease will always be made up (or more than made up) by an increase elsewhere. The second law states that the total entropy of the universe may not decrease.

In calculating the change in the entropy of a system, you must always use the absolute temperature scale.

IV. True or False Questions and Responses

True or False

_____ 1. Work can never be converted completely into internal energy.

_____ 2. When a wooden block is pushed across a wooden desktop at constant speed, energy is transferred as heat to the block-desktop system.

_____ 3. Work can never be converted completely into heat.

_____ 4. Heat can never be converted completely into work.

_____ 5. The transfer of heat across a finite temperature difference is an irreversible process.

_____ 6. According to the second law of thermodynamics, a heat engine cannot have an efficiency of 100%.

_____ 7. According to the second law of thermodynamics, a refrigerator cannot have a COP of 1 or greater.

_____ 8. All heat engines operating between the same two heat reservoirs have the same efficiency.

_____ 9. All quasi-static processes are reversible.

_____ 10. All reversible processes are quasi-static.

_____ 11. In a reversible heat engine, heat must be absorbed or rejected isothermally.

_____ 12. The entropy of a system depends only upon the state of the system.

_____ 13. In any adiabatic process, the entropy change of the system is zero.

_____ 14. The entropy of an isolated system cannot decrease.

_____ 15. When an irreversible process takes place, the universe becomes more disordered.

_____ 16. Disorder is more probable than order.

Responses to True or False

1. False. In an adiabatic compression no heat is transferred into or out of the system so the change in internal energy of the system is equal to the total work done on the system.

2. False. Energy is transferred as *work* to the block-desktop system. The work done by the pushing agent equals the increase in internal energy of the block-desktop system plus any energy that is transferred to the surroundings (such as acoustic energy). Heat is the energy transferred due to a temperature difference; however, there is no temperature difference involved in this transfer.

3. True. In a reversible, isothermal (constant temperature) compression of an ideal gas, the internal energy of the gas remains constant and the work done on the gas equals the heat transferred out of the gas. To be truly reversible the heat transfer must occur with only an infinitesimal temperature difference. Because no real process is 100% reversible, this does not occur.

4. False. In a reversible, isothermal (constant temperature) expansion of an ideal gas, the internal energy of the gas remains constant and the heat transferred to the gas is equal to the total work without some other change. In our case, we are not contradicting the second law by saying that the statement "Heat can never converted completely into work" is false because some other change does occur. That "other change" is the increase in the volume of the gas. For a cyclic process, there is no "other change" and the heat absorbed cannot be converted completely into work.

5. True. Heat flows from a region of high temperature to a region of low temperature, and never the other way around.

6. True. For an engine operating at 100% efficiency, no heat is rejected to the lower-temperature reservoir. Thus after an integral number of cycles, there are no other effects than that of extracting heat from the high-temperature reservoir and performing an equivalent amount of work. This contradicts the heat-engine statement of the second law of thermodynamics.

7. False. The COP $\left(\text{COP} = Q_c / W \right)$ of a refrigerator is typically greater than 1.

8. False. All *reversible* heat engines operating between the same two heat reservoirs have the same efficiency. Actual heat engines are not reversible and have lower efficiencies.

9. False. For example, the quasi-static free expansion of a gas is not a reversible process.

10. True.

11. True. Any heat flow through a finite temperature difference is irreversible. Heat never flows from a lower temperature to a higher temperature.

12. True.

13. False. $\Delta S = \int dQ_{rev} / T$ for a reversible process, but not all adiabatic processes are reversible. For example, an adiabatic free (unstrained) expansion of a gas is an irreversible process which results in an increase in entropy.

14. True.

15. True.

16. True. If all states are equally probable, there are many more disordered states than ordered states. Just look at your desk. Most of the time, unless you continuously straighten it up, it will become more disordered as time goes on.

V. Questions and Answers

Questions

1. When we say "engine," we think of something mechanical with moving parts. In such an engine, friction always reduces the engine's efficiency. Why is this?

2. There are people who try to keep cool on a hot summer day by leaving the refrigerator doors open, but you can't cool your kitchen this way! Why not?

3. Why do engineers designing a steam-electric generating plant always try to design for as high a feed-steam temperature as possible?

4. The conduction of heat across a temperature difference is an irreversible process, but the object that lost heat can always be rewarmed, and the one that gained it can be recooled. The dissipation of mechanical energy, as in the case of an object sliding across a rough table and slowing down, is irreversible, but the object can be cooled and set moving again at its original speed. So in just what sense are these processes "irreversible"?

5. In a slow, steady isothermal expansion of an ideal gas against a piston, the work done is equal to the heat input. Is this consistent with the first law?

6. If a gas expands freely into a larger volume in an insulated container so that no heat is added to the gas, its entropy increases. In view of the definition of ΔS, how can this be?

7. Heat flows from a hotter to a colder object. By how much is the entropy of the universe changed? In what sense does this correspond to energy becoming unavailable for doing work?

8. In discussing the Carnot cycle, we say that extracting heat from a reservoir isothermally does not change the entropy of the universe. In a real process, this is a limiting situation that can never quite be reached. Why not? What is the effect on the entropy of the universe?

Answers

1. The force of kinetic friction always transforms mechanical energy into thermal energy. The engine is rated by how well it transfers mechanical energy to an external agent as work, so if some of the mechanical energy is dissipated via kinetic friction, the engine has less mechanical energy to transfer to the external agent. Thus the efficiency of the engine is reduced.

2. Energy is transferred to the refrigerator's working substance both as heat from the things inside the box and as work done on it by the compressor. The refrigerator then transfers all of this energy as heat to the room air. With the refrigerator door kept open, the compressor has to work even harder. This extra energy is transferred by the refrigerator to the room air as heat, causing the room to grow even hotter.

3. High efficiency is almost always a primary objective for a steam-electric generating plant. In such a plant, the steam is the working fluid for the heat engine that drives the electric generator. The temperature of the steam is the temperature of the high-temperature heat source. The temperature of the low-temperature reservoir is usually fixed by circumstances, such as the temperature of a nearby lake. Thus, increasing the feed-steam temperature is the only way to increase the Carnot efficiency limit for the generator.

4. To say that a process is irreversible means, essentially, that the universe as a whole won't go in the other direction. In the examples given, we can put each "system" back where it started, but not without making a permanent change in its "surroundings." That is, the agents used to return each "system" back to where it started are changed in the process, so the *universe* is not restored to its original state.

5. This is consistent with the first law, which states that the heat absorbed by the gas equals the change in the internal energy of the gas plus the work done by the gas. The internal energy of an ideal gas depends only upon its temperature. Thus, in an isothermal expansion the change in internal energy is zero and the heat absorbed equals the work done.

6. For the adiabatic, free expansion of an ideal gas no work is done by the gas. The first law tells us that the internal energy, and thus the temperature, does not change. Since the expansion is adiabatic, no heat is exchanged by the gas, which might tempt you to think that the change in entropy $\left(\Delta S = \int dQ_{rev} / T \right)$ is zero. However, this is incorrect. To calculate the change in entropy, we must consider a *reversible* process that brings the gas from its initial to its final state. Thus, we consider a reversible process that consists of two segments: a reversible adiabatic expansion to the final volume (during which work is done by the gas and so its temperature decreases), followed by a reversible, constant-volume warming where heat is absorbed by the gas until it reaches its final temperature. No heat is absorbed (or released) during the reversible adiabatic expansion. Therefore, during that segment the entropy of the gas does not change. During the reversible, constant-volume warming segment, heat is absorbed by the gas and its entropy increases. Therefore, in an adiabatic, free expansion, the entropy of an ideal gas must increase.

7. Let Q be the heat transferred and let T_1 be the higher and T_2 be the lower temperature. The change in entropy of the high-temperature object is $-Q/T_1$ and the change in entropy of the low-temperature object is Q/T_2. Therefore, the change in entropy of the two objects, taken together, is

$$\Delta S = \frac{Q}{T_2} - \frac{Q}{T_1}$$

If this is all that happens, then this is the total change in entropy of the universe. If, instead of just letting the heat flow, we had run a Carnot engine between these two objects as temperature "reservoirs," the work we could have gotten from it is

$$W = \varepsilon_c Q = \left(1 - T_2 / T_1\right) Q = T_2 \, \Delta S$$

If, instead of running a Carnot engine, we just let the heat flow, this work would not have been done and the energy that could have done this work would no longer be available.

8. In any real process, the entropy of the universe will increase. In reality, if everything is at exactly the same temperature, it's all at thermal equilibrium and there can be no heat transfer; there must be some small temperature difference to have a heat transfer. However, if there is a temperature difference, the entropy loss Q/T_h of the warmer object is less than the entropy gain Q/T_c of the cooler object. The net entropy change of everything put together is always an increase.

VI. Problems, Solutions, and Answers

Example #1. A certain engine absorbs 150 J of heat and rejects 88 J in each cycle. (*a*) What is its efficiency? (*b*) If it runs at 200 cycles/min, what is its power output?

Picture the Problem. In any complete cycle, the heat the engine absorbs equals the work it does plus the heat it rejects.

1. Find the amount of work the engine does in one complete cycle.	$Q_h = W + Q_c$ $W = Q_h - Q_c = 150\,\text{J} - 88\,\text{J} = 62\,\text{J}$
2. The efficiency is the work done by the engine divided by the amount of heat it absorbs.	$\varepsilon = \dfrac{W}{Q_h} = \dfrac{62\,\text{J}}{150\,\text{J}} = 0.41 = 41\%$
3. The power is the work done per unit time. In one minute, the engine goes through 200 cycles.	$P = \dfrac{W}{t} = \dfrac{200(62\,\text{J})}{60\,\text{s}} = 200\,\text{W}$

Example #2—Interactive. An electric refrigerator removes 13 MJ of heat from its interior for each kilowatt-hour of electric energy used. What is its coefficient of performance?

Picture the Problem. The coefficient of performance is the ratio of the heat extracted by the refrigerator to the work done on the refrigerator by an external agent. **Try it yourself.** Work the problem on your own, in the spaces provided, to get the final answer.

1. Convert the work done on the refrigerator (one kilowatt-hour) to joules.	
2. Solve for the coefficient of performance (COP).	COP = 3.6

Example #3. A certain refrigerator has a power rating of 88.0 W. Consider it to be a reversible refrigerator. If the temperature of the room is 26.0°C, how long will the refrigerator take to freeze 2.50 kg of water that is put into it at 0°C?

Picture the Problem. The power rating is the work input per unit time. Heat is extracted from the interior of the refrigerator at 0°C since it is extracted from freezing water. To freeze the water, the total latent heat of fusion will have to be extracted from the water. Since the cycle is reversible, the refrigerator is ideal and the heats exhausted to and extracted from the hot and cold reservoirs are proportional to their absolute temperatures.

1. Write an expression for the COP of the refrigerator. This will allow us to solve for the time. The work done is the power multiplied by the time. The heat extracted is the mass times the latent heat of fusion.	$COP = \dfrac{Q_c}{W} = \dfrac{T_c}{\Delta T}$ $\dfrac{mL_f}{Pt} = \dfrac{T_c}{T_h - T_c}$
2. Solve for time.	$t = \dfrac{(T_h - T) mL_f}{PT_c} = \dfrac{(26.0\,\text{K})(2.50\,\text{kg})(334\,\text{kJ/kg})}{(88.0\,\text{W})(273\,\text{K})}$ $= 904\,\text{s}$

Example #4—Interactive. A certain refrigerator requires 35.0 J of work to remove 45.0 cal of heat from its interior. (*a*) What is its coefficient of performance? (*b*) How much heat is rejected to the surroundings at 22.0°C? (*c*) If the refrigerator cycle is reversible, what is the temperature inside the refrigerator?

Picture the Problem. You will need to convert calories to joules to maintain consistency of units. **Try it yourself.** Work the problem on your own, in the spaces provided, to get the final answer.

1. The coefficient of performance is the ratio of the heat removed to the work done.	COP = 5.37
2. The heat rejected equals the work done plus the heat extracted from the interior.	$Q_h = 223\,\text{J}$

3. Since the cycle is thermodynamically reversible, the heats exhausted to and extracted from the hot and cold reservoirs are proportional to their absolute temperatures.	
	$T_c = 249\,\text{K} = -24°\text{C}$

Example #5. A not very clever idea for a ship's engine goes as follows: A Carnot engine extracts heat from seawater at 18°C and exhausts it to evaporating dry ice, which the ship carries with it, at −78°C. If the ship's engines are to run at 8000 horsepower, what is the minimum amount of dry ice it must carry for a single day's running?

Picture the Problem. The efficiency of the engine can be determined from the temperatures involved. Determine how much work, in joules, must be done by the engines in one day. The heat rejected to the low-temperature reservoir determines how much dry ice must be sublimated. The latent heat of sublimation of dry ice (carbon dioxide) can be found in Table 18-2 on page 563 of the text.

1. Begin by converting all values to SI units.	$8000\,\text{hp} = (8000\,\text{hp})(746\,\text{W/hp}) = 5.97\,\text{MW}$ $1\,\text{day} = (1\,\text{day})(24\,\text{hrs/day})(3600\,\text{s/hr}) = 8.64\times10^4\,\text{s}$ $18°\text{C} = 273\,\text{K} + 18\,\text{K} = 291\,\text{K}$ $\text{-}78°\text{C} = 273\,\text{K} - 73\,\text{K} = 2195\,\text{K}$
2. Determine the efficiency of the Carnot engine.	$\mathcal{E}_C = \dfrac{W}{Q_h} = 1 - \dfrac{T_c}{T_h}$
3. Work is power multiplied by time, and the heat the engine absorbs equals the work it does plus the heat it rejects. The heat rejected to the dry ice is equal to the product of the mass of the ice and the heat of sublimation.	$W = Pt$ $Q_h = W + Q_c = Pt + Q_c$ $Q_c = mL_s$ $Q_h = Pt + mL_s$
4. Substitute in and solve for the required mass of dry ice. That's about 2000 tons of dry ice. We said it wasn't a very good scheme.	$\dfrac{W}{Q_h} = 1 - \dfrac{T_c}{T_h}$ $\dfrac{Pt}{Pt + mL_s} = 1 - \dfrac{T_c}{T_h}$ $m = \dfrac{Pt}{L_s}\left(\dfrac{T_c}{T_h} - 1\right)^{-1}$ $= \dfrac{(5.97\times10^6\,\text{W})(8.64\times10^4\,\text{s})}{573\,\text{kJ/kg}}\left(\dfrac{291\,\text{K}}{195\,\text{K}} - 1\right)^{-1}$ $= 1.83\times10^6\,\text{kg}$

Example #6—Interactive. A certain electric generating plant produces electrical energy by using steam that enters its turbine at a temperature of 320.0°C and leaves it at 40.0°C. Over the course of a year, the plant consumes 4.40×10^{16} J of heat and produces an average electric power output of 600.0 MW. What is its second-law efficiency?

Picture the Problem. The second-law efficiency is the ratio of the actual efficiency to the Carnot efficiency. **Try it yourself.** Work the problem on your own, in the spaces provided, to get the final answer.

1. Calculate the Carnot efficiency from the temperatures involved.	
2. The actual efficiency of the plant is the ratio of the work done (electrical energy produced) to the heat absorbed.	$\mathcal{E} = 0.911$

Example #7. A certain engine has a second-law efficiency of 85%. In each cycle it absorbs 480 J of heat from a reservoir at 300°C and rejects 300 J of heat to a cold-temperature reservoir. (*a*) What is the temperature of the cold reservoir? (*b*) How much more work could be done by a Carnot engine working between the same two reservoirs and extracting the same 480 J of heat in each cycle?

Picture the Problem. The actual efficiency can be calculated using the information provided. The Carnot efficiency can then be calculated, which may be used to determine the temperature of the low-temperature reservoir. This engine has an efficiency equal to 85% of a Carnot engine.

1. Write an expression for the actual efficiency of the engine.	$\mathcal{E} = \dfrac{W}{Q_h} = 1 - \dfrac{Q_c}{Q_h}$
2. Write an expression for the Carnot efficiency.	$\mathcal{E}_C = 1 - \dfrac{T_c}{T_h}$
3. These two efficiencies are related by the second law efficiency. We can use the result to solve for the temperature of the low-temperature bath.	$\mathcal{E} = \mathcal{E}_{sl}\mathcal{E}_C$ $1 - \dfrac{Q_c}{Q_h} = \mathcal{E}_{sl}\left(1 - \dfrac{T_c}{T_h}\right)$ $T_c = T_h\left[1 - \dfrac{1}{\mathcal{E}_{sl}}\left(1 - \dfrac{Q_c}{Q_h}\right)\right]$ $= (573\,\mathrm{K})\left[1 - \dfrac{1}{0.85}\left(1 - \dfrac{300\,\mathrm{J}}{480\,\mathrm{J}}\right)\right]$ $= 320\,\mathrm{K} = 47°\mathrm{C}$

4. We can now find the work done by a Carnot engine operating between the same temperatures.	$\varepsilon_{\mathrm{C}} = \dfrac{W}{Q_{\mathrm{h}}} = 1 - \dfrac{T_{\mathrm{c}}}{T_{\mathrm{h}}}$ $W = Q_{\mathrm{h}}\left(1 - \dfrac{T_{\mathrm{c}}}{T_{\mathrm{h}}}\right) = \left(480\,\mathrm{J}\right)\left(1 - \dfrac{320\,\mathrm{K}}{573\,\mathrm{K}}\right) = 212\,\mathrm{J}$
5. The work done by the engine in this problem is 85% of the work done by a Carnot engine, so we can calculate the extra work done by the Carnot Engine.	$W_{\text{additional}} = W_{\mathrm{C}} - W_{\text{actual}} = \left(212\,\mathrm{J}\right) - 0.85\left(212\,\mathrm{J}\right)$ $= 32\,\mathrm{J}$

Example #8—Interactive. A Carnot engine moves 1200 J of heat from a high-temperature reservoir and rejects 600 J to the atmosphere at a temperature of 20°C. (*a*) What is the efficiency of this engine? (*b*) What is the temperature of the high-temperature reservoir? **Try it yourself.** Work the problem on your own, in the spaces provided, to get the final answer.

1. Determine the efficiency of the engine from the given values of heat input and the work done.	
	$\varepsilon = 0.500$
2. This is a Carnot engine. Relate its efficiency to the operating temperatures to solve for the temperature of the high-temperature reservoir.	
	$T_{\mathrm{h}} = 586\,\mathrm{K}$

Example #9. When 1 kg of water vaporizes at a pressure of 1 atm and a temperature of 100°C, by how much has its entropy increased?

1. The change in entropy of the water is equal to the heat it absorbs divided by the temperature.	$dS = \dfrac{Q}{T}$
2. The heat absorbed is the mass of the water times its latent heat of vaporization.	$Q = mL_{\mathrm{v}}$
3. Solve for the change in entropy.	$dS = \dfrac{mL_{\mathrm{v}}}{T} = \dfrac{\left(1\,\mathrm{kg}\right)\left(2257\,\mathrm{J/kg}\right)}{373\,\mathrm{k}} = 6.05\,\mathrm{kJ/K}$

Example #10—Interactive. When 1 kg of water is frozen under standard conditions, by how much does its entropy change? **Try it yourself.** Work the problem on your own, in the spaces provided, to get the final answer.

1. The change in entropy of the water is equal to the heat it absorbs divided by the temperature. When the water is frozen, it loses heat, so the gain in heat is negative.	
2. The heat extracted is the mass of the water times its latent heat of fusion.	
3. Solve for the change in entropy. "Standard conditions" means that the heat is extracted from the water at $0°C$. The entropy of the water decreases as heat is extracted from it.	$dS = -1221\,\text{J/K}$

Example #11. A heat engine works in a cycle between reservoirs at 273 K and 490 K. In each cycle the engine absorbs 1250 J of heat from the high-temperature reservoir and does 475 J of work. (*a*) What is its efficiency? (*b*) What is the change in entropy of the universe when the engine goes through one complete cycle? (*c*) How much energy becomes unavailable for doing work when this engine goes through one complete cycle?

1. Find the efficiency of the engine from the work done and the heats given.	$\varepsilon = \dfrac{W}{Q_h} = \dfrac{475\,\text{J}}{1250\,\text{J}} = 0.380$
2. The change in entropy of the universe is the sum of the entropy change of the high-temperature reservoir, the entropy change of the low-temperature reservoir, and the change in entropy of the engine. The change in entropy of the working substance of the engine is zero because the engine has completed one entire cycle, and is thus unchanged. The change in entropy of the high-temperature reservoir is negative because heat is extracted from it. The change in entropy of the low-temperature reservoir is positive.	$dS_U = dS_h + dS_c + dS_{\text{engine}}$ $= -\dfrac{Q_h}{T_h} + \dfrac{Q_c}{T_c} + 0 = -\dfrac{Q_h}{T_h} + \dfrac{Q_h - W}{T_c}$ $= -\dfrac{1250\,\text{J}}{490\,\text{K}} + \dfrac{(1250\,\text{J}) - (475\,\text{J})}{273\,\text{K}} = 0.288\,\text{J/K}$
3. The "lost work" is the additional work that a Carnot engine operating between the same temperatures would do after extracting the same amount of heat from the high-temperature reservoir.	$W_{\text{lost}} = W_C - W = \varepsilon_C Q_h - \varepsilon Q_h = (\varepsilon_C - \varepsilon) Q_h$ $= \left[\left(1 - \dfrac{T_c}{T_h}\right) - \varepsilon\right] Q_h$ $= \left[\left(1 - \dfrac{273\,\text{K}}{490\,\text{K}}\right) - 0.380\right](1250\,\text{J}) = 78.6\,\text{J}$

Example #12—Interactive. The interior of a refrigerator's freezing compartment is at 10°F. The kitchen is at 78°F. Suppose that heat leaks through the walls into the freezing compartment at a rate of 70 cal/min. (*a*) In one hour, how much has the entropy of the universe been increased by this heat leakage? (*b*) How much energy becomes unavailable for doing work when this heat leaks into the freezer compartment? **Try it yourself.** Work the problem on your own, in the spaces provided, and check your answer.

1. Convert the temperatures to the Kelvin scale and the rate of heat leakage to Joules per second.	
2. The entropy change of the freezing compartment increases by an amount equal to the heat delivered to it divided by its absolute temperature.	
3. In the same way, obtain an expression for the entropy decrease of the kitchen outside the refrigerator.	
4. The entropy change of the universe is that of the freezing compartment plus that of the outside world.	$dS = 8.6\,\text{J/K}$
6. Relate the "lost" work to the entropy change and the temperature of the low-temperature reservoir.	$W_{\text{lost}} = 2.24\,\text{kJ}$

Example #13. In a vacuum bottle, 350 g of water and 150 g of ice are initially in equilibrium at 0°C. The bottle is not a perfect insulator. Over time, its contents come to thermal equilibrium with the outside air at 25°C. How much does the entropy of universe increase in this process?

Picture the Problem. The total change in entropy of the universe will equal the increase in entropy of the ice and water as they warm, plus the decrease in entropy of the surrounding air as it loses heat.

1. The total increase in entropy of the ice and water equals its increase during the melting of the ice at 273 K plus the increase of all the water as it warms. Because the temperature is not constant during the warming of the water, we need to integrate to find the total change	$dS_{\text{iw}} = dS_{\text{m}} + dS_{\text{w}} = \dfrac{Q_{\text{m}}}{T_1} + \displaystyle\int \dfrac{dQ_{\text{w}}}{T}$ $= \dfrac{m_i L_f}{T_1} + \displaystyle\int_{T_1}^{T_2} \dfrac{mc\,dT}{T} = \dfrac{m_i L_f}{T_1} + mc\displaystyle\int_{T_1}^{T_2} \dfrac{dT}{T}$ $= \dfrac{m_i L_f}{T_1} + mc\ln\dfrac{T_2}{T_1}$

in entropy. The heat gained during warming is equal to the mass of the water times its specific heat times the change in temperature.	
2. The change in entropy of the air is equal to the heat lost in melting the ice plus the heat lost in warming the water, divided by the constant temperature of the air.	$dS_{air} = -\dfrac{Q}{T_2} = -\dfrac{m_i L_f + mc\,\Delta T}{T_2}$
3. The change in entropy of the universe is the sum of these changes in entropy.	$dS_u = dS_{iw} + dS_{air}$ $= \dfrac{m_i L_f}{T_1} + mc\ln\dfrac{T_2}{T_1} - \dfrac{m_i L_f + mc\,\Delta T}{T_2}$ $= m_i L_f\left(\dfrac{1}{T_1} - \dfrac{1}{T_2}\right) - mc\left(\dfrac{\Delta T}{T_2} - \ln\dfrac{T_2}{T_1}\right)$ $= (0.15\,\text{kg})(334\,\text{kJ/kg})\left(\dfrac{1}{273\,\text{K}} - \dfrac{1}{298\,\text{K}}\right)$ $\quad -(0.5\,\text{kg})(4.18\,\text{kJ/}(\text{kg}\cdot\text{K}))\left(\dfrac{25\,\text{K}}{298\,\text{K}} - \ln\dfrac{298\,\text{K}}{273\,\text{K}}\right)$ $= 23.2\,\text{J/K}$

Example #14—Interactive. I have a cup containing 220 g of hot water at 75°C. To cool it I pour 60 g of tap water at 26°C into it. How much does the entropy of the universe change during this cooling?

Picture the Problem. The total change in entropy of the universe will equal the increase in entropy of the cool water as it warms, plus the decrease in entropy of the hot water as it cools. **Try it yourself.** Work the problem on your own, in the spaces provided, to get the final answer.

1. Calculate the final temperature of 280 g of water.	
2. Calculate the increase in entropy of the cool water. You will have to integrate because the temperature of the water is not constant.	
3. In the same manner, calculate the decrease in entropy of the warm water as it cools.	

| 4. The net entropy change of the universe is the change in entropy of the tap water plus the change in entropy of the hot water. | $dS_u = 2.21\,\text{J/K}$ |

Example #15. Consider an engine in which the working substance is 1.23 mol of an ideal gas for which γ is 1.41. The engine runs reversibly in the cycle shown on the *PV* diagram in Figure 19-1. The cycle consists of an isobaric (constant pressure) expansion a at a pressure of 15 atm, during which the temperature of the gas increases from 300 K to 600 K, followed by an isothermal expansion b until its pressure becomes 3 atm. Next is an isobaric compression c at a pressure of 3 atm, during which the temperature decreases from 600 K to 300 K, followed by an isothermal compression d until its pressure returns to 15 atm. Find the work done by the gas, the heat absorbed by the gas, the internal energy change, and the entropy change of the gas, first for each part of the cycle, and then for the complete cycle.

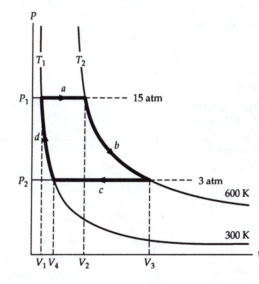

Figure 19-1

| 1. Find the work done by the gas during the isobaric process of stage a. We use the ideal-gas law to relate the product of pressure and volume to the temperature. | $W_a = \int P\,dV = P_1\,\Delta V = P_1V_2 - P_1V_1$
 $= nRT_2 - nRT_1 = nR\,\Delta T$
 $= (1.23\,\text{mol})(8.31\,\text{J/(mol·K)})(300\,\text{K}) = 3070\,\text{J}$ |
| 2. Determine the heat absorbed by the gas during this expansion.

We calculate c_p' explicitly, because this value will be used extensively throughout the rest of the problem. | $Q_a = nc_p'\,\Delta T$
 $c_p' = c_v' + R = \dfrac{c_p'}{\gamma} + R$
 $= \dfrac{\gamma R}{\gamma-1} = \dfrac{1.41(8.31\,\text{J/(mol·K)})}{0.41} = 28.6\,\text{J/(mol·K)}$
 $Q_a = (1.23\,\text{mol})(28.6\,\text{J/(mol·K)})(300\,\text{K}) = 10{,}550\,\text{J}$ |

3. Find the change in the internal energy of the engine during this stage.	$\Delta U_a = Q_a - W_a = 10,550\,\text{J} - 3070\,\text{J} = 7480\,\text{J}$
4. Find the change in entropy of the gas during this stage.	$dS_a = \int \dfrac{dQ}{T} = nc'_p \int_{T_1}^{T_2} \dfrac{dT}{T} = nc'_p \ln \dfrac{T_2}{T_1}$ $= (1.23\,\text{mol})(28.6\,\text{J}/(\text{mol}\cdot\text{K})) \ln \dfrac{600\,\text{K}}{300\,\text{K}} = 24.4\,\text{J/K}$
5. Find the work done as the gas expands isothermally along path b. The ideal-gas law can be used to express the pressure in terms of the volume and temperature. Because this is an isothermal expansion, the ratio of the final and initial volumes is equal to the ratio of the initial and final pressures.	$W_b = \int_{V_2}^{V_3} P\,dV = nRT_2 \int_{V_2}^{V_3} \dfrac{dV}{V} = nRT_2 \ln \dfrac{V_3}{V_2}$ $= nRT_2 \ln \dfrac{P_1}{P_2}$ $= (1.23\,\text{mol})(8.31\,\text{J}/(\text{mol}\cdot\text{K}))(600\,\text{K}) \ln \dfrac{15\,\text{atm}}{3\,\text{atm}}$ $= 9870\,\text{J}$
6. The internal energy of the gas depends only upon its temperature. Therefore during an isothermal process there is no change in the internal energy.	$\Delta U_b = 0$
7. Because there is no change in internal energy, the heat absorbed is equal to the work done.	$Q_b = W_b = 9870\,\text{J}$
8. Find the change in entropy.	$dS_b = \int \dfrac{dQ}{T} = \dfrac{1}{T_2} \int dQ = \dfrac{Q_b}{T_2} = \dfrac{9870\,\text{J}}{600\,\text{K}} = 16.5\,\text{J/K}$
9. Using the same process as in steps 1 through 5, we can find the work done and the heat absorbed by the gas, and the internal energy change and the change in entropy of the gas along path c.	$W_c = nR\,\Delta T$ $= (1.23\,\text{mol})(8.31\,\text{J}/(\text{mol}\cdot\text{K}))(-300\,\text{K})$ $= -3070\,\text{J}$ $Q_c = nc'_p\,\Delta T$ $= (1.23\,\text{mol})(28.6\,\text{J}/(\text{mol}\cdot\text{K}))(-300\,\text{K})$ $= -10,550\,\text{J}$ $\Delta U_c = Q_c - W_c = -10,600\,\text{J} + 3070\,\text{J} = -7480\,\text{J}$ $dS_c = nc'_p \ln \dfrac{T_1}{T_2}$ $= (1.23\,\text{mol})(28.6\,\text{J}/(\text{mol}\cdot\text{K})) \ln \dfrac{300\,\text{K}}{600\,\text{K}}$ $= -24.4\,\text{J/K}$

10. Along path *d*, we can use the same procedures we use for the isothermal expansion along path *b*.	$$W_d = nRT_1 \ln \frac{P_1}{P_2}$$ $$= (1.23\,\text{mol})(8.31\,\text{J}/(\text{mol}\cdot\text{K}))(300\,\text{K})\ln\frac{3\,\text{atm}}{15\,\text{atm}}$$ $$= -4940\,\text{J}$$ $$\Delta U_d = 0$$ $$Q_d = W_d = -4940\,\text{J}$$ $$dS_d = \frac{Q_d}{T_1} = \frac{-4940\,\text{J}}{300\,\text{K}} = -16.5\,\text{J/K}$$
11. To find the net values of the work done by the gas, the heat absorbed by the gas, the net internal energy change of the gas, and the net entropy change during a complete cycle, simply add the results from each segment of the cycle.	$$W_{net} = W_a + W_b + W_c + W_d$$ $$= 3070\,\text{J} + 9870\,\text{J} - 3070\,\text{J} - 4940\,\text{J} = 4930\,\text{J}$$ $$Q_{net} = Q_a + Q_b + Q_c + Q_d$$ $$= 10,600\,\text{J} + 9870\,\text{J} - 10,600\,\text{J} - 4940\,\text{J} = 4930\,\text{J}$$ $$\Delta U_{net} = \Delta U_a + \Delta U_b + \Delta U_c + \Delta U_d$$ $$= 7480\,\text{J} + 0 - 7480\,\text{J} - 0 = 0$$ $$dS_{net} = 24.4\,\text{J/K} + 16.5\,\text{J/K} - 24.4\,\text{J/K} - 16.5\,\text{J/K} = 0$$

Chapter **20**

Thermal Properties and Processes

I. Key Ideas

Section 20-1. Thermal Expansion. When the temperature of a solid or liquid increases, it usually expands. When a bar of length L is heated, raising its temperature by ΔT, the length of the bar changes by ΔL. The fractional change in the length is $\Delta L / L$. The ratio of the fractional change in the length of an object to the change in temperature is called the **coefficient of linear expansion** α:

$$\alpha = \lim_{\Delta T \to 0} \frac{\Delta L / L}{\Delta T} = \frac{1}{L}\frac{dL}{dT} \qquad \qquad \text{Coefficient of linear expansion}$$

Similarly, the **coefficient of volume expansion** β is defined as the ratio of the fractional change in volume to the change in temperature (at constant pressure):

$$\beta = \lim_{\Delta T \to 0} \frac{\Delta V / V}{\Delta T} = \frac{1}{V}\frac{dV}{dT} \qquad \qquad \text{Coefficient of volume expansion}$$

For a given material, the relation between the coefficient of volume expansion and the coefficient of linear expansion is $\beta = 3\alpha$.

Section 20-2. The van der Waals Equation and Liquid–Vapor Isotherms. The van der Waals equation of state describes the behavior of gases over a wide range of pressures more accurately than does the ideal-gas equation of state $(PV = nRT)$. The van der Waals equation of state for n moles of a gas is

$$\left(p + \frac{an^2}{V^2} \right)(V - bn) = nRT \qquad \qquad \text{The van der Waals equation of state}$$

The constants a and b provide for the attractive forces between molecules and the finite size of the molecules, respectively.

The pressure at which a liquid coexists in equilibrium with its own vapor is called the **vapor pressure.** The temperature at which the vapor pressure equals one atmosphere is called the **normal boiling point.**

Section 20-3. Phase Diagrams. The equilibrium state of a substance held at constant volume depends on its temperature and pressure. A plot of pressure versus temperature for such a substance is called a **phase diagram** where the state of a substance can be represented by a point. On a phase diagram, one region represents states of solid phase, another region represents states of liquid phase, and yet another region represents states of gaseous phase.

The point on a phase diagram where the liquid and vapor phases have the same density is called the **critical point,** and the temperature at this point is called the critical-point temperature T_c. At this point and above it, there is no distinction between the liquid and gas phases. Every substance has a unique triple point, the point at which the vapor, liquid, and solid phases can coexist in equilibrium. The triple-point temperature for water is $273.16 \text{ K} = 0.01°C$ and the triple-point pressure is 4.58 mmHg.

Section 20-4. The Transfer of Thermal Energy. Temperature differences result in the transfer of energy from one place to another by three processes: conduction, convection, and radiation. In **conduction,** heat is transferred by interactions among atoms or molecules, though there is no transport of the atoms or molecules themselves. In **convection,** heat is transferred via mass transport. This occurs, for example, in a flowing fluid. In **radiation,** heat is transported via electromagnetic radiation that is emitted and absorbed. In all three processes the net flow of heat (that is, the transfer of energy) is from a region of higher temperature to a region of lower temperature.

When heat flows by conduction along a solid bar, the rate at which heat flows is called the **thermal current** I, measured in units such as joules per second. The thermal current is related to the temperature gradient $\Delta T / \Delta x$ by

$$I = \frac{\Delta Q}{\Delta t} = kA \frac{\Delta T}{\Delta x} \qquad\qquad \text{Thermal conduction}$$

where ΔQ is the net flow of heat during time ΔT, A is the cross-sectional area of the bar, and the constant k is the **coefficient of thermal conductivity** of the substance. Solving this equation for the temperature difference ΔT we obtain

$$\Delta T = I \frac{\Delta x}{kT} = IR$$

where R is the **thermal resistance** of the material of thickness Δx :

$$R = \frac{\Delta x}{kA} \qquad\qquad \text{Thermal resistance}$$

The equation $\Delta T = IR$ is analogous to the equation for the viscous flow of a fluid through a pipe, $\Delta P = I_v R$, except that I stands for the flow of heat rather than the volume flow rate of a fluid, R is the thermal resistance, and ΔT has replaced the pressure difference ΔP.

If heat flows from a warmer region to a cooler region through a series of slabs, the equivalent thermal resistance R_{eq} of the series is the sum of the thermal resistances of the individual slabs:

$$R_{eq} = R_1 + R_2 + \dots$$ Resistance in series

Similarly, if heat is conducted from a warmer region to a cooler region through two or more parallel paths, the reciprocal of the equivalent thermal resistance of the paths is the sum of the reciprocals of the thermal resistances of the individual paths. That is,

$$\frac{1}{R_{eq}} = \frac{1}{R_1} + \frac{1}{R_2} + \dots$$ Resistances in parallel

The heat transferred to or from an object by convection is approximately proportional to the surface area of the object and to the difference in temperature between the object and the surrounding fluid.

The rate at which a surface radiates heat is given by the **Stefan–Boltzmann law**:

$$P_r = e\sigma A T^4$$ Stefan–Boltzmann law

where e is the **emissivity** of the surface ($0 \le e \le 1$), A is the area of the surface, T is the surface temperature in kelvins, and σ is Stefan's constant:

$$\sigma = 5.6703 \times 10^{-8}\ \text{W/}(\text{m}^2 \cdot \text{K}^4)$$ Stefan's constant

The rate P_a at which a body absorbs radiant heat from its surroundings is

$$P_a = e\sigma A T_0^4$$ Absorption of radiation

where T_0 is the temperature of its surroundings. Therefore the net power radiated by an object at temperature T in an environment at temperature T_0 is given by

$$P_{net} = e\sigma A\left(T^4 - T_0^4\right)$$ Net radiative thermal current

Note that a good emitter (an object having a high emissivity e) is also a good absorber.

An object that absorbs all the radiation incident upon it has an emissivity equal to 1 and is called a **blackbody**. A blackbody is also an ideal radiator. When a body emits thermal radiant energy, it does so over a continuum of wavelengths. The wavelength λ_{max} at which a blackbody emits radiant energy at the greatest rate is inversely proportional to the absolute temperature of the body. This result is known as Wien's displacement law:

$$\lambda_{max} = \frac{2.898\ \text{mm} \cdot \text{K}}{T}$$ Wien's displacement law

All three mechanisms of heat flow are driven by a difference in temperature. Independent of which of the mechanisms are at work, the rate of cooling of a warm body is approximately proportional to the temperature difference between the body and its surroundings. This result, called **Newton's law of cooling,** holds whether heat is being transferred by conduction, convection, radiation, or some combination of the three. It is established by applying the differential approximation to the various heat transfer mechanisms and is most accurate when the temperature differences are small.

II.   Physical Quantities and Key Equations

Physical Quantities

Stefan's constant $\sigma = 5.6703 \times 10^{-8}$ $\text{W/m}^2 \cdot \text{K}^4$

Key Equations

Coefficient of linear expansion $\alpha = \lim_{\Delta T \to 0} \dfrac{\Delta L / L}{\Delta T} = \dfrac{1}{L}\dfrac{dL}{dT}$

Coefficient of volume expansion $\beta = \lim_{\Delta T \to 0} \dfrac{\Delta V / V}{\Delta T} = \dfrac{1}{V}\dfrac{dV}{dT}$

The van der Waals equation of state $\left(p + \dfrac{an^2}{V^2}\right)(V - bn) = nRT$

Thermal conduction $I = \dfrac{\Delta Q}{\Delta t} = kA\dfrac{\Delta T}{\Delta x}$

Thermal resistance $R = \dfrac{\Delta x}{kA}$

so $\Delta T = I\dfrac{\Delta x}{kT} = IR$

Resistance in series $R_{eq} = R_1 + R_2 + \ldots$

Resistances in parallel $\dfrac{1}{R_{eq}} = \dfrac{1}{R_1} + \dfrac{1}{R_2} + \ldots$

Stefan–Boltzmann law $P = e\sigma A T^4$

Absorption of radiation $P_a = e\sigma A T_0^4$

Net radiative thermal current $P_{net} = P - P_a = e\sigma A\left(T^4 - T_0^4\right)$

Wien's displacement law $\lambda_{max} = \dfrac{2.898 \text{ mm} \cdot \text{K}}{T}$

III. Potential Pitfalls

In problems involving thermal expansion, the units of temperature must be consistent with those of α or β. On the other hand, any units may be used for L and ΔL (or V and ΔV) as long as they are the same.

There are three and only three mechanisms for the transfer of energy via heat. They are conduction, convection, and radiation. Remember, by definition, heat is the transfer of energy due to a difference in temperature.

In conduction and convection problems, only temperature *differences* matter, so Celsius degrees and kelvins may be used interchangeably. In radiation problems this is not true; you must use absolute temperatures. Remember, heat transfers are always from warmer regions to cooler regions.

IV. True or False Questions and Responses

True or False

_____ 1. The van der Waals equation is a more accurate equation of state for real gases than the ideal-gas equation.

_____ 2. The constant a that appears in the van der Waals equation is related to the volume of the molecules themselves.

Responses to True or False

1. True.

2. False. The constant b is related to the volume of the molecules. The constant a is related to the weak attractive forces between the gas molecules.

V. Questions and Answers

Questions

1. A metal plate with a circular hole drilled through it is uniformly heated. As the plate gets hotter, it expands. Does the hole get bigger or smaller?

2. The ordinary thermometers we see every day are mostly alcohol in glass. How would the use of such a thermometer be affected if alcohol and glass had the same thermal coefficient of volume expansion?

3. If an ordinary (liquid-in-glass) thermometer is placed in something quite hot, the liquid column may actually drop a little before it starts to rise. What is going on here?

4. When you are trying to open a glass jar of food with a stuck lid, it often helps to run hot water over the metal lid for a little while. Why?

5. Under what conditions does the van der Waals equation of state reduce to the ideal-gas law?

6. Vessels like vacuum bottles or dewar flasks, which are designed to keep fluids very cold, are made with double-glass walls. The inner surfaces are silvered and there is a vacuum between the walls. Discuss how this design minimizes heat losses.

7. Materials used commercially for building insulation tend to have relatively little mass and occupy a lot of space: they are foamy, porous materials or masses of compacted fibers or some such. Why?

Answers

1. The hole enlarges with the plate. Every linear dimension of the metal plate undergoes the same fractional increase in length, including the diameter of the hole.

2. This would make them useless, of course. The reason the top of the fluid column rises against the scale marked on the glass is that the volume of the fluid expands more for a given temperature change than does the volume of the glass.

3. If this happens, it is because the glass envelope holding the liquid gets hot first and expands a little before the temperature of the liquid inside has had time to rise. Then, as the glass, the liquid, and the surroundings all come to thermal equilibrium, the liquid rises with respect to the glass tube.

4. You may just be dissolving away sticky goo that is gluing the lid to the jar. More relevant to our discussion here is the fact that the metal lid will expand more as you increase its temperature than the glass jar will, so the lid loosens.

5. The van der Waals equation of state reduces to the ideal-gas law when the volume per mole is very large and thus the density of the gas is very low.

6. The vacuum between the double-glass walls prevents heat transfer through the walls by conduction and convection. Silvering the inner surfaces of the double walls cuts down on heat loss by radiation by reducing their emissivity. The double walls must join at the neck, and there will be some heat loss by conduction there. This is reduced by using glass that is a relatively poor conductor.

7. The molecules of a gas spend most of their time between collisions. Thus, gases are good insulators with regard to the transfer of heat by conduction. They are poor insulators with regard to convection, however. By using materials that prevent convection currents from building up—materials that confine the gas in numerous small pockets—convective transfers of heat are all but eliminated. Thus, products used for insulation are light because they consist of numerous pockets of trapped gas. They occupy lots of space in order to minimize the temperature gradient, which in turn minimizes the thermal current.

VI. Problems, Solutions, and Answers

Example #1. A brass pin is exactly 5.00 cm long when it is at a temperature of 140.0°C. What is its length when it cools to 20.0°C?

Picture the Problem. Look up the coefficient of thermal expansion for brass in Table 20-1 on page 629 of the text, and use it to determine the new length.

1. The coefficient of thermal expansion is the fractional change in length per kelvin. Use this to solve for the change in length. The final length will be equal to the initial length plus the change in length	$\alpha = \dfrac{\Delta L/L}{\Delta T}$ $L' = L + \Delta L = L + \alpha L\,\Delta T = L(1 + \alpha\,\Delta T)$ $\quad = (0.0500\,\text{m})\left[1 + \left(19\times10^{-6}\ \text{K}^{-1}\right)(-120\,\text{K})\right]$ $\quad = 0.0499\,\text{m}$

Example #2—Interactive. A surveyor's steel tape is manufactured to be accurate at 20°C. On a day when the temperature is 38°C it is used to mark off a 100-m-long track. If the length as measured with this tape is 100 m, what is the error in the length of the track due to the thermal expansion of the steel tape?

Picture the Problem. Look up the coefficient of thermal expansion for steel in Table 20-1 on page 629 of the text, and use it to determine the new length. **Try it yourself.** Work the problem on your own, in the space provided, to get the final answer.

1. The coefficient of thermal expansion is the fractional change in length per kelvin. Use this to solve for the change in length of the 100-m tape, which is the error in the length of the tape. The final length will be equal to the initial length plus the change in length.	 $\Delta L = 2\,\text{cm}$, or a 0.02% error

Example #3. A 10.0-L Pyrex flask is filled to the brim with acetone at 12.4°C. To what temperature must the flask and its contents be heated, or cooled, so that 85.0 cm³ of acetone overflow the flask?

Picture the Problem. Obtain an expression for the change in the volume of the flask and the acetone in terms of the temperature change. You will need the coefficients of thermal expansion for Pyrex glass and acetone found in Table 20-1 on page 629 of the text. Remember the volume coefficient of thermal expansion is three times the linear coefficient of thermal expansion. You want the change in these volumes to be 85.0 cm³.

1. Obtain an expression for the change in volume due to the temperature change.	$\beta = 3\alpha = \dfrac{\Delta V/V}{\Delta T}$ $\Delta V = 3\alpha V_0\,\Delta T = \beta V_0\,\Delta T$

2. The difference of the change in volume of the flask and the acetone must be $85\,\text{cm}^3$.	$\Delta V_a - \Delta V_f = 85.0\,\text{cm}^3$ $\beta_a V_0 \Delta T - 3\alpha_p V_0 \Delta T = 85.0\,\text{cm}^3$ $\Delta T = \dfrac{85.0\,\text{cm}^3}{\left(\beta_a - 3\alpha_f\right)V_0}$ $= \dfrac{\left(85.0\,\text{cm}^3\right)\left(10^{-3}\,\text{L/cm}^3\right)}{\left(\left(1.5\times10^{-3}\,\text{K}^{-1}\right)-3\left(3.2\times10^{-6}\,\text{K}^{-1}\right)\right)\left(10\,\text{L}\right)}$ $= +5.7\,\text{K}$

Example #4—Interactive. My car has a 40-L gasoline tank. If I fill it completely full of gasoline at a temperature of 12°C and then let the car sit in the sun until the temperatures of the gasoline and the tank reach 30°C, how much gasoline spills out of the tank? The thermal coefficient of volume expansion for gasoline is $9\times10^{-4}\ \text{K}^{-1}$ and the tank is made of steel.

Picture the Problem. Obtain an expression for the change in the volume of the gas tank and the gasoline in terms of the temperature change. You will need the coefficient of thermal expansion for steel found in Table 20-1 on page 629 of the text. Remember the volume coefficient of thermal expansion is three times the linear coefficient of thermal expansion. The difference in the volume increase of the two substances is the overflow volume. **Try it yourself.** Work the problem on your own, in the spaces provided, to get the final answer.

1. Obtain an expression for the change in volume due to the temperature change.	
2. The difference of the change in volume of the gasoline and the tank is the overflow volume.	overflow $= 0.624\,\text{L}$

Example #5. One mole of steam is confined in a 30.0-L container at 600°C. (*a*) Calculate the pressure using the ideal-gas equation. (*b*) The van der Waals constants for steam are $a = 5.43\ \text{atm}\cdot\text{L}^2/\text{mol}^2$ and $b = 0.03$ L/mol. Calculate the pressure using the van der Waals equation. (*c*) Actual measurements show the pressure to be 2.39 atm. Compare the pressures calculated using the ideal-gas equation and the van der Waals equation with the measured pressure.

1. Calculate the pressure using the ideal-gas equation.	$P_1 = \dfrac{nRT}{V}$ $= \dfrac{(1\,\text{mol})(0.0821\,\text{L·atm}/(\text{mol·K}))(873\,\text{K})}{30.0\,\text{L}} = 2.39\,\text{atm}$
2. Calculate the pressure using the van der Waals equation.	$\left(P + \dfrac{an^2}{V^2}\right)(V - bn) = nRT$ $P_2 = \dfrac{nRT}{V - bn} - \dfrac{an^2}{V^2}$ $= \dfrac{(1\,\text{mol})(0.0821\,\text{L·atm}/(\text{mol·K}))(837\,\text{K})}{(30.0\,\text{L}) - (0.03\,\text{L/mol})(1\,\text{mol})}$ $- \dfrac{(5.43\,\text{atm·L}^2/\text{mol}^2)(1\,\text{mol})^2}{(30.0\,\text{L})^2} = 2.39\,\text{atm}$

The actual, ideal-gas, and van der Waals pressures all agree to better than one part in 239.

Example #6—Interactive. Fifty moles of steam are confined in a 30.0-L container at 600°C. (*a*) Calculate the pressure using the ideal-gas equation. (*b*) The van der Waals constants for steam are $a = 5.43\,\text{atm} \cdot \text{L}^2/\text{mol}^2$ and $b = 0.03\,\text{L/mol}$. Calculate the pressure using the van der Waals equation. (*c*) Actual measurements show the pressure to be 113 atm. Compare the pressures calculated using the ideal-gas equation and the van der Waals equation with the measured pressure. **Try it yourself.** Work the problem on your own, in the spaces provided, to get the final answer.

1. Calculate the pressure using the ideal-gas equation.	$P_1 = 119\,\text{atm}$
2. Calculate the pressure using the van der Waals equation.	$P_2 = 111\,\text{atm}$

The ideal-gas pressure is about 5% higher than the measured pressure. The van der Waals pressure is about 2% lower than the measured pressure.

The pressures calculated using the van der Waals equation are closer to the experimentally measured values than are the pressures calculated using the ideal-gas law. However, the results from both of these equations are in close agreement with the measured pressure when the density of the gas is low. This is evidenced by the results from Example #5, where both calculated pressures differed from the measured pressure by less than 0.4%.

The constant a that appears in the van der Waals equation is a measure of the forces exerted by the gas molecules on each other when the molecules are between collisions. Water molecules are electrically polarized. Therefore the forces they exert on each other are large in comparison with the forces exerted by many other molecules. The constant a is least for the inert gases such as helium, neon, and argon.

Example #7. Figure 20-1 shows a system holding liquid helium at 4.20 K (the boiling point of helium). A cylindrical can 5.00 cm in diameter and 7.00 cm high is supported by two stainless steel pins from the walls, which are at 77.0 K (the boiling point of nitrogen). The space between the can and the walls is evacuated. The steel pins are each 1.00 mm in diameter and 6.00 cm long and have a thermal conductivity of 13.4 W/(m·K). If the latent heat of vaporization of helium is 21.0 kJ/kg, at what rate does the helium boil off? (Assume that the emissivity of the can's exterior is 0.250.)

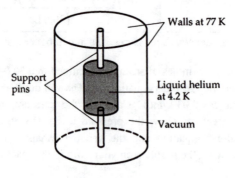

Figure 20-1

Picture the Problem. The heat that evaporates the helium is delivered to it through the support pins and by radiation from the 77-K walls. Thus the rate at which the helium must absorb heat is equal to the conductive thermal currents through the pins plus the radiative thermal current.

1. Equate the rate required to boil off the helium to the rate at which heat is delivered to the helium. We can then solve for the rate at which the helium is boiled off.	$\dfrac{dQ_{evap}}{dt} = \dfrac{d(mL_v)}{dt} = P_{conductive} + P_{radiative}$ $\dfrac{dm}{dt} = \dfrac{P_{conductive} + P_{radiative}}{L_v}$
2. Determine the rate at which heat is transferred via conduction through the pins. The factor of two appears because there are two pins.	$P_{conductive} = 2kA_{pin}\dfrac{\Delta T}{\Delta x}$ $= 2\big(13.4 \text{ W/(m·K)}\big)\big(7.85\times10^{-7}\text{ m}^2\big)\dfrac{77.0\text{ K} - 4.20\text{ K}}{0.0600\text{ m}}$ $= 0.0255 \text{ W}$

3. Determine the rate at which heat is transferred via radiation from the surface area of the can. When calculating the surface area, remember to use the area of the side of the can, as well as the top and bottom.	$P_{radiative} = e\sigma A_{can}\left(T^4 - T_0^4\right)$ $= 0.25\left(5.67\times10^{-8}\ \text{W}/\left(\text{m}^2\cdot\text{K}\right)\right)\left(0.0149\ \text{m}^2\right)$ $\times\left[\left(77.0\ \text{K}\right)^4 - \left(4.20\ \text{K}\right)^4\right]$ $= 0.00743\ \text{W}$
4. Now we can calculate the rate of boiloff by substituting into the equation of step 1.	$\dfrac{dm}{dt} = \dfrac{0.0255\ \text{W} - 0.00743\ \text{W}}{2.1\times10^4\ \text{J/kg}} = 1.57\times10^{-6}\ \text{kg/s}$

Example #8—Interactive. A vacuum bottle with an inside diameter of 6.00 cm contains 150.0 g of water and 75.0 g of ice at 0.00°C (see Figure 20-2). Heat leakage through the walls of the bottle is negligible, but it is closed with a cork stopper 2.00 cm thick whose thermal conductivity is $0.500\ \text{W}/\left(\text{m}\cdot\text{K}\right)$. If the surroundings are at 28.0°C, how long does it take for all the ice to melt?

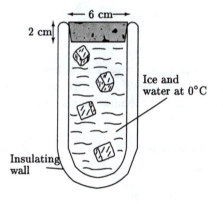

Figure 20-2

Picture the Problem. The heat required to melt the water comes from the cork, providing thermal conductivity from the warmer surroundings to the ice water. **Try it yourself.** Work the problem on your own, in the spaces provided, to get the final answer.

1. Equate an expression for the product of the time and the thermal current through the cork stopper with an expression for the heat absorbed by the melting ice.	
2. Solve for the time at which all the ice is melted.	$t = 3.50\ \text{hr}$

Example #9. A home has a window area of 263 ft^2 of single-pane glass 0.135 in. thick. (*a*) What is the rate of conductive heat loss through the glass when the temperature inside is 72°F and the temperature outside is 10°F? (*b*) For comparison, calculate the rate of conductive heat loss that would occur if the entire temperature difference of 62F° were applied across the glass. (Use $5.6\,\text{Btu}\cdot\text{in}/(\text{hr}\cdot\text{ft}^2\cdot\text{F}°)$ as the thermal conductivity of the glass.)

Picture the Problem. For part (*a*), use the *R* factor of glass from Table 20-4 on page 639 of the text. Note that this is independent of the thickness of the glass. This is because the thermal resistance of a pane of glass is dominated by the resistances of the layers of air on either side of the window. The thermal resistance equals the *R* factor divided by the area. The thermal current is the temperature difference divided by the thermal resistance.

1. Find the thermal current for part (*a*), using the *R* value for glass.	$I = \dfrac{\Delta T}{R} = \dfrac{A\,\Delta T}{R_f} = \dfrac{\left(263\,\text{ft}^2\right)\left(72°\,\text{F}-10°\,\text{F}\right)}{0.9\,\text{h}\cdot\text{ft}^2\cdot\text{F}° \,/\,\text{Btu}}$ $= 1.81\times10^4\ \text{Btu/hr}$
2. Find the thermal current assuming the entire temperature gradient were across the glass.	$I = \dfrac{kA\,\Delta T}{\Delta x} = \dfrac{\left(5.6\,\text{Btu}\cdot\text{in}/\left(\text{hr}\cdot\text{ft}^2\cdot\text{F}°\right)\right)\left(263\,\text{ft}^2\right)\left(72°\,\text{F}-10°\,\text{F}\right)}{0.135\,\text{in}}$ $= 6.76\times10^5\ \text{Btu/hr}$

Example #10—Interactive. The walls of a certain house consist of 160 m^2 of brick, 10.0 cm thick, with a thermal conductivity of $0.8\,\text{W}/(\text{m}\cdot\text{K})$. At what rate is the heat lost by conduction through the walls if the temperature of the inside surface of the wall is 16°C and the outside temperature is 5°C?

Picture the Problem. See part (*b*) of Example #9 for an example of how to work this problem. **Try it yourself.** Work the problem on your own, in the space provided, to get the final answer.

1. Find the thermal current assuming the entire temperature gradient is across the brick. The result of Example #9, however, should make you wonder if this is an overestimate.	$I = 14.1\,\text{kW}$

Example #11. Assume that the surface of a human body can be considered a perfect blackbody at infrared wavelengths. Take the surface area to be 2.2 m^2 and the surface temperature to be 33°C. (*a*) Calculate the peak wavelength of the body's radiated spectrum. (*b*) If the body's surroundings are at a temperature of 22°C, calculate the body's total rate of heat loss by radiation.

1. Calculate the peak wavelength.	$\lambda_{\text{max}} = \dfrac{2.898\ \text{mm}\cdot\text{K}}{T} = \dfrac{2.898\times10^{-3}\ \text{m}\cdot\text{K}}{306\,\text{K}}$ $= 9.48\times10^{-6}\ \text{m}=9.48\,\mu\text{m}$

2. Calculate the net rate of radiative heat loss of the body, assuming the emissivity is 1.	$P_{rad} = e\sigma A\left(T^4 - T_0^4\right)$
This result, which estimates only radiative heat-loss rate, is larger than the body's entire resting heat-loss rate, so some of the assumptions made must be incorrect. In fact, the emissivity of the human body is not quite 1, and it is not ordinarily considered socially acceptable to expose our entire surface area directly to the surroundings.	$= 1\left(5.67 \times 10^{-8}\ \text{W/}\left(\text{m}^2 \cdot \text{K}^4\right)\right)\left(2.2\,\text{m}^2\right)$ $\times \left[\left(306\,\text{K}\right)^4 - \left(295\,\text{K}\right)^4\right]$ $= 149\,\text{W}$

Example #12—Interactive. In a certain experiment, heat is transferred from a source at 227°C to water at 30.0°C through a rod 3.00 cm in diameter as shown in Figure 20-3. The rod consists of a 12.0-cm-long aluminum rod butted end to end with a 5.00-cm-long copper rod. What is the temperature of the aluminum–copper junction if the aluminum rod is in thermal contact with the 227°C heat source and the copper rod is in contact with the 30.0°C water bath? (The thermal conductivities can be found in Table 20-3 on page 636 of the text.)

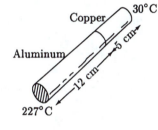

Figure 20-3

Picture the Problem. The thermal current is related to the thermal conductivities, cross-sectional areas, and the temperature gradients of each rod. Equilibrium will be reached when the thermal currents from each rod are equal. **Try it yourself.** Work the problem on your own, in the space provided, to get the final answer.

1. Obtain expressions for the thermal currents of each rod.	
2. Equate the two expressions for the thermal current and solve for the temperature at the junction.	$T = 70°\text{C}$